BY ED YONG

*I Contain Multitudes: The Microbes
Within Us and a Grander View of Life*

*An Immense World: How Animal Senses
Reveal the Hidden Realms Around Us*

AN IMMENSE WORLD

AN
IMMENSE
WORLD

How Animal Senses
Reveal the Hidden Realms
Around Us

ED YONG

RANDOM HOUSE / NEW YORK

Published in the United States by Random House,
an imprint and division of Penguin Random House LLC, New York.

RANDOM HOUSE and the HOUSE colophon are registered
trademarks of Penguin Random House LLC.

LIBRARY OF CONGRESS CATALOGING-IN-PUBLICATION DATA
NAMES: Yong, Ed, author.
TITLE: An immense world : how animal senses reveal the hidden
realms around us / Ed Yong.
DESCRIPTION: New York : Random House, 2022. |
Includes bibliographical references and index.
IDENTIFIERS: LCCN 2021046048 (print) | LCCN 2021046049 (ebook) |
ISBN 9780593133231 (hardcover) | ISBN 9780593133248 (ebook)
SUBJECTS: LCSH: Senses and sensation. | Animal behavior. | Physiology. | Neurosciences.
CLASSIFICATION: LCC QP431 .Y68 2022 (print) | LCC QP431 (ebook) |
DDC 591.5—dc23/eng/20211221
LC record available at https://lccn.loc.gov/2021046048
LC ebook record available at https://lccn.loc.gov/2021046049

Printed in the United States of America on acid-free paper

randomhousebooks.com

14 16 18 19 17 15 13

Book design by Barbara M. Bachman

Illustrations redrawn by Mapping Specialists Ltd.

For Liz Neeley, who sees me

Contents

How do you know but ev'ry Bird
that cuts the airy way,
Is an immense world of delight,
clos'd by your senses five?

—WILLIAM BLAKE

AN IMMENSE WORLD

Introduction

The Only
True Voyage

I MAGINE AN ELEPHANT IN A ROOM. THIS ELEPHANT IS NOT THE proverbial weighty issue but an actual weighty mammal. Imagine the room is spacious enough to accommodate it; make it a school gym. Now imagine a mouse has scurried in, too. A robin hops alongside it. An owl perches on an overhead beam. A bat hangs upside down from the ceiling. A rattlesnake slithers along the floor. A spider has spun a web in a corner. A mosquito buzzes through the air. A bumblebee sits upon a potted sunflower. Finally, in the midst of this increasingly crowded hypothetical space, add a human. Let's call her Rebecca. She's sighted, curious, and (thankfully) fond of animals. Don't worry about how she got herself into this mess. Never mind what all these animals are doing in a gym. Consider, instead, how Rebecca and the rest of this imaginary menagerie might perceive one another.

The elephant raises its trunk like a periscope, the rattlesnake flicks out its tongue, and the mosquito cuts through the air with its antennae. All three are smelling the space around them, taking in the floating scents. The elephant sniffs nothing of note. The rattlesnake detects the trail of the mouse, and coils its body in ambush. The mosquito smells the alluring carbon dioxide on Rebecca's breath and the aroma of her skin. It lands on her arm, ready for a meal, but before it can bite, she swats it away—and her slap disturbs the mouse. It squeaks in alarm, at a pitch that is audible to the bat but too high for the elephant to hear. The elephant, meanwhile, unleashes a deep, thunderous rumble too low-pitched for the mouse's ears or the bat's but felt by the vibration-sensitive belly of the rattlesnake. Rebecca, who is oblivious to both the ultrasonic mouse squeaks and the infrasonic elephant rumbles, listens instead to the robin, which is singing at frequencies better suited to her ears. But her hearing is too slow to pick out all the complexities that the bird encodes within its tune.

The robin's chest looks red to Rebecca but not to the elephant, whose eyes are limited to shades of blue and yellow. The bumblebee can't see red, either, but it *is* sensitive to the ultraviolet hues that lie beyond the opposite end of the rainbow. The sunflower it sits upon has at its center an ultraviolet bullseye, which grabs the attention of both the bird and the bee. The bullseye is invisible to Rebecca, who thinks the flower is only yellow. Her eyes are the sharpest in the room; unlike the elephant or the bee, she can spot the small spider sitting upon its web. But she stops seeing much of anything when the lights in the room go out.

Plunged into darkness, Rebecca walks slowly forward, arms outstretched, hoping to feel obstacles in her way. The mouse does the same but with the whiskers on its face, which it sweeps back and forth several times a second to map its surroundings. As it skitters between Rebecca's feet, its footsteps are too faint for her to hear, but they are easily audible to the owl perched overhead. The disc of stiff feathers on the owl's face funnels sounds toward its sensitive ears, one of which is slightly higher than the other. Thanks to this asymmetry, the owl can pinpoint the source of the mouse's skittering in both the vertical and horizontal planes. It swoops in, just as the mouse blunders within range of the waiting rattlesnake. Using two pits on its snout, the snake can sense the infrared radiation that emanates from warm objects. It effectively sees in heat, and the mouse's body blazes like a beacon. The snake strikes . . . and collides with the swooping owl.

All of this commotion goes unnoticed by the spider, which barely hears or sees the participants. Its world is almost entirely defined by the vibrations coursing through its web—a self-made trap that acts as an extension of its senses. When the mosquito strays into the silken strands, the spider detects the telltale vibrations of struggling prey and moves in for the kill. But as it attacks, it is unaware of the high-frequency sound waves that are hitting its body and bouncing back to the creature that sent them—the bat. The bat's sonar is so acute that it not only finds the spider in the dark but pinpoints it precisely enough to pluck it from its web.

As the bat feeds, the robin feels a familiar attraction that most of the other animals cannot sense. The days are getting colder, and it is time to migrate to warmer southern climes. Even within the enclosed gym, the robin can feel Earth's magnetic field, and, guided by its internal compass, it points due south and escapes through a window. It leaves behind one elephant, one bat, one bumblebee, one rattlesnake, one slightly ruffled owl, one extremely fortunate mouse, and one Rebecca. These seven creatures share the same physical space but experience it in wildly and wondrously different ways. The same is true for the billions of other animal species on the planet and the countless individuals within those species.* Earth teems with sights and textures, sounds and vibrations, smells and tastes, electric and magnetic fields. But every animal can only tap into a small fraction of reality's fullness. Each is enclosed within its own unique sensory bubble, perceiving but a tiny sliver of an immense world.

THERE IS A WONDERFUL word for this sensory bubble—*Umwelt*. It was defined and popularized by the Baltic-German zoologist Jakob von Uexküll in 1909. Umwelt comes from the German word for "environment," but Uexküll didn't use it simply to refer to an animal's surroundings. Instead, an Umwelt is specifically the part of those surroundings that an animal can sense and experience—its *perceptual* world. Like the occupants of our imaginary room, a multitude of creatures could be standing in the same physical space and have completely different Umwelten. A tick, questing for mammalian blood, cares about body heat, the touch of hair, and the odor of butyric acid that

---

* To understand how varied senses can be in a single species, just look at humans. For some people, red and green look identical. For others, body odor smells like vanilla. For yet others, coriander (cilantro) tastes of soap.

emanates from skin. These three things constitute its Umwelt. Trees of green, red roses too, skies of blue, and clouds of white—these are not part of its wonderful world. The tick doesn't willfully ignore them. It simply cannot sense them and doesn't know they exist.

Uexküll compared an animal's body to a house. "Each house has a number of windows," he wrote, "which open onto a garden: a light window, a sound window, an olfactory window, a taste window, and a great number of tactile windows. Depending on the manner in which these windows are built, the garden changes as it is seen from the house. By no means does it appear as a section of a larger world. Rather, it is the only world that belongs to the house—its [Umwelt]. The garden that appears to our eye is fundamentally different from that which presents itself to the inhabitants of the house."

This was a radical notion at the time—and in some circles, it might still be. Unlike many of his contemporaries, Uexküll saw animals not as mere machines but as sentient entities, whose inner worlds not only existed but were worth contemplating. Uexküll didn't exalt the inner worlds of humans over those of other species. Rather, he treated the Umwelt concept as a unifying and leveling force. The human's house might be bigger than the tick's, with more windows overlooking a wider garden, but we are still stuck inside one, looking out. Our Umwelt is still limited; it just doesn't *feel* that way. To us, it feels all-encompassing. It is all that we know, and so we easily mistake it for all there is *to* know. This is an illusion, and one that every animal shares.

We cannot sense the faint electric fields that sharks and platypuses can. We are not privy to the magnetic fields that robins and sea turtles detect. We can't trace the invisible trail of a swimming fish the way a seal can. We can't feel the air currents created by a buzzing fly the way a wandering spider does. Our ears cannot hear the ultrasonic calls of rodents and hummingbirds or the infrasonic calls of elephants and whales. Our eyes cannot see the infrared radiation that rattlesnakes detect or the ultraviolet light that the birds and the bees can sense.

Even when animals share the same senses with us, their Umwelten can be very different. There are animals that can hear sounds in what seems to us like perfect silence, see colors in what looks to us like total darkness, and sense vibrations in what feels to us like complete stillness. There are animals with eyes on their genitals, ears on their knees, noses on their limbs, and tongues all over their skin. Starfish see with the tips of their arms, and sea urchins with their entire bodies. The star-nosed mole feels around with its nose, while the manatee uses its lips. We are no sensory slouches, either. Our hearing is decent, and certainly better than that of the millions of insects that have no ears at all. Our eyes are unusually sharp, and can discern patterns on animal bodies that the animals themselves cannot see. Each species is constrained in some ways and liberated in others. For that reason, this is not a book of lists, in which we childishly rank animals according to the sharpness of their senses and value them only when their abilities surpass our own. This is a book not about superiority but about diversity.

This is also a book about animals as animals. Some scientists study the senses of other animals to better understand ourselves, using exceptional creatures like electric fish, bats, and owls as "model organisms" for exploring how our own sensory systems work. Others reverse-engineer animal senses to create new technologies: Lobster eyes have inspired space telescopes, the ears of a parasitic fly have influenced hearing aids, and military sonar has been honed by work on dolphin sonar. These are both reasonable motivations. I'm not interested in either. Animals are not just stand-ins for humans or fodder for brainstorming sessions. They have worth in themselves. We'll explore their senses to better understand *their* lives. "They move finished and complete, gifted with extensions of the senses we have lost or never attained, living by voices we shall never hear," wrote the American naturalist Henry Beston. "They are not brethren, they are not underlings; they are other nations, caught with ourselves in the net of life and time, fellow prisoners of the splendour and travail of the earth."

A FEW TERMS WILL act as guideposts on our journey. To sense the world, animals detect *stimuli*—quantities like light, sound, or chemicals—and convert them into electrical signals, which travel along neurons toward the brain. The cells that are responsible for detecting stimuli are called *receptors:* Photoreceptors detect light, chemo-receptors detect molecules, and mechanoreceptors detect pressure or movement. These receptor cells are often concentrated in *sense organs,* like eyes, noses, and ears. And sense organs, together with the neurons that transmit their signals and the parts of the brain that process those signals, are collectively called *sensory systems.* The visual system, for example, includes the eyes, the photoreceptors inside them, the optic nerve, and the visual cortex of the brain. Together, these structures give most of us the sense of sight.

The preceding paragraph could have been pulled from a high school textbook. But take a moment to consider the miracle of what it describes. Light is just electromagnetic radiation. Sound is just waves of pressure. Smells are just small molecules. It's not obvious that we should be able to detect *any* of those things, let alone convert them into electrical signals or derive from those signals the spectacle of a sunrise, or the sound of a voice, or the scent of baking bread. The senses transform the coursing chaos of the world into perceptions and experiences—things we can react to and act upon. They allow biology to tame physics. They turn stimuli into *information.* They pull relevance from randomness, and weave meaning from miscellany. They connect animals to their surroundings. And they connect animals to each other via expressions, displays, gestures, calls, and currents.

The senses constrain an animal's life, restricting what it can detect and do. But they also define a species' future, and the evolutionary possibilities ahead of it. For example, around 400 million years ago, some fish began leaving the water and adapting to life on land. In open air, these pioneers—our ancestors—could see over much longer distances than they could in water. The neuroscientist Malcolm MacIver thinks that this change spurred the evolution of advanced mental abilities, like planning and strategic thinking. Instead of simply reacting to whatever was directly in front of them, they could be proactive. By seeing farther, they could think ahead. As their Umwelten expanded, so did their minds.

An Umwelt cannot expand indefinitely, though. Senses always come at a cost. Animals have to keep the neurons of their sensory sys-

tems in a perpetual state of readiness so that they can fire when necessary. This is tiring work, like drawing a bow and holding it in place so
that when the moment comes, an arrow can be shot. Even when your
eyelids are closed, your visual system is a monumental drain on your
reserves. For that reason, no animal can sense everything well.

Nor would any animal want to. It would be overwhelmed by the
flood of stimuli, most of which would be irrelevant. Evolving according to their owner's needs, the senses sort through an infinity of stimuli,
filtering out what's irrelevant and capturing signals for food, shelter,
threats, allies, or mates. They are like discerning personal assistants who
come to the brain with only the most important information.* Writing
about the tick, Uexküll noted that the rich world around it is "con-
stricted and transformed into an impoverished structure" of just three
stimuli. "However, the poverty of this environment is needful for the
certainty of action, and certainty is more important than riches." Nothing can sense everything, and nothing needs to. That is why Umwelten
exist at all. It is also why the act of contemplating the Umwelt of another creature is so deeply human and so utterly profound. Our senses
filter in what we need. We must choose to learn about the rest.

THE SENSES OF ANIMALS have fascinated people for millennia,
but mysteries still abound. Many of the animals whose Umwelten
are most different from ours live in habitats that are inaccessible or
impenetrable—murky rivers, dark caves, open oceans, abyssal depths,
and subterranean realms. Their natural behavior is hard to observe, let
alone to interpret. Many scientists are limited to studying creatures
that can be kept in captivity, with all the strangeness that entails. Even
in labs, animals are challenging to work with. Experiments that might
reveal how they use their senses are hard to design, especially when
those senses are drastically different from ours.

Amazing new details—and, sometimes, entirely new senses—are

---

* In 1987, German scientist Rüdiger Wehner described these as "matched filters"—aspects of
an animal's sensory systems that are tuned to the sensory stimuli that it most needs to detect.

being discovered regularly. Giant whales have a volleyball-sized sensor at the tip of their lower jaw, which was only discovered in 2012 and whose function is still unclear. Some of the stories in these pages are decades or centuries old; others emerged as I was writing. And there's still so much we can't explain. "My dad, who is an atomic physicist, once asked me a bunch of questions," Sonke Johnsen, a sensory biologist, tells me. "After a few *I don't know*s, he said: *You guys really don't know anything*." Inspired by that conversation, Johnsen published a paper in 2017 entitled "We Don't Really Know Anything, Do We? Open Questions in Sensory Biology."

Consider the seemingly simple question *How many senses are there?* Around 2,370 years ago, Aristotle wrote that there are five, in both humans and other animals—sight, hearing, smell, taste, and touch. This tally persists today. But according to the philosopher Fiona Macpherson, there are reasons to doubt it. For a start, Aristotle missed a few in humans: proprioception, the awareness of your own body, which is distinct from touch; and equilibrioception, the sense of balance, which has links to both touch and vision.

Other animals have senses that are even harder to categorize. Many vertebrates (animals with backbones) have a second sensory system for detecting odors, governed by a structure called the vomeronasal organ; is this part of their main sense of smell, or something separate? Rattlesnakes can detect the body heat of their prey, but their heat sensors are wired to their brain's visual center; is their heat sense simply part of vision, or something distinct? The platypus's bill is loaded with sensors that detect electric fields and sensors that are sensitive to pressure; does the platypus's brain treat these streams of information differently, or does it wield a single sense of electrotouch?

These examples tell us that "senses cannot be clearly divided into a limited number of discrete kinds," Macpherson wrote in *The Senses*. Instead of trying to shove animal senses into Aristotelian buckets, we should instead study them for what they are.* Though I have orga-

---

* If you were being maximally reductive, you could reasonably argue that there are really only two senses—chemical and mechanical. Chemical senses include smell, taste, and vision. Mechanical senses include touch, hearing, and electrical senses. The magnetic sense might be-

nized this book into chapters that revolve around specific stimuli, like light or sound, that's largely for convenience. Each chapter is a gateway into the varied things that animals do with each stimulus. We will not concern ourselves with counting senses, nor talk nonsensically about a "sixth sense." We will instead ask how animals use their senses, and attempt to step inside their Umwelten.

It won't be easy. In his classic 1974 essay, "What Is It Like to Be a Bat?," the American philosopher Thomas Nagel argued that other animals have conscious experiences that are inherently subjective and hard to describe. Bats, for example, perceive the world through sonar, and since this is a sense that the majority of humans lack, "there is no reason to suppose that it is subjectively like anything we can experience or imagine," Nagel wrote. You could envision yourself with webbing on your arms or insects in your mouth, but you'd still be creating a mental caricature of *you* as a bat. "I want to know what it is like for a *bat* to be a bat," Nagel wrote. "Yet if I try to imagine this, I am restricted to the resources of my own mind, and those resources are inadequate to the task."

In thinking about other animals, we are biased by our own senses and by vision in particular. Our species and our culture are so driven by sight that even people who are blind from birth will describe the world using visual words and metaphors.* You agree with people if you *see* their point, or share their *view*. You are oblivious to things in your *blind spots*. Hopeful futures are *bright* and *gleaming;* dystopias are *dark* and *shadowy*. Even when scientists describe senses that humans lack altogether, like the ability to detect electric fields, they talk about *images* and *shadows*. Language, for us, is both blessing and curse. It gives us the tools for describing another animal's Umwelt even as it insinuates our own sensory world into those descriptions.

---

long to either category or both. This framework will probably make absolutely no sense right now, but should become clearer as you continue in the book. I'm not especially wedded to it, but it is one possible way of thinking about the senses—and one that might appeal to the lumpers among you.

* Let me just say that avoiding visual metaphors when describing other senses is extremely difficult over the length of an entire book. I have tried to do so, or at least to be judicious and explicit whenever I have to resort to visual terms.

Scholars of animal behavior often discuss the perils of anthropomorphism—the tendency to inappropriately attribute human emotions or mental abilities to other animals. But perhaps the most common, and least recognized, manifestation of anthropomorphism is the tendency to forget about other Umwelten—to frame animals' lives in terms of *our* senses rather than *theirs*. This bias has consequences. We harm animals by filling the world with stimuli that overwhelm or befuddle their senses, including coastal lights that lure newly hatched turtles away from the oceans, underwater noises that drown out the calls of whales, and glass panes that seem like bodies of water to bat sonar. We misinterpret the needs of animals closest to us, stopping smell-oriented dogs from sniffing their environments and imposing the visual world of humans upon them. And we underestimate what animals are capable of to our own detriment, missing out on the chance to understand how expansive and wondrous nature truly is—the delights that, as William Blake wrote, are "clos'd by your senses five."

Throughout this book, we'll encounter animal abilities that others had long thought impossible or absurd. Zoologist Donald Griffin, who co-discovered the sonar of bats, once wrote that biologists have been overly swayed by what he called "simplicity filters." That is, they seemed reluctant to even consider that the senses they were studying might be more complex and refined than whatever data they had collected could suggest. This lament contradicts Occam's razor, the principle that states that the simplest explanation is usually the best. But this principle is only true *if you have all the necessary information to hand.* And Griffin's point was that you might not. A scientist's explanations about other animals are dictated by the data she collects, which are influenced by the questions she asks, which are steered by her imagination, which is delimited by her senses. The boundaries of the human Umwelt often make the Umwelten of others opaque to us.

Griffin's words are not carte blanche to put forward convoluted or paranormal explanations for animal behavior. I see them, and Nagel's essay, as a call for humility. They remind us that other animals are sophisticated, and that, for all our vaunted intelligence, it is very hard for us to understand other creatures, or to resist the tendency to view their senses through our own. We can study the physics of an animal's environment, look at what they respond to or ignore, and trace the web of neurons that connects their sense organs to their brains. But the ultimate feats of understanding—working out what it's like to be a bat, or an elephant, or a spider—always require what psychologist Alexandra Horowitz calls "an informed imaginative leap."

Many sensory biologists have backgrounds in the arts, which may enable them to see past the perceptual worlds that our brains automatically create. Sonke Johnsen, for example, studied painting, sculpture, and modern dance well before he studied animal vision. To represent the world around us, he says, artists already have to push against the limits of their Umwelt and "look under the hood." That capacity helps him "think about animals having different perceptual worlds." He also notes that many sensory biologists are perceptually divergent. Sarah Zylinski studies the vision of cuttlefish and other cephalopods; she has prosopagnosia and can't recognize even familiar faces, including her mother's. Kentaro Arikawa studies color vision in butterflies; he is red-green color-blind. Suzanne Amador Kane studies the visual and vibrational signals of peacocks; she has slight differences in her color vision in each eye, so that one gives her a slightly reddish tint. Johnsen suspects that these differences, which some might bill as "disorders," actually predispose people to step outside their Umwelten and embrace those of other creatures. Perhaps people who experience the world in ways that are considered atypical have an intuitive feeling for the limits of typicality.

We can all do this. I began this book by asking you to conjure a room full of hypothetical animals, and I'm asking you to perform similar feats of imagination over the next 13 chapters. The task will be hard, as Nagel predicted. But there is value and glory in the striving. On this journey through nature's Umwelten, our intuitions will be our biggest liabilities, and our imaginations will be our greatest assets.

ONE LATE MORNING IN June 1998, Mike Ryan hiked into the Panamanian rainforest to search for animals with his former student Rex

Cocroft. Usually, Ryan would have looked for frogs. But Cocroft had taken a liking to sap-sucking insects called treehoppers, and he had something cool to show his friend. Heading out from their research station, the duo pulled off a road and walked along a river. Once Cocroft spotted the right kind of shrub, he turned over a few leaves and quickly found a family of tiny treehoppers of the species *Calloconophora pinguis*. Cocroft had found a mother surrounded by babies, their black backs capped with forward-pointing domes that looked like Elvis's hair.

Treehoppers communicate by sending vibrations through the plants on which they stand. These vibrations are not audible but can be easily converted into sounds. Cocroft clipped a simple microphone to the plant, handed Ryan some headphones, and told him to listen. Then he flicked the leaf. Immediately the baby treehoppers ran away, while producing vibrations by contracting muscles in their abdomens. "I figured it was probably going to be some kind of scurrying noise," Ryan recalls. "And what I heard instead was like cows mooing." The sound was deep, resonant, and unlike anything you'd expect from an insect. As the babies settled down and returned to their mother, their cacophony of vibrational moos turned into a synchronized chorus.

Still watching them, Ryan took the headphones off. All around him, he heard birds singing, howler monkeys roaring, and insects chirping. The treehoppers were quiet. Ryan put the headphones back on, "and I was transported into a totally different world," he tells me. Once more, the jungle noises dropped out of his Umwelt, and the mooing treehoppers returned. "It was the coolest experience," he says. "It was sensory travel. I was in the same place, but stepping between these two really cool environments. It was such a stark demonstration of Uexküll's idea."

The Umwelt concept can feel constrictive because it implies that every creature is trapped within the house of its senses. But to me, the idea is wonderfully expansive. It tells us that all is not as it seems and that everything we experience is but a filtered version of everything that we *could* experience. It reminds us that there is light in darkness, noise in silence, richness in nothingness. It hints at flickers of the unfamiliar in the familiar, of the extraordinary in the everyday, of magnificence in mundanity. It shows us that clipping a microphone onto a plant can be an intrepid act of exploration. Stepping between Umwelten, or at least trying to, is like setting foot upon an alien planet. Uexküll even billed his work as a "travelogue."

When we pay attention to other animals, our own world expands and deepens. Listen to treehoppers, and you realize that plants are thrumming with silent vibrational songs. Watch a dog on a walk, and you see that cities are crisscrossed with skeins of scent that carry the biographies and histories of their residents. Watch a swimming seal, and you understand that water is full of tracks and trails. "When you look at an animal's behavior through the lens of that animal, suddenly all of this salient information becomes available that you would otherwise miss," Colleen Reichmuth, a sensory biologist who works with seals and sea lions, tells me. "It's like a magic magnifying glass, to have that knowledge."

Malcolm MacIver argues that when animals moved onto land, the greater range of their vision spurred the evolution of planning and advanced cognition: Their Umwelten expanded, and so did their minds. Similarly, the act of delving into other Umwelten allows us to see further and think more deeply. I'm reminded of Hamlet's plea to Horatio that "there are more things in heaven and Earth . . . than are dreamt of in your philosophy." The quote is often taken as an appeal to embrace the supernatural. I see it rather as a call to better understand the natural. Senses that seem paranormal to us only appear this way because we are so limited and so painfully unaware of our limitations. Philosophers have long pitied the goldfish in its bowl, unaware of what lies beyond, but our senses create a bowl around us too—one that we generally fail to penetrate.

But we can try. Science-fiction authors like to conjure up parallel universes and alternate realities, where things are similar to this one but slightly different. Those exist! We will visit them one at a time, beginning with the most ancient and universal of senses—the chemical ones, like smell and taste. From there, via an unexpected route, we'll visit the realm of vision, the sense that dominates the Umwelt of most people but that still holds surprises galore. We'll stop to savor the delightful world of color before heading into the harsher territories of pain and heat. We'll sail smoothly through the various mechanical senses that respond to pressure and movement—touch, vibration, hearing, and the most impressive use of hearing, echolocation. Then, as experienced sensory travelers whose imaginations have been fully primed, we'll make our most difficult imaginative leaps yet, through the strange senses that animals use to detect the electric and magnetic fields that we cannot. Finally, at journey's end, we'll see how animals unify the information from their senses, how humans are polluting and distorting that information, and where our responsibilities to nature now lie.

As the writer Marcel Proust once said, "The only true voyage . . . would be not to visit strange lands but to possess other eyes . . . to see the hundred universes that each of them sees." Let us begin.

Leaking Sacks of Chemicals

Smells and Tastes

"I DON'T THINK HE'S BEEN IN HERE BEFORE," ALEXANDRA Horowitz tells me. "So it should be very smelly."

By "he," she means Finnegan—her ink-black Labrador mix, who also goes by Finn. By "here," she means the small, windowless room in New York City in which she runs psychological experiments on dogs. By "smelly," she means that the room should be bursting with unfamiliar aromas, and thus should prove interesting to Finn's inquisitive nose. And so it does. As I look around, Finn smells around. He explores nostrils-first, intently sniffing the foam mats on the floor, the keyboard and mouse on the desk, the curtain draped over a corner, and the space beneath my chair. Compared to humans, who can explore new scenes by subtly moving our heads and eyes, a dog's nasal explorations are so meandering that it's easy to see them as random and thus aimless. Horowitz thinks of them differently. Finn, she notes, is interested in objects that people have touched and interacted with. He follows trails and checks out spots where other dogs have been. He examines vents, door cracks, and other places where moving air imports new odorants—scented molecules.* He sniffs different parts of the same object, and he'll sniff them at different distances, "like he's

---

* In the official parlance, an odorant is the molecule itself, and an odor is the sensation that said molecule produces; isoamyl acetate, an odorant, has the odor of bananas.

approaching the Van Gogh and seeing what the brushstrokes look like up close," says Horowitz. "They're in that state of olfactory exploration all the time."

Horowitz is an expert on dog olfaction—their sense of smell—and I'm here to talk with her about all things sniffy and nasal. And yet, I'm so relentlessly visual that when Finn finishes nosing around and approaches me, I'm instantly drawn to his eyes, which are captivating and brown like the darkest chocolate.* It takes concerted effort to refocus on what's right in front of them—his nose, prominent and moist, with two apostrophe-shaped nostrils curving to the side. This is Finn's main interface with the world. Here's how it works.

Take a deep breath, both as demonstration and to gird yourself for some necessary terminology. When you inhale, you create a single airstream that allows you to both smell and breathe. But when a dog sniffs, structures within its nose split that airstream in two. Most of the air heads down into the lungs, but a smaller tributary, which is for smell and smell alone, zooms to the back of the snout. There it enters a labyrinth of thin, bony walls that are plastered with a sticky sheet called the olfactory epithelium. This is where smells are first detected. The epithelium is full of long neurons. One end of each neuron is exposed to the incoming airstream and snags passing odorants using specially shaped proteins called odorant receptors. The other end is plugged directly into a part of the brain called the olfactory bulb. When the odorant receptors successfully grab their targets, the neurons notify the brain, and the dog perceives a smell. You can breathe out now.

Humans share the same basic machinery, but dogs just have more of everything: a more extensive olfactory epithelium, dozens of times more neurons in that epithelium, almost twice as many kinds of olfactory receptors, and a relatively larger olfactory bulb.† And their

---

* It's no coincidence that I'm drawn to Finn's eyes. Dogs have a facial muscle that can raise their inner eyebrows, giving them a soulful, plaintive expression. This muscle doesn't exist in wolves. It's the result of centuries of domestication, in which dog faces were inadvertently reshaped to look a bit more like ours. Those faces are now easier to read, and better at triggering a nurturing response.

† I've deliberately avoided putting hard numbers on the scale of these differences. It is easy to find estimates, and very hard to find primary sources for them; after an hours-long search

hardware is packed off into a separate compartment, while ours is exposed to the main flow of air through our noses. This difference is crucial. It means that whenever we exhale, we purge the odorants from our noses, causing our experience of smell to strobe and flicker. Dogs, by contrast, get a smoother experience, because odorants that enter their noses tend to stay there, and are merely replenished by every sniff.

The shape of their nostrils adds to this effect. If a dog is sniffing a patch of ground, you might imagine that every exhalation would blow odorants on the surface *away* from the nose. But that's not what happens. The next time you look at a dog's nose, notice that the front-facing holes taper off into side-facing slits. When the animal exhales while sniffing, air exits through those slits and creates rotating vortices that waft fresh odors *into* the nose. Even when breathing out, a dog is *still* sucking air in. In one experiment, an English pointer (who was curiously named Sir Satan) created an uninterrupted inward airstream for 40 seconds, despite exhaling 30 times during that period.

With such hardware, it's no wonder that dog noses are incredibly sensitive. But how sensitive? Scientists have tried to find the thresholds at which dogs can no longer smell certain chemicals, but their answers are all over the place, varying by factors of 10,000 from one experiment to another.* Rather than focusing on these dubious statistics, it's more instructive to look at what dogs can actually do. In past experiments, they have been able to tell identical twins apart by smell. They could detect a single fingerprint that had been dabbed onto a microscope slide, then left on a rooftop and exposed to the elements for a week. They could work out which direction a person had walked in after smelling just five footsteps. They've been trained to detect bombs, drugs, landmines, missing people, bodies, smuggled cash, truffles, in-

---

that included a scientific paper that sourced a factoid to a book in the For Dummies series, I fell into an existential void and questioned the very nature of knowledge. Regardless, the differences are there, and they're substantial; it's only a question of exactly how substantial they are.

* In one study, two dogs could detect amyl acetate—think bananas—at just 1 or 2 parts per trillion, which would make them 10,000 to 100,000 times better than humans. But it also makes them 30 to 20,000 times better than six beagles that were tested on the same chemical 26 years earlier, using different methods.

vasive weeds, agricultural diseases, low blood sugar, bedbugs, oil pipe-
line leaks, and tumors.

Migaloo can find buried bones at archeological sites. Pepper uncov-
ers lingering oil pollution on beaches. Captain Ron detects turtle nests
so that the eggs can be collected and protected. Bear can pinpoint hidden electronics, while Elvis specializes in pregnant polar bears. Train,
who flunked out of drug detection school for being too energetic, now
uses his nose to track the scat of jaguars and mountain lions. Tucker
used to hang off the bow of boats and sniff for orca poop; he has since
retired, and his duties now fall to Eba. If it has a scent, a dog can be
trained to detect it. We redirect their Umwelten in service of our
needs, to compensate for our olfactory shortcomings. These feats of
detection are worth marveling at, but they are also parlor tricks. They
allow us to abstractly appreciate that dogs have a great sense of smell,
without truly appreciating what that means for their inner lives or
how their olfactory world differs from a visual one.

Unlike light, which always moves in a straight line, smells diffuse
and seep, flood and swirl. When Horowitz observes Finn sniffing a
new space, she tries to ignore the clear edges that her vision affords,
and instead pictures "a shimmering environment, where nothing has a
hard boundary," she says. "There are focal areas, but everything is sort
of seeping together." Smells travel through darkness, around corners,
and in other conditions that vex vision. Horowitz can't see into the bag
slung over the back of my chair, but Finn can *smell* into it, picking up
molecules drifting from the sandwich within. Smells linger in a way
that light does not, revealing history.* The past occupants of Horowitz's room have left no ghostly visual traces, but their chemical imprint is there for Finn to detect. Smells can arrive before their sources,
foretelling what's to come. The scents unleashed by distant rain can
clue people in to advancing storms; the odorants emitted by humans
arriving home can send their dogs running to a door. These skills are
sometimes billed as extrasensory, but they are simply sensory. It's just

---

* I can think of one exception: Some marine worms release glowing "bombs" full of lumi-
nescent chemicals, whose persistent light distracts predators from the escaping worms.

that things often become apparent to the nose before they appear to the eyes. When Finn sniffs, he is not merely assessing the present but also reading the past and divining the future. And he is reading biographies. Animals are leaking sacks of chemicals, filling the air with great clouds of odorants.* While some species deliberately send messages by releasing smells, all of us inadvertently do so, giving away our presence, position, identity, health, and recent meals to creatures with the right noses.†

"I never thought much about the nose at all," says Horowitz. "It didn't occur to me."‡ When she started studying dogs, she focused on things like their attitudes to unfairness—the kind of topic that's interesting to psychologists. But after reading Uexküll and thinking about the Umwelt concept, she shifted her attention to smell—the kind of topic that's interesting to *dogs*.

She notes, for example, that many dog owners deny their animals the joys of sniffing. To a dog, a simple walk is an odyssey of olfactory exploration. But if an owner doesn't understand that and instead sees a walk as simply a means of exercise or a route to a destination, then every sniffy act becomes an annoyance. When the dog pauses to examine some invisible trace, it must be hurried along. When the dog sniffs at poop, a carcass, or something the owner's senses find displeasing, it must be pulled away. When the dog sticks its nose in the crotch of an-

---

* Leopard urine smells of popcorn. Yellow ants smell of lemons. Depending on the species, stressed frogs can smell of peanut butter, curry, or cashew nuts, according to scientists who painstakingly sniffed 131 species and won an Ig Nobel Prize for their efforts. Crested auklets—comical seabirds that have tufted heads—roost in massive colonies that, quite delightfully, smell of tangerines.

† One possible exception is the puff adder, a venomous African snake. It sits in ambush for weeks at a time, and protects itself by visually blending into its environment. But somehow, it seems to blend in chemically, too. In 2015, Ashadee Kay Miller found that keen-nosed animals, including dogs, mongooses, and meerkats, can't detect a puff adder, even when they walk over one. Dogs can detect the scent of shed skin, but for reasons that no one understands, the living snakes are undetectable to their noses.

‡ Scientists fall prey to this, too. When Horowitz tallied every study of dog behavior published in the last decade, she found that only 4 percent focused on smell. Just 17 percent described the odor environment in which experiments were done—including airflow, temperature, humidity, or the previous presence of people or food. It's as if vision researchers hadn't thought to mention if their laboratory lights were on or not.

other dog, it's being indecorous: Bad dog! After all, in Western cultures at least, humans don't smell each other.* "You could give someone a hug, but if you actually sniffed them, that would be very weird," says Horowitz. "I could say that your hair smells great, but I can't say that *you* smell great, unless we're intimate." Time and again, people impose their values—and their Umwelt—onto their dogs, forcing them to look instead of sniff, dimming their olfactory worlds and suppressing an essential part of their caninehood. That was never clearer to Horowitz than when she took Finn to a nosework class.

Oddly billed as a sport, these classes simply train dogs to find hidden scents, under increasingly difficult conditions. That should come naturally, but it didn't to many of the animals in Finn's class. Several seemed to lack any agency: They had to be pulled from box to box by their owners, or were completely unsure what to do. Others became agitated in the presence of other dogs and barked at them. But after a summer of sniffing, those behavioral quirks diminished. The reticent dogs regained their volition. The reactive dogs became tolerant. All seemed more easygoing. Fascinated, Horowitz and her colleague Charlotte Duranton ran their own experiment with 20 dogs. In front of each animal, Duranton placed a bowl in one of three locations: one where the bowl always contained food, a second where it was always empty, and a third where the outcome was ambiguous. The dogs quickly learned to approach the food-filled bowl and ignore the empty one. What about the ambiguous one? A dog's willingness to approach *that* bowl indicates what a cognitive psychologist might call *positive judgment bias* and what everyone else might call *optimism*. Horowitz found that dogs became more optimistic after just two weeks of nosework. As their sense of smell brightened, so did their outlook. (By contrast, dogs didn't change after two weeks of heelwork—an owner-led obedience activity that involves neither olfaction nor autonomy.)

For Horowitz, the implications are clear: Let dogs be dogs. Appreciate that their Umwelt is different, and lean into that difference. She

---

* At the Oscars ceremony in 2021, a journalist asked South Korean actor Yuh-Jung Youn what Brad Pitt smells like. Youn replied, "I didn't smell him! I'm not a dog!"

does this by taking Finn on dedicated smell walks, when he's allowed to sniff to his olfactory bulb's content. If he stops, she stops. His nose sets the pace. The walks are slower, but she has no destination in mind. We go on such a walk together, heading a few blocks west of her office and into Manhattan's Riverside Park. It's a hot summer's day, and the air is redolent with garbage, urine, and exhaust—and that's only what I can smell. Finn detects more. He runs his nose along the cracks in the pavement. He investigates a traffic sign. He pauses to sniff a hydrant "because it's been visited by all the other dogs of Columbia University," Horowitz says. Sometimes she'll see Finn sniff a fresh patch of urine, raise his head, look around (or smell around), and find the dog that just left it. The smell isn't just an object unto itself but a reference point, and the walk isn't just an intermediate state between points A and B but a tour of Manhattan's layered, unseen stories.

Once we're inside the park, the air fills with greenery, cut grass, mulch, and barbecues. Another dog walks past and Finn turns to breathe in an odor sample, puffing his cheeks out like a cigar smoker. Two large poodles approach, but before they can get close, their owner pulls them away and body-checks them against a fence. Horowitz looks sad. She's happier when a female Australian shepherd arrives and circles Finn, both enthusiastically sniffing each other's genitals, while we make small talk with the owner. We glean the other dog's sex through pronouns; Finn worked it out through smell. We ask about her age; Finn can guess. We don't ask about her health or readiness to mate; Finn doesn't need to ask. "There was a time when I would try to smell what he's smelling, but I do that less often simply because I know I'm not getting what he's getting," Horowitz says. But there's room for improvement. Though the human nose lacks the anatomical complexity of a dog's and is unhelpfully farther from the ground, it is also underused. By taking more sniffs herself, and paying closer attention to odors, Horowitz says that she has become a better smeller (and a more socially awkward one). "We have perfectly good noses. We just don't use them as well as the dog."

A FUNNY THING HAPPENS when you mention dogs to neuroscientists who study olfaction in humans, as Horowitz learned while writing her book *Being a Dog*. They get a little territorial, a little . . . well . . . sniffy. Some dislike that dogs get treated like special olfactory paragons when many other mammals are excellent smellers, including rats (which can also detect landmines), pigs (whose olfactory epithelium can be twice as large as a German shepherd's), and elephants (which we'll get to later). Others point to massive discrepancies in studies that test dogs' ability to detect specific odors. These have variously claimed that dogs are a billion times more sensitive than humans, or a million times, or just ten thousand times. In some cases, humans do *better*: Of 15 odorants where both species have been tested, humans outperformed our canine companions on five, including beta-ionone (cedar wood) and amyl acetate (bananas). People also excel at discriminating between smells. While it's easy to find two colors that humans can't tell apart, it's very hard to find indistinguishable pairs of odors. Neuroscientist John McGann has tried, and tells me, "We tried odors that *mice* can't tell apart and humans were like: No, we've got this."

Yet textbooks still claim our sense of smell is terrible. McGann traced the origin of this pernicious myth to the nineteenth century. In 1879, neuroscientist Paul Broca noted that our olfactory bulbs are relatively puny compared to those of other mammals. He reasoned that smell is a base and animalistic sense, and the loss of it was necessary for us to have higher thought and free will. He then classified us (along with other primates and whales) as non-smellers. The label stuck, even though Broca never actually measured how well animals smell, relying instead on sketchy inferences based on the dimensions of their brains. Compared to a mouse, a human has an olfactory bulb smaller relative to other parts of the brain, but also physically bigger, with roughly as many neurons. It's not clear what any one of these metrics says about an animal's experience of smell.*

The textbook perspective is also a Western one, based on cultures

---

* The olfactory bulb might not even be necessary for smell. In 2019, Tali Weiss identified several women who seem to lack this structure altogether and could smell just fine. How they do it is anyone's guess.

where smell has long been undervalued. Plato and Aristotle argued that olfaction was too vague and ill-formed to produce anything other than emotional impressions. Darwin deemed it to be "of extremely slight service." Kant said that "smell does not allow itself to be described, but only compared through similarity with another sense." The English language confirms his view with just three dedicated smell words: *stinky, fragrant,* and *musty*. Everything else is a synonym (*aromatic, foul*), a very loose metaphor (*decadent, unctuous*), a loan from another sense (*sweet, spicy*), or the name of a source (*rose, lemon*). Of the five Aristotelian senses, four have vast and specific lexicons. Smell, as Diane Ackerman wrote, "is the one without words."

The Jahai people of Malaysia would disagree, as would the Semaq Beri, the Maniq, and the many other hunter-gatherer groups who have dedicated smell vocabularies. The Jahai use a dozen words for smells and smell alone. One describes the scent in gasoline, bat droppings, and millipedes. Another is for some quality shared by shrimp paste, rubber tree sap, tigers, and rotten meat. Yet another refers to soap, the pungent durian fruit, and the popcorn-like twang of the binturong.* They "have this ease of talking about smells," says psychologist Asifa Majid, who found that the Jahai can name smells as easily as English-speakers can name colors. Just as tomatoes are red, the binturong is *ltpit*. Smell is also a fundamental part of their culture. Once, Majid was told off by Jahai friends for sitting too close to her research partner and allowing their smells to mingle. Another time, she tried to name the smell of a wild ginger plant; children mocked her not only for failing but also for treating the whole plant as a single object, when the stem and flowers *obviously* had distinct smells. The myth of poor human olfaction "might have been overridden much earlier if the humans under consideration had been Jahai instead of Brits and Americans," Majid tells me.

Even Westerners can pull off surprising olfactory feats when given the chance. In 2006, neuroscientist Jess Porter took blindfolded stu-

---

* The binturong is a black, shaggy, 2-meter-long creature that looks like a cross between a cat, weasel, and bear. It's also known as a bearcat, and makes a cameo appearance in my first book, *I Contain Multitudes*.

dents to a park in Berkeley and asked them to follow a 10-meter trail of chocolate oil that she had drizzled on the grass. The students got down on all fours, snuffled about like dogs, and looked ridiculous. But they succeeded, and got better with practice.

When I visit Alexandra Horowitz, she challenges me to the same test and lays some chocolate-scented string on the floor. Eyes closed and nostrils open, I kneel down and sniff away. I quickly pick up the smell of chocolate and follow it. When I lose the scent, I cast my head from side to side, exactly like a dog would. But there end the similarities. A dog can sniff six times a second, wafting a steady conveyor of air over its olfactory receptors. I start to hyperventilate after several consecutive sniffs, and when I pause to exhale, I lose the trail. I succeed in tracking the string, but it takes me a minute to do what Finn manages in half a second. Even if I practiced regularly, I couldn't come close; I don't have the hardware. And crucially, Horowitz adds after whipping away the string, a dog can still follow a trail once the odor source is gone. We both try, bending down to sniff. "I don't smell anything left," she says. We humans underestimate our sense of smell, but it's also clear that we simply don't live in the same olfactory world as a dog. And that world is so complicated that it's a wonder we can make sense of it at all.

MANY LIVING THINGS CAN sense light. Some can respond to sound. A select few can detect electric and magnetic fields. But every thing, perhaps without exception, can detect chemicals. Even a bacterium, which consists of just one cell, can find food and avoid danger by picking up on molecular clues from the outside world. Bacteria can also release their own chemical signals to communicate with each other, launching infections and performing other coordinated actions only when their numbers are large enough. Their signals can then be detected and exploited by bacteria-killing viruses, which have a chemical sense even though they are such simple entities that scientists disagree about whether they're even alive. Chemicals, then, are the most ancient and universal source of sensory information. They've been part of Umwelten for as long as Umwelten have existed. They're also among the hardest parts of it to understand.

Scientists who work on vision and hearing have it comparatively easy. Light and sound waves can be defined by clear and measurable properties like brightness and wavelength, or loudness and frequency. Shine wavelengths of 480 nanometers into my eyes, and I'll see blue. Sing a note with a frequency of 261 hertz (Hz), and I'll hear middle C. Such predictability simply doesn't exist in the realm of smells. The variation among possible odorants is so wide that it might as well be infinite. To classify them, scientists use subjective concepts like intensity and pleasantness, which can only be measured by asking people. Even worse, there are no good ways of predicting what a molecule smells like—or even if it smells at all—from its chemical structure.* And yet, many animals naturally grapple with the intricacy of olfaction, without any training in chemistry or neuroscience. Their noses are kings of infinite space. How do they work?

The basics became clearer after Linda Buck and Richard Axel made a pivotal discovery in 1991. In work that would earn them a Nobel Prize, the duo identified a large group of genes that produce odorant receptors—the proteins that initially recognize smelly molecules.† We

---

* Unless you actually stuck your nose over some benzaldehyde, you couldn't guess that it smells like almonds. If you saw dimethyl sulfide drawn on a page, you couldn't foresee that it carries the scent of the sea. Even similar molecules can produce immensely different smells. Heptanol, with a backbone of seven carbon atoms, smells green and leafy. Add another carbon atom to the chain and you get octanol, which smells more like citrus. Carvone exists in two forms that contain exactly the same atoms but are mirror images of each other: One smells of caraway seeds and the other of spearmint. Mixtures are even more confusing. When mixed, some pairs of odors still smell distinct, while others produce a third smell that's unlike the two parents. Meanwhile, perfumes that contain hundreds of chemicals don't smell any more complex than individual odorants, and people typically struggle to name more than three ingredients in a blend. Noam Sobel, a neurobiologist who studies olfaction, has come closer than anyone else to wrangling this complexity. While I was writing this book, he and his team developed a measure that analyzes 21 features of odorant molecules and collapses these into a single number. The closer this smell metric is for any two molecules, the more similar their odors. This isn't quite the same as predicting scent from structure, but it's the next best thing—predicting scent from similarity to other scents.

† The terminology is confusing. In sensory biology, the word *receptor* is usually used to describe a sensory cell, like a photoreceptor or a chemoreceptor. In this case, the odorant receptors are proteins on the surface of those cells. Don't blame me; I didn't make the rules.

encountered them earlier in this chapter while discussing dogs, but they underlie the sense of smell throughout the animal kingdom. The odorant receptors probably recognize their target molecules, like electric sockets accepting certain cables.* When this happens, the neurons that harbor these receptors send signals to the smell centers of the brain, and the animal perceives a scent. But the details of this process are still murky. There aren't enough receptors to account for the huge range of possible odorants, so the perception of scent must depend on the combination of olfactory neurons that are firing. If one group goes off, you delight at the scent of a rose. If another group activates, you wince at the whiff of vomit. Such a code must exist, but its nature is still mostly mysterious.

Odorant receptors can also vary from one individual to another in dramatic ways. For example, the OR7D4 gene creates a receptor that responds to androstenone, the chemical behind the stench of sweaty socks and body odor. To most people, it's repulsive. But to a lucky few who inherit a slightly different version of OR7D4, androstenone smells like *vanilla*. That's just one receptor out of hundreds, and all exist in varied forms, bestowing every individual with their own subtly personalized Umwelt. Everyone likely smells the world in a slightly different way. And if it's that hard to appreciate the olfactory Umwelt of another human, imagine how hard the task becomes for another species.

We should be skeptical of any claim that pits one animal's sense of smell against another's. I have repeatedly read that an elephant's sense of smell is five times more sensitive than a bloodhound's, but that's an utterly meaningless statement. Does that mean the elephant detects five times more chemicals? Does it sense certain chemicals at a fifth the concentration, or from five times the distance? Does it remember smells for five times as long? Such comparisons will always be flawed because smell is diverse and often unquantifiable. We need to stop asking "How good is an animal's sense of smell?" Better questions would

---

* One widely popularized theory, which says that smells are encoded in the vibrations of different molecules, has been thoroughly debunked.

be "How important is smell to that animal?" and "What does it use its sense of smell for?"

Male moths, for example, are tuned to sexual chemicals released by females. They pick up these odorants from miles away using feathery antennae, and slowly flutter over to the source. Smell is so important to them that when scientists transplanted the antennae of female sphinx moths onto males, the recipients behaved like females, seeking out the scent of egg-laying sites instead of mates. Their sense of smell is clearly amazing, as evidenced by the continued existence of moths. But they only put this amazing sense toward a few specific tasks. Moths have been described as "odor-guided drones," and that's not an exaggeration. Many males don't even have mouthparts when they reach adulthood. Freed from the need to feed, their short lives are devoted to flying, finding, and . . . mating. Their behaviors are simple enough that they can be easily diverted. By mimicking female moth odors, bolas spiders can lure male moths into fatal ambushes, while farmers can lure them into traps. Other insects, however, process smells in more sophisticated ways.

IN A LAB IN New York City, Leonora Olivos Cisneros pulls out a large Tupperware container and lifts the lid to reveal a writhing sea of dark-red dots. They're ants. Specifically, they're clonal raiders—an obscure species that's stockier than most ants and, unusually, has neither queens nor males. Every individual is female and every one can reproduce by cloning herself. About 10,000 of them are scurrying around the container. Most have formed a makeshift nest from their own bodies and are tending to their young grubs. The rest are wandering around in search of food. Olivos Cisneros feeds them on other ants, including escamoles—the larvae of a much larger species that she brings over from Mexico.

The clonal raiders are so small that it's hard to focus on any one of them. Under the microscope, they're much easier to see, not just because they've been magnified but also because Olivos Cisneros has painted them. With practiced hands, she uses insect pins to dab splotches of yellow, orange, magenta, blue, and green onto the insects' backs, giving each individual a unique color code that can be tracked by an automated camera system. The colors also make them easier to observe by eye. Every now and then, I notice one of them tapping at another with the tips of its clubby antennae. This action, delightfully known as antennating, is the ant equivalent of a sniff. It's the means through which they inspect the chemicals on each other's bodies and discern colony-mates from interlopers. These ants normally live underground and are completely blind. "There's nothing visual going on," Daniel Kronauer, who leads the lab, tells me. "In terms of their communication, everything is chemical."

The chemicals they use are pheromones—an important term that is frequently misunderstood. It refers to chemical signals that carry messages between members of *the same species*. Bombykol, which female moths use to attract males, is a pheromone; the carbon dioxide that draws mosquitoes to my body is not. Pheromones are also *standardized messages,* whose use and meaning do not vary between individuals of a given species. All female silk moths use bombykol and all males are attracted to it; by contrast, the smells that distinguish one person's scent from another's are *not* pheromones. Indeed, despite the existence of pheromone parties where singletons sniff each other's clothes, or pheromone sprays that are marketed as aphrodisiacs, it's still unclear if human pheromones even exist. Despite decades of searching, none have been identified.*

Ant pheromones are another story. There are many, and ants put them to different uses depending on their properties. Lightweight chemicals that easily rise into the air are used to summon mobs of workers that can rapidly overwhelm prey, or to raise fast-spreading alarms. Crush the head of an ant, and within seconds, nearby colony-

---

* Human pheromones likely exist, but finding them is a chore. In animals, researchers typically look for stereotyped behaviors or physiological reactions that reveal the reaction to a pheromone—a flaring of the lips, a fluttering of antennae, or a rise in testosterone. Humans are so annoyingly varied and complex that few of our actions fit the bill. Some researchers once suspected that women synchronize their menstrual cycles because of some unidentified pheromone, but such synchronicity is itself a myth. Others now think that breasts might release a pheromone that prompts babies to suckle, but again, no chemical has been isolated.

mates will sense the aerosolized pheromones and charge into battle. Medium-weight chemicals that become airborne more slowly are used to mark trails. Workers lay these down when they find food, leading other colony-mates to foraging hotspots. As more workers arrive, the trail is strengthened. As the food runs out, the trail decays. Leafcutter ants are so sensitive to their trail pheromone that a milligram is enough to lay a path around the planet three times over. Finally, the heaviest chemicals, which barely aerosolize, are found on the surface of the ants' bodies. Known as cuticular hydrocarbons, they act as identity badges. Ants use them to discern their own species from other kinds of ants, nestmates from other colonies, and queens from workers. Queens also use these substances to stop workers from breeding or to mark unruly subjects for punishment.

Pheromones hold such sway over ants that they can force the insects to behave in bizarre and detrimental ways, in disregard of other pertinent sensory cues. Red ants will look after the caterpillars of blue butterflies, which look nothing like ant grubs but *smell* exactly like them. Army ants are so committed to following their pheromone trails that if those paths should accidentally loop back onto themselves, hundreds of workers will walk in an endless "death spiral" until they die from exhaustion.* Many ants use pheromones to discern dead individuals: When the biologist E. O. Wilson daubed oleic acid onto the bodies of living ants, their sisters treated them as corpses and carried them to the colony's garbage piles. It didn't matter that the ant was alive and visibly kicking. What mattered was that it *smelled* dead.

"The ant world is a tumult, a noisy world of pheromones being passed back and forth," Wilson said. "We don't see it, of course. We don't see anything more than these little ruddy creatures scurrying around on the ground, but there's a huge amount of activity, of coordination and communication going on." That's all based on pheromones. These smelly substances allow ants to transcend the limits of

---

* In September 2020, I noted that the army ant death spiral was the perfect metaphor for the United States' response to the COVID-19 pandemic: "The ants can sense no picture bigger than what's immediately ahead. They have no coordinating force to guide them to safety. They are imprisoned by a wall of their own instincts."

individuality and act as a superorganism, producing complex and transcendent behaviors from the unknowing actions of simple individuals. They allow army ants to act as unstoppable predators, Argentine ants to create supercolonies that extend for miles, and leafcutter ants to develop their own agriculture by gardening fungi. Ant civilizations are among the most impressive on Earth, and as ant researcher Patrizia d'Ettorre once wrote, their "genius is definitely in their antennae."

Kronauer's research with the clonal raider ant shows how that genius might have evolved. Ants are essentially a group of highly specialized wasps that evolved between 140 and 168 million years ago and rapidly transitioned from a solitary existence to an extremely social one. Along the way, their repertoire of odorant receptor genes—the ones that allow them to sense smelly chemicals—ballooned in size. While fruit flies have 60 of these genes and honeybees have 140, most ants have between 300 and 400, and the clonal raider has a record-breaking 500.* Why? Here are three clues. First, a third of the clonal raiders' odorant receptors are only produced on the underside of their antennae—the parts that they pat each other with during antennation. Second, these receptors specifically detect the heavyweight pheromones that ants wear as identity badges. Third, these 180 or so receptors all arose from just one gene, which was repeatedly duplicated at roughly the time that ancestral ants went from living alone to living in colonies. Putting these clues together, Kronauer reasons that all that extra olfactory hardware might have helped ants to better recognize their nestmates. After all, they are not only looking for the presence or absence of *one* pheromone but weighing up the relative proportions of a few dozen of them. That's a challenging computation, but one that undergirds everything else that ants do. By expanding their powers of smell, they gained the means of regulating their sophisticated societies.

It becomes especially obvious how much ants rely on smell when they are disconnected from that sense. When Kronauer deprived his clonal raiders of a gene called *orco,* which odorant receptors need to

---

* A word of caution: It is dangerous to assess an animal's sensory abilities by counting its genes. Dogs have twice the number of working odorant receptor genes as humans, but that doesn't mean that their sense of smell is twice as good.

detect their target molecules, the mutant ants behaved in entirely un-ant-like ways. "Right from the beginning, there was something wrong with those ants," Olivos Cisneros tells me. "It was super-easy to spot." They wouldn't follow pheromone trails. They ignored barriers whose intense smells would ward off normal ants, like lines drawn by Sharpies. They ignored the grubs that they're normally duty-bound to care for. They ignored their colonies altogether, and went walkabout on their own for days at a time. If they accidentally found themselves within a colony, their presence was disruptive. Sometimes they'd release alarm pheromones without provocation, sending their nestmates into an unnecessary panic. "They can't tell that there are other ants there," Kronauer says. "They just can't sense them at all." It's hard not to feel sorry for them. An ant without olfaction is an ant without a colony, and an ant without a colony is barely an ant at all.*

Ants are perhaps the most dramatic example of the power of pheromones, but they're hardly the only ones. Female lobsters urinate into the faces of males to tempt them with a sex pheromone. Male mice produce a pheromone in their urine that makes females especially attracted to other components in their odor; this substance is called darcin, after *Pride and Prejudice*'s male hero. The early spider-orchid deceives male bees into carrying its pollen by mimicking their sexual pheromones. "We live, all the time, especially in nature, in great clouds of pheromones," E. O. Wilson once said. "They're coming out in spumes in millionths of a gram that can travel for maybe a kilometer." These tailored messages drive the entire animal kingdom, from the smallest of creatures to the very biggest.

IN 2005, LUCY BATES arrived in Kenya's Amboseli National Park to study its elephants. On her first day out, her experienced field assistants told her that these animals, which had been observed by scientists since the 1970s, would almost certainly realize that a fresh face had

---

* There's precedent for this. Back in 1874, the Swiss scientist Auguste Forel showed that an ant's antennae are its main organs of smell. When he removed those antennae, ants wouldn't build their nests, care for their young, or attack interlopers from other colonies.

joined the research group. Bates was skeptical. How would they know? Why would they care? But as soon as the team found one of the herds and switched off their vehicle's engine, the elephants immediately turned toward them. "One of them came up, stuck her trunk in my window, and had a good sniff," Bates tells me. "They knew someone new was inside."

Over the next few years, Bates came to realize what anyone who spends time with elephants knows: Their lives are dominated by smell. You don't need to know about an elephant's record-breaking catalog of 2,000 olfactory receptor genes, or the size of its olfactory bulb. Just watch the trunk. No other animal has a nose so mobile and conspicuous, and so no other animal is as easy to watch in the act of smelling. Whether an elephant is walking or feeding, alarmed or relaxed, its trunk is constantly in motion, swinging, coiling, twisting, scanning, sensing. Sometimes the entire 6-foot organ periscopes dramatically to inspect an object. Sometimes its movements are subtle. "You can approach a feeding elephant who's heard you coming, and without turning its head, it'll flick just the tip of its trunk toward you," says Bates.

African elephants can use their trunks to detect their favorite plants, even when obscured in lidded boxes, and even when hidden among a messy botanical buffet. They can learn unfamiliar smells: After being briefly taught to detect TNT, which is supposedly odorless to humans, three African elephants could identify the substance more skillfully than highly trained detection dogs. Two of those same elephants, Chishuru and Mussina, could sniff a human and identify the matching scent from a row of nine jars laced with the odors of different people. Asian elephants are no slouches, either. In one study, they could correctly identify which of two covered buckets contained more food through smell alone—a feat that humans can't duplicate and that (in one of Alexandra Horowitz's experiments) even dogs struggled with.* "We could tell the difference if we looked, but if we were just smelling it, there's no way," says Bates. "The level of information they can get is just so far beyond what we can comprehend."

---

* Horowitz thinks that the dogs might just not have been motivated to do it.

Elephants can also smell danger. Some time after Bates arrived in Amboseli, one of her colleagues gave a ride to a couple of Maasai men in a jeep that the team had used for decades. The next day, when the team drove out, the elephants were unexpectedly cautious around the familiar vehicle. Young Maasai men will sometimes spear elephants, and Bates reasoned that the creatures were disconcerted by the lingering scents in the jeep—some combination of the cows that the Maasai raise, the dairy products they eat, and the ochre they daub on their bodies. To test this idea, she hid various bundles of clothes in elephant country. When the animals approached washed garments or those worn by the Kamba, who pose no threat to them, they were curious but unconcerned. But every time they got wind of clothes worn by the Maasai, their reactions were unmistakable. "Once the first trunk went up, the whole group *ran* away as fast as they could, and almost always into long grass," Bates tells me. "It was incredibly stark—every group, every time."

Food and foes aside, few sources of odor are as pertinent to an elephant as other elephants. They'll regularly inspect each other with their trunks, probing away at glands, genitals, and mouths. When African elephants reunite after a prolonged separation, they go through intense greeting rituals. Human observers can see their flapping ears and hear their throaty rumbles, but for the elephants themselves, the experience must also be olfactory pandemonium. They vigorously urinate and defecate, while aromatic liquid pours forth from glands behind their eyes, filling the air around them with scents.

Few people have done more to study elephant odors than Bets Rasmussen,* a biochemist who was once crowned "the queen of elephant secretions, excretions and exhalations." If an elephant produced it, Rasmussen likely sniffed it and possibly tasted it. Those secretions, she realized, are full of pheromones, and thus full of meaning. In 1996, after 15 years of work, she isolated a chemical called Z-7-dodecen-1-yl

---

* Given that elephants live in matriarchal societies that are led by females, it's fitting that the study of elephant senses has been led by women: Bets Rasmussen for olfaction; Katy Payne, Joyce Poole, and Cynthia Moss for hearing; and Caitlin O'Connell for seismic senses. We'll meet the others in later chapters.

acetate, which females release in their urine to inform bulls that they're ready to mate. It was astonishing that just one compound could so greatly affect the sex lives of so complex an animal. It was even more astonishing that female moths attract males with the same substance. Fortunately, male moths aren't drawn to female elephants, because the attractant is just one of several compounds on their search list. Luckier still, male elephants don't try to mate with female moths, because the latter produce piddling amounts of the pheromone. Other elephants, however, shine like odorous beacons. Rasmussen eventually discovered that elephants can tell, through smell, when females are at different parts of their estrus cycles, or when bulls are in the hyperaggressive sexual state called musth. They can also identify individuals. As they walk the time-worn trails that connect their home ranges, they leave dung and urine behind—not waste, but personal stories to be read by the trunks of others around them.

In 2007, Lucy Bates found a clever way of testing this idea. She followed family groups of elephants and waited for one to urinate. Once the herd had left, she drove over, scooped up the urine-soaked soil with a trowel, and placed it in an ice cream tub. She then drove around the savannah until she found either the same herd of elephants or a different one. Cutting them off, she emptied the container of soil onto the path ahead of them, sped off to a distant vantage point, and waited. "It was not the most pleasant experiment," she tells me. "Often, you'd think you know where they were going and put the sample out, and they'd change direction. That was quite soul-destroying." When she got it right, the elephants would always inspect the urine as they approached. If it came from a different family group, they quickly ignored it. If it came from a family member who wasn't part of the current unit, they showed more interest. But if it came from an elephant who was part of the same group *and walking behind them,* they were especially curious. They knew exactly who had left the urine, and since that individual couldn't possibly have teleported ahead, they seemed confused and carefully investigated the displaced scent. Elephants move in large family groups, and it seems they know not only who's around but where those individuals are. Scent cements that awareness. "The amount of information that they must be picking up all the time as they're walking along, from all the different smells they're taking in . . . I think it just must be overwhelming," says Bates.

The exact nature of that information is hard to discern. Smells aren't easily captured, so while scientists can photograph an animal's displays and record its calls, those who care about olfaction have to do things like scoop up urine-soaked soil. Smells aren't easily reproduced, either: You can't play back an odor through a speaker or a screen, so researchers have to do things like drive piss-soaked soil in front of elephant herds. And that's if they think about olfaction at all. In many cases, elephant researchers have tested the brains of these animals through experiments that are implicitly visual and involve objects like mirrors. How much have we missed about an elephant's mind because we've ignored its primary senses?

When they walk their favorite routes and encounter the smelly deposits of other elephants, what are they getting besides identity? Do they know the emotional states of those previous passers-by? Can they sense stress or diagnose illnesses? What of their wider environment? Elephants that have returned to postwar Angola seem to skirt around the millions of landmines that still dot the land—unsurprising, perhaps, given how quickly they can be trained to detect TNT. They've been known to dig wells in times of drought, and George Wittemyer, who has also worked in Amboseli, is sure that they're using the smell of buried water to do so. He also thinks that they can detect approaching rain from the smells it unleashes as it splashes onto faraway soils. "That smell is exhilarating," he tells me. "It makes me feel excited and alive, and you'll also see elephants rising up to it."

Rasmussen once speculated that elephants might guide their long migrations using "chemical memories of landscapes, terrain, pathways, mineral and salt sources, waterholes, the scenting of rain or flooding rivers, and tree odors signifying seasons." No one has tested these claims, but they make sense. After all, dogs, humans, and ants can all track trails of scent. Salmon can return to the very streams in which they were born by homing in on the distinctive scents of those natal waters.* Whip spiders use the smell sensors on the tips of their extremely long, thread-like front legs to find their way back to their shelters amid the clutter of a rainforest. Polar bears might be able to navigate across thousands of miles of indistinct ice because glands in their paws leave scent behind with every step. These examples are so common that some scientists believe the main purpose of animal olfaction isn't to detect chemicals but to use them in navigating through the world. With the right noses, landscapes can be mapped as odorscapes, and fragrant landmarks can show the way to food and shelter. Ironically, the best evidence for such feats comes from animals that, until recently, were thought to be unable to smell.

JOHN JAMES AUDUBON, the avid naturalist and artist, was best known for painting North America's birds, and compiling those pieces into a seminal ornithological tome. But he was also responsible for seeding a centuries-long falsehood about birds through some truly abysmal experiments involving vultures.

Since Aristotle, scholars believed that vultures had a keen sense of smell. Audubon thought differently. When he left a putrefying pig carcass in the open, no vultures came to eat. By contrast, when he put out a deerskin stuffed with straw, a turkey vulture swooped in and pecked away. These birds, he claimed in 1826, find their food with sight, not smell. His supporters bolstered that claim with equally dodgy evidence. One noted that vultures would attack a painting of an eviscerated sheep, and that captive vultures refused to eat after being blinded. Another showed that a turkey—not a turkey vulture, mind you; an actual turkey—would still eat food that was tainted with sulfuric acid and potassium cyanide, a strong-smelling concoction that proved violently fatal. These bizarre studies struck a chord. Never mind that vultures prefer fresh carcasses and ignore overly stinky meat like the kind Audubon used. Forget that Audubon confused black vultures (which

* Arthur Hasler confirmed this ability in the 1950s after having his own olfactory epiphany. While hiking near a waterfall, the familiar smells brought back long-buried childhood memories, and he wondered if migrating salmon experience something similar.

are less reliant on smell) with turkey vultures, or that oil paints at the time gave off certain chemicals also found in decaying flesh. Disregard the many reasons a mutilated animal might not feel very peckish. The idea that turkey vultures—and by dubious extension, *all birds*—can't smell became textbook wisdom. Evidence to the contrary was ignored for decades, and the study of avian olfaction lapsed into neglect.*

Betsy Bang revitalized it. An amateur ornithologist and medical illustrator, she dissected the nasal passages of bird after bird and sketched what she saw. And what she saw—large cavities filled with convoluted scrolls of thin bone, much like what lurks within a dog's snout—convinced her that birds must be able to smell. Why else would they have all that hardware? Concerned that the textbooks were spouting misinformation, Bang spent the 1960s carefully examining the brains of more than a hundred species and measuring their olfactory bulbs. She showed that these smell centers were especially large in turkey vultures, the kiwis of New Zealand, and the tubenoses—a group of seabirds that includes albatrosses, petrels, shearwaters, and fulmars. Tubenoses are named for the obvious nostrils on their beaks, which were originally thought to be channels for expelling salt. Bang's work suggested another purpose: The tubes draw air into the nose, allowing the birds to catch the scent of food while soaring over the ocean. For them, "olfaction is of primary importance," Bang wrote.† ("She didn't mind taking on a fight, even if it meant taking on Audubon," her son Axel later said.)

Elsewhere in California, Bernice Wenzel had come to the same conclusion. A physiology professor (and one of the few women in the

---

* Ornithologist Kenneth Stager ran much better versions of Audubon's studies and showed that turkey vultures do indeed home in on the scent of hidden carcasses. He also learned that an oil company had begun tracking leaks in its pipelines by adding ethyl mercaptan—a gas that smells of farts and decay—and scanning the skies for circling vultures. Intrigued, Stager fashioned his own mercaptan dispenser and deployed it at various sites in California. Whenever he did, vultures arrived. Audubon was wrong: Turkey vultures not only can smell, but can smell well enough to detect the faintest plumes of odorants from miles overhead.

† Birds evolved from the same group of small, predatory dinosaurs that included celebrities like *Velociraptor*. By scanning the skulls of these animals, paleontologist Darla Zelenitsky showed that they had large olfactory bulbs for their size—as did larger cousins like *Tyrannosaurus*. These dinosaurs likely used their sense of smell to hunt, and birds are the modern inheritors of that ancient Umwelt.

United States to hold such a position in the 1950s), Wenzel showed that when homing pigeons catch a whiff of scented air, their hearts beat faster and the neurons in their olfactory bulbs buzz excitedly. She repeated that test with other birds—turkey vultures, quails, penguins, ravens, ducks—and all reacted similarly. She proved what Bang deduced: Birds can smell. Both Bang and Wenzel, who have since passed away, have been described as "mavericks of their generation" who pushed against incorrect dogma and allowed others to explore a sensory world that was deemed nonexistent. And because of the examples they set and the mentorship they offered, many of the scientists who followed in their footsteps were also women.

One, Gabrielle Nevitt, was in the audience when Wenzel discussed her seabird studies in one of her final pre-retirement talks. Inspired, Nevitt began a career-long quest to find out how tubenoses make use of smell. Beginning in 1991, she would get onto any Antarctic voyage that she could, while trying "to figure out how to test birds from the deck of an icebreaker without getting killed," she tells me. She'd soak tampons in fish oils and fly them from kites. She'd release slicks of pungent oils from the sterns of ships. And every time, tubenoses arrived quickly. Nevitt suspected that the birds were drawn to a specific chemical within the pungent glop, but she didn't know what it might be, or how the birds found it across featureless water. She only learned the answer on a later Antarctic voyage, and in unexpected circumstances.

During the trip, a fierce storm rocked Nevitt's ship, throwing her across her room and slamming her into a tool chest. She tore her kidney and was confined to her bunk, even after her ship had docked and a fresh crew had come on board. Still recuperating, Nevitt chatted with the new chief scientist—an atmospheric chemist named Tim Bates, who had come to study a gas called dimethyl sulfide, or DMS. In the oceans, plankton release DMS when they're eaten by krill—shrimp-like animals that are, in turn, eaten by whales, fish, and seabirds. DMS doesn't dissolve easily in water, and eventually makes its way into the air. If it rises high enough, it seeds clouds. If it enters the nose of a sailor, it evokes an odor that Nevitt describes as "a lot like oysters" or "kind of seaweed-y." It's the scent of the sea.

In particular, DMS is the scent of *bountiful* seas, where huge blooms of plankton feed equally huge swarms of krill. As Nevitt talked to Bates, it dawned on her that DMS was exactly the chemical she had envisioned—an olfactory dinner bell that alerted seabirds when waters were teeming with prey. Bates cemented this impression by giving Nevitt a map that showed DMS levels across parts of Antarctica. In the varying levels of the chemical, Nevitt saw a seascape of odorous mountains and unscented valleys. She realized that the ocean wasn't as featureless as she had once imagined; rather, it had a secret topography that was invisible to the eye but evident to the nose. She began to perceive the sea the way a seabird might.

Once back on her feet, Nevitt carried out a string of studies that confirmed the DMS hypothesis. She found that tubenoses will flock to slicks of the chemical. She calculated that they can detect it at the kind of low, feeble traces that might realistically drift on the wind. She showed that some tubenoses are drawn to DMS before they can even fly.* Many species nest in deep burrows, and their chicks, which resemble grapefruit-sized balls of lint, hatch into a world of darkness. Their early Umwelt is bereft of light but awash in odor, wafting in from the burrow entrance or carried in on the beaks and feathers of their parents. These hatchlings have no knowledge of the ocean, but they know to head toward DMS. And even after they emerge into the light, trading their claustrophobic nurseries for the immensity of the sky, smells remain their north star. They soar for thousands of miles, searching for diffuse plumes of scent that might betray the presence of krill beneath the surface.†

---

* Tubenoses aren't the only animals that track DMS. Penguins, reef fish, and sea turtles can all detect the chemical, and are all drawn to it.

† Tracking such plumes is harder than following a straight line of sight. A bird's best option is to fly across the wind to maximize its chances of blundering into a stray odor molecule, and then follow it upwind on a zigzagging path. That's how male moths find the pheromones released by their females, and it's how albatrosses find the odors released by their prey. Henri Weimerskirch fitted wandering albatrosses—the birds with the world's greatest wingspan—with GPS loggers to track their whereabouts and stomach temperature recorders to log when they ate. By analyzing that data, Gabrielle Nevitt shows that the birds use zigzagging, smell-tracking flights to capture at least half of their food.

But smells are more than dinner bells. In the ocean, they're also signposts. Geological features, like submerged mountains or slopes in the seafloor, affect the levels of nutrients in the water, which in turn influence concentrations of plankton, krill, and DMS. The smellscapes that seabirds track are intimately tied to actual landscapes, and so are surprisingly predictable. Over time, Nevitt suspects, seabirds build up a map of these features, using their noses to learn the locations of the richest feeding spots and their home nests.

This is a hard idea to test, but Anna Gagliardo found compelling evidence for it. She transported a few Cory's shearwaters—a kind of tubenose—to locations 500 miles from their nesting colonies and temporarily shut down their sense of smell with a nasal wash. When released, these birds struggled to travel home, taking weeks or months to do what normal shearwaters did in mere days. Without smell, they lost their way. Without smell, the ocean was stripped of landmarks. As the writer Adam Nicolson described in *The Seabird's Cry,* "What may be featureless to us, a waste of undifferentiated ocean, is for them rich with distinction and variety, a fissured and wrinkled landscape, dense in patches, thin in others, a rolling olfactory prairie of the desired and the desirable, mottled and unreliable, speckled with life, streaky with pleasures and dangers, marbled and flecked, its riches often hidden and always mobile, but filled with places that are pregnant with life and possibility."

SHEARWATERS, DOGS, ELEPHANTS, and ants all smell with different organs, but they all smell in stereo, using a pair of nostrils or antennae. By comparing the odorants that land on each side, they can track the source of a scent. Even humans can do this: The string-tracking task that Alexandra Horowitz asked me to try is much harder if one nostril is blocked. Directionality comes more easily to a paired detector, which also explains the distinctive shape of one of nature's least likely but most effective smell organs—the forked tongue of snakes.

Snake tongues come in shades of lipstick red, electric blue, and inky black. Outstretched and splayed, they can be longer and wider than their owners' heads. Kurt Schwenk has been fascinated by them for decades, and he often finds that he's alone in that. In the second year of his PhD, he told a fellow student what he was working on, eager to revel in the joys of scientific pursuits with a like-minded soul. The student (who is now a famous ecologist) burst out laughing. "That would have been enough to hurt my feelings, but this was a guy who studied the mites that hang out in the nostrils of hummingbirds," Schwenk tells me, still slightly outraged. "Someone who studied hummingbird nostril mites thought that what *I* did was funny! For some reason, people find tongues funny."

Perhaps there's something unseemly about studying organs that are linked to carnal delights like sex and food. Perhaps it's weird to seriously investigate things that we protrude in jest or defiance. Or perhaps it's that the forked tongue has become a symbol of malevolence and duplicity. Whatever the case, serious scholars have put forward some very strange hypotheses for how snakes use their tongues, or for why those tongues are forked. Some have described them as venomous stingers, or fly-catching forceps, or tactile organs akin to hands, or even nostril-cleaning tools. Aristotle suggested that the fork doubled the pleasure that a snake gets from its food—but the snake's tongue has no taste buds and conveys no sensory information on its own. Instead, as scientists finally discovered in the 1920s, it's a chemical collector. When it darts into the world, its tips snag odor molecules that lie on the ground or drift through the air. When it retracts, saliva sweeps the chemical bounty into a pair of chambers—the vomeronasal organ— that connect to the brain's smell centers.* With the aid of its tongue, a snake *smells* the world. Each flick is the equivalent of a sniff. Indeed, the very first thing that a hatchling serpent does upon breaking out of its egg is to flick its tongue. "That tells you something about the primacy of the sense," Schwenk says.

---

* For the longest time, researchers have claimed that the tongue delivers chemicals to the snake's vomeronasal organ, also known as Jacobson's organ, by threading its tips through two holes in the roof of a snake's mouth. This is a myth. X-ray movies show that they do nothing of the sort, and the tongue simply nestles into the roof of the mouth. But to Schwenk's eternal annoyance, the misconception still persists and abounds in textbooks.

Using its tongue, a male garter snake can track a slithering female by following the trail of pheromones she leaves behind. By comparing what she deposited on different sides of objects she pushed against, he can work out her direction. Once he finds her, he can gauge her size and health, possibly with just one or two flicks. He can do this all in the dark. A male can even be fooled into vigorously mating with a paper towel that has been imbued with a female's scent. But all of these feats could be just as easily accomplished with a paddle-shaped, human-esque tongue. So why do snakes have forked ones? Schwenk reasoned that the fork allows snakes to smell in stereo, by comparing chemical traces at two points in space. If both tips detect trail pheromones, the snake stays on course. If the right tip gets a hit but the left one doesn't, the snake veers right. If both come up empty, it swings its head from side to side until it regains the trail. The fork allows the snake to pre-cisely define the edges of the path.

As a timber rattlesnake slithers over the forest floor, its tongue turns the world into both map and menu, revealing the crisscrossing tracks of scurrying rodents and discerning the scents of different species. Amid the tangled trails, it can pick out those of its favorite prey* and find sites where those tracks are common and fresh. It hides nearby, coiled in ambush. When a rodent runs past, the snake explodes out-ward four times faster than a human can blink. It stabs the rodent with its fangs and injects venom. The toxins usually take a while to work, and since rodents have sharp teeth, the snake avoids injury by releasing its prey and letting it run off. After several minutes, it starts flicking its tongue to track down the now-dead victim. The venom helps. Aside from lethal toxins, rattlesnake venom also includes compounds called disintegrins, which aren't toxic but react with a rodent's tissues to re-lease odorants. The snakes can use these aromas to distinguish envenom-ated rodents from healthy ones and to tell rodents envenomated by *their own species* from those bitten by other kinds of rattlesnakes. They

---

* Rulon Clark, whom we'll meet in a later chapter, showed that even inexperienced lab-born rattlesnakes can distinguish the smells of favored prey like chipmunks and white-footed mice from those of unfamiliar lab rats. He also found, rather sinisterly, that rosy boas are specifically drawn to the odors of female mice who have litters of young.

can even track the specific individual that they attacked because they instantly learn the victim's scent at the moment of a bite. "There are presumably odors of multiple mice around, but they know which trail to follow," Schwenk says.

Snakes can also catch trails of scent on the breeze. Chuck Smith, one of Schwenk's former students, demonstrated this by implanting copperheads with radio transmitters and tracking their movements. Twice, he released a female snake into the wild and watched as she stayed in exactly the same place. She couldn't have left a scent trail, but she still managed to attract males who were randomly wandering hundreds of yards away, then suddenly crawled directly to her in a straight line.

Schwenk guessed that their secret lies in the way they flick. Lizards, the group from which snakes evolved, also smell with their tongues, which are also sometimes forked. But when lizards stick their tongues out, they usually flick once. The tips extend, scrape the ground, and retract. Snakes, without exception, flick repeatedly and rapidly, often never touching the ground. The tongues bend in the middle as if moving on a hinge, and the tips carve out a wide circular arc, 10 to 20 times a second. Bill Ryerson, another of Schwenk's students, analyzed those movements by getting snakes to tongue-flick into clouds of cornstarch. He illuminated the clouds with laser light, and filmed the swirling particles with high-speed cameras. When Schwenk saw the footage, "my brain nearly exploded," he says.

It turns out that the tongue's tips splay out at the ends of each flick and get closer at the midpoint. This motion creates two donut-shaped rings of continuously moving air that draw in odorants from the left and right sides of the snake. It's as if the snake temporarily conjures up two large fans that suck in odors from either side, concentrating diffuse odor molecules onto the tips of its tongue. And since the odors come in from left and right, the fork can still provide a sense of direction, even when flicking in air.

This style of smelling is unusual in two ways. First, it involves a tongue, which is traditionally an organ of taste—a sense that snakes barely use, for reasons I'll get to. Second, it involves an organ that, in most other animals, is either nonexistent or of secondary importance. Many backboned animals have *two* distinct systems for detecting odors. The main one includes all the structures, receptors, and neurons that I described in the head of a dog at the start of this chapter. The vomeronasal organ is its sidekick; it has its own kinds of odor-sensing cells, its own sensory neurons, and its own connections to the brain. It's usually found inside the nasal cavity, just above the roof of the mouth. Don't bother trying to feel around for yours, though. For some reason, humans lost our vomeronasal organ during our evolution, as did other apes, along with whales, birds, crocodiles, and some bats.

Most other mammals, reptiles, and amphibians have kept theirs. When one elephant touches another with its trunk and brings the pheromone-coated tip into its mouth, those molecules head to the vomeronasal. When horses or cats curl back their upper lip to expose their teeth, they're cutting off their nostrils and sending inhaled odorants to the vomeronasal. And when a snake retracts its tongue and squeezes the tips between the floor and roof of its mouth, the collected molecules are squirted to the vomeronasal. In snakes, this sidekick is the star. Without it, garter snakes stop following trails and stop eating, while rattlesnakes botch half their strikes and fail to capture what they hit. These snakes can still inhale odorants through their nostrils, but their "main" olfactory system can't seem to do much with that information. It has been relegated to a passive role, informing the brain if there's something interesting around to tongue-flick at.

Snakes are unusual not just because their vomeronasal organ is so important but also because we actually understand what it does. In other animals, the organ is a mystery, albeit one that seems to attract confident claims.* For the moment, no one really knows why some species have two separate systems for smelling. Nor is it entirely clear

---

* It's often mythologized as a specialized pheromone detector, but that can't be true, since it also responds to other odorants, while the main smell system also picks up pheromones. It might detect molecules too heavy to float through the airways of the main olfactory system, but this idea hasn't been adequately tested. It might control instinctive responses to smells while the main system governs responses that animals learn through experience; this idea hasn't been thoroughly tested, either.

why most animals have another distinct chemical sense. I'm talking, of course, about taste.

EVERY APRIL, THE ASSOCIATION for Chemoreception Sciences holds its annual meeting in Florida, and, per tradition, scientists who study smell square off against those who study taste in a heated softball game. "Smell usually wins," smell scientist Leslie Vosshall tells me, "because the field is vastly larger. It's like four or five to one." Like smell, taste—or gustation, in the fancy scientific parlance—is a means of detecting chemicals in the environment. But beyond that, the two senses are distinct. Put your nose next to vanilla oil, and you'll inhale a pleasing odor; drop that same oil on your tongue, and you'll likely flinch in disgust.

The difference between smell and taste is surprisingly complicated. You might reasonably say that animals smell with noses and taste with tongues, but snakes use their tongues to collect odors, and other animals (which we'll meet shortly) taste with unusual body parts. You could also argue (and many scientists do) that we smell molecules that drift through the air, but taste those that stay in liquid or solid form. Smell works at a distance; taste works through contact. That's a better distinction, but it has several problems. First, the receptors that are responsible for recognizing smells are always covered in a thin layer of liquid, so odorant molecules must first dissolve to be detected. So smell—like taste—always involves a liquid step and always involves close contact even if those smells have traveled from afar. Second, as we've seen, ants and other insects can smell by contact, using their antennae to pick up pheromones that are too heavy to go airborne. Third, fish can smell even though everything they're smelling is dissolved in water. For creatures like these that are constantly immersed in liquid, the distinction between taste and smell can be so confusing that one neuroscientist just told me, "I avoid thinking about it."

But John Caprio, a physiologist who studies catfish, says the difference between smell and taste couldn't be clearer. Taste is reflexive and innate, while smell is not.* From birth, we recoil from bitter substances, and while we can learn to override those responses and appreciate beer, coffee, or dark chocolate, the fact remains that there's something instinctive to override. Odors, by contrast, "don't carry meaning until you associate them with experiences," Caprio says. Human infants aren't disgusted by the smell of sweat or poop until they get older. Adults vary so much in their olfactory likes and dislikes that when the U.S. Army tried to develop a stink bomb for crowd control purposes, they couldn't find a smell that was universally disgusting to all cultures. Even animal pheromones, which are traditionally thought to trigger hardwired responses, are surprisingly flexible in their effects, which can be sculpted through experience.

Taste, then, is the simpler sense. As we've seen, smell covers a practically infinite selection of molecules with an indescribably vast range of characteristics, which the nervous system represents through a combinatorial code so fiendish that scientists have barely begun to crack it. Taste, by contrast, boils down to just five basic qualities in humans—salt, sweet, bitter, sour, and umami (savory)—and perhaps a few more in other animals, which are detected through a small number of receptors. And while smell can be put to complex uses—navigating the open oceans, finding prey, and coordinating herds or colonies—taste is almost always used to make binary decisions about food. Yes or no? Good or bad? Consume or spit?

It's ironic that we associate taste with connoisseurship, subtlety, and fine discrimination when it is among the coarsest of senses. Even our ability to taste bitter, which warns us of hundreds of potentially toxic compounds, isn't built to *distinguish* between them. There's only one sensation of bitter because you don't need to know which bitter thing you're tasting—you just need to know to stop tasting it. Taste is mostly a final check before consumption: Should I eat this? That's why snakes barely bother with taste. With their flickering tongues, they can make

---

* The two senses use different receptors and different neurons, which connect to different parts of the brain. In vertebrates, the taste system is mostly wired to the hindbrain, which controls basic vital functions. The smell system is hooked up to the forebrain, which controls more advanced abilities like learning.

decisions about whether something is worth eating through *smell* well before their mouths make contact.* It's almost unheard of for a snake to strike a prey animal and then spit it out. (We tend to wrongly equate taste with flavor, when the latter is more dominated by smell. That's why food seems bland when you have a cold: Its taste is the same, but the flavor dims because you can't smell it.)

Reptiles, birds, and mammals taste with their tongues. Other animals aren't so restricted. If you're very small, food isn't just something you put in your mouth, but something you can walk upon. As such, most insects can taste with their feet and legs. Bees can detect the sweetness of nectar just by standing on a flower. Flies can taste the apple you're about to eat by landing on it. Parasitic wasps can use taste sensors on the tips of their stings to carefully implant their eggs in the bodies of other insects. One species can even taste the difference between hosts that have already been parasitized by other wasps and those that are currently vacant.†

If a mosquito lands on a human arm, "it's a delight of the senses," says Leslie Vosshall. "Human skin has a taste to it, which gives them more confirmation that they made it to the right place." But if that arm is covered with bitter-tasting DEET, the receptors on their feet force them to take off before they get a chance to bite. Vosshall has videos in which a mosquito lands on a gloved hand and walks over to a small patch of exposed but DEET-covered skin. Its leg touches the skin, and immediately withdraws. It circles, tries again, and retreats again. "It's poignant," she tells me, in a strange display of sympathy for a mosquito. "It's also really psychedelic. We have no idea what it'd be like to taste with our fingers." Insects can taste with other body parts, too, which expands the uses to which they can put this typically lim-

---

* Schwenk thinks that's because snakes eat infrequently, but in bulk. They'll often tackle much larger prey, and then remodel their innards to digest their meals. When a python swallows a pig or deer, its guts and liver double in size and its heart swells by 40 percent over just a few days. For them, every meal soaks up a lot of energy, and they need to know as early as possible whether to pay that cost.

† The stinger of a parasitic wasp is like a Swiss army knife. Aside from taste sensors, it can also carry smell sensors, touch sensors, and bits of metal. It's a drill, a nose, a tongue, and a hand.

ited sense. Some can find good sites for laying their eggs using taste receptors on their egg-laying tubes. Some have taste receptors on their wings, which might alert them to traces of food as they fly. Flies will start grooming themselves if they taste the presence of bacteria on their wings. Even decapitated flies will do this.

The most extensive sense of taste in nature surely belongs to catfish. These fish are swimming tongues. They have taste buds spread all over their scale-free bodies, from the tips of their whisker-like barbels to their tails. There's hardly a place you can touch a catfish without brushing thousands of taste buds. If you lick one of them, you'll both simultaneously taste each other.* "If I were a catfish, I'd love to jump into a vat of chocolate," John Caprio tells me. "You could taste it with your butt." With their body-wide buds, catfish have turned taste into an omnidirectional sense—albeit one that's still devoted to evaluating food. They eat meat, and if you put a piece anywhere on their skin (or add meat juices to the water around them), they'll turn and snap at the right place. They're exquisitely sensitive to amino acids—the building blocks of proteins and flesh.† They aren't great at detecting sugars, though: Unfortunately for Caprio, his chocolate fantasy would be underwhelming.

This inability to sense sugar and other classic tastes is surprisingly common, and varies according to an animal's diet. Cats, spotted hyenas, and many other mammals that eat meat and nothing else similarly lack a sweet tooth. Vampire bats, which drink only blood, have also lost their taste for sweetness, and for umami. Pandas have no need to sense umami either, since they only eat bamboo, but they gained an

---

* Some catfish have venomous spines and (as we'll discover in a later chapter) others can create electricity, so animal welfare issues aside, I would highly recommend not licking one, except as part of a thought experiment.

† Amino acids come in two forms that are mirror images of each other, called L and D. Nature relies primarily on the L forms, and the D forms are incredibly rare in animals. So in the mid-1990s, when Caprio tested the marine hardhead catfish, he was shocked to learn that almost half its taste buds react to D-amino acids. "I thought that's got to be a mistake," he says. "Where are there D-amino acids in the environment that would be important to catfish?" He eventually learned that several marine worms and clams can flip L-amino acids into their mirrored D opposites. Scientists only discovered that marine animals make D-amino acids in the 1970s. "The catfish knew it hundreds of millions of years ago," Caprio says.

expanded set of bitter-sensing genes to warn them of the myriad possible toxins in their mouthfuls.* Other leaf-eating specialists, like koalas, have also gained more bitter detectors, while mammals that swallow their prey whole, including sea lions and dolphins, have lost most of theirs. Repeatedly and predictably, the gustatory Umwelten of animals have expanded and contracted to make sense of the foods they most often encounter. And sometimes those changes altered their destinies.

Like cats and other modern carnivores, small predatory dinosaurs probably lost the ability to taste sugar. They passed their restricted palate on to their descendants, the birds, many of which still have no sense for sweetness. Songbirds—the vocal and hugely successful group that includes robins, jays, cardinals, tits, sparrows, finches, and starlings—are an exception. In 2014, evolutionary biologist Maude Baldwin showed that some of the earliest songbirds regained their sweet tooth by tweaking a taste receptor that normally senses umami into one that also senses sugar. This change occurred in Australia, a land whose plants produce so much sugar that its flowers overflow with nectar and its eucalyptus trees exude a syrupy substance from their bark. Perhaps these abundant sources of energy allowed the newly sweet-toothed songbirds to thrive in Australia, to endure marathon migrations to other continents, to find nectar-rich flowers wherever they arrived, and to diversify into a massive dynasty that now includes half the world's bird species. This story is unproven but nonetheless beguiling. It's possible that if a random Australian bird hadn't expanded its Umwelt tens of millions of years ago, none of us would be waking up to the melodic sounds of birdsong today.†

YOU CAN SPLIT THE SENSES into different groups depending on the stimuli that they detect. Smell, its vomeronasal variant, and taste are

---

* Remember, though, that taste is more about coarse detection than fine discrimination: Compared to a dog, a panda might recognize more things as being bitter, but it likely experiences those things in the same consistent way.

† Baldwin also showed that hummingbirds repurposed their umami receptor into a sugar one. They changed the same gene as the songbirds, but independently, and in an almost com-

chemical senses, which detect the presence of molecules. They are ancient, universal, and seem to sit apart from the others, which is partly why I chose them as the first stop on our journey. But they aren't entirely distinct. On closer inspection, they share common ground with at least one other sense, in an unexpected way.

At the start of this chapter, we saw that dogs and other animals detect smells using proteins called odorant receptors. These are part of a much larger group of proteins called G-protein-coupled receptors, or GPCRs. Ignore the convoluted name; it doesn't matter. What matters is that they are chemical sensors. They sit on the surface of cells, grabbing specific molecules that float past. Through their actions, cells can detect and react to the substances around them. This process is temporary: After the GPCRs are done, they either release or destroy the molecules that they've grabbed. But one group of them bucks this trend: opsins. They are special because they keep hold of their target molecules, and because those molecules absorb light. This is the entire basis of vision. This is how all animals see—using light-sensitive proteins that are actually modified chemical sensors.

In a way, we see by smelling light.

---

pletely different way. She tells me that in some species, the altered receptor can still detect umami, which means "they may not be able to distinguish between sweet and savory." Imagine being unable to tell the difference between soy sauce and apple juice.

2.

Endless Ways of Seeing

Light

I AM STARING AT A JUMPING SPIDER, AND EVEN THOUGH ITS body is pointing away from me, it is staring back. Four pairs of eyes encircle its turret-like head, two pointing forward and two pointing sideways and backward. The spider has close to wraparound vision, and its only blind spot is immediately behind it. When I waggle my finger in its five o'clock, it sees my vibrating digit and turns around. As I move the finger, the spider follows. Jumping spiders "are the only spiders that will turn and look at you routinely," says Elizabeth Jakob, whose lab in Amherst, Massachusetts, I am currently visiting. "A lot of spiders spend a lot of time just sitting motionless on a web and waiting for something to happen. But these are active."

Humans are such a visual species that those of us with sight instinctively equate active eyes with an active intellect. In their flitting, darting movements, we see another curious mind investigating the world. In the case of jumping spiders, this is not unwarranted anthropomorphism. Despite their poppy-seed-sized brains, they really are surprisingly smart.* The *Portia* species are famed for planning out strategic

---

* I ask Jakob how much of a jumping spider's above-average intelligence (for a spider) is baked into its senses. Spiders that mostly sense vibrations along their webs don't have a huge amount of information to interpret, she says. "For the really visual spiders, the complexity of information they have to deal with is so much higher," she says. "I can't help but think it's valuable for them to be able to interpret it, and that seems like a good opening for evolution to

routes when stalking prey, or flexibly switching between sophisticated hunting tactics. The bold jumping spiders (*Phidippus audax*) that Jakob studies are less ingenious, but she still houses them in the company of stimulating objects—the kind of environmental enrichment that zookeepers might provide for captive mammals. Some have brightly colored sticks in their terraria. One individual, I note, has a red Lego brick. We joke about what it might build when our backs are turned.

Barely bigger than my smallest fingernail, the bold jumping spider is mostly black, except for white fuzz on its knees and vibrant turquoise splotches on the appendages that hold its fangs. It is unexpectedly cute. Its stocky body, short limbs, large head, and wide eyes are all rather childlike, and stir the same deep psychological bias that makes babies and puppies adorable. But its proportions didn't evolve to engender empathy. The short limbs power great leaps: Unlike other spiders that sit in ambush, jumping spiders stalk and pounce upon their prey. And unlike other spiders that mostly sense the world through vibrations and touch, jumping spiders rely on vision. That's why the eight eyes occupy up to half the volume of their large heads. They are the spiders whose Umwelten are closest to ours. In that similarity, I find affinity. I watch the spider, and it watches me back, two starkly different species connected by our dominant sense.

The late British neurobiologist Mike Land, described to me by one of his colleagues as "the god of eyes," pioneered the study of jumping spider vision. In 1968, he developed an ophthalmoscope for spiders, which he could use to observe the creatures' retinas as they, in turn, gazed at images. Jakob and her colleagues have refined Land's design; during my visit, they've placed a jumping spider in their device, which is currently trained upon the creature's central eyes. These point straight ahead and are the largest of the four pairs. They are also the sharpest. Despite being just a few millimeters long, they can see as clearly as the eyes of pigeons, elephants, or small dogs. Each eye is a

---

push them toward higher and higher cognitive skills. But I don't know. We have to factor in our own human bias toward being visual."

long tube, with a lens at the front and a retina at the back.* The lens is fixed in place, but the spider can look around by swiveling the rest of the tube inside its head. (Imagine gripping a flashlight by its head, and then aiming its beam by moving the tube.)† The female spider in the eye tracker is doing exactly that. Her body is still. Her eyes look still, too. But on the monitor, we can see that her retinas are moving. "She's really looking around," Jakob says.

For reasons that no one fully understands, the retinas of her central eyes are shaped like boomerangs. At first, on Jakob's screen, they seem separate (> <). But when she shows the spider a black square, the two retinas converge upon it, forming crosshairs (><). As the square moves, the retinas follow. After a while, though, the spider loses interest, and the retinas diverge. Jakob replaces the square with the silhouette of a cricket, and the retinas converge again. This time, they dance over the image, flitting between the antennae, body, and legs with the same jerky hops that our eyes make when taking in a scene. The retinas also rotate together, twisting clockwise and anticlockwise, perhaps because the spider is searching for specific angles that might help it identify what it's looking at. Mike Land once wrote that it is "an exhilarating but very weird experience to look into the moving eyes of another sentient creature, particularly one so far removed in its evolution from oneself." I couldn't agree more. At least 730 million years of evolution separate humans from jumping spiders, and it is hard to interpret the behavior of such a different creature. But on Jakob's monitor, I can watch a spider paying attention and losing interest. I can observe it observing. By watching its gaze, I can get as close as possible to glimpsing its mind. And, despite many similarities, I can see just how different its vision is from mine.

---

* Each central eye actually has two lenses, one at the top and one at the bottom. The top lens collects and focuses light, while the bottom one spreads it out. This arrangement enlarges images before they hit the spider's retina, which is why these tiny animals can see as sharply as small dogs. The telescopes that Galileo started using in 1609 work in the same way, using tubes with lenses at both ends to peer at distant objects. Unbeknownst to him, he was unwittingly plagiarizing a structure that jumping spiders had evolved millions of years prior, and which, on clear nights, they can use to see the moon.

† Baby jumping spiders are transparent. With good lighting, you can see their eye tubes moving about inside their heads.

For a start, it has more eyes. The central pair may be sharp and mobile, but their field of view is very narrow. If they were all the spider had, its vision would be like two flashlights sweeping around a dark room. The secondary eyes on either side of the central pair compensate for this shortcoming with a much broader field of view. And though they are themselves immobile, they are highly sensitive to motion. If a fly buzzes in front of the spider, the secondary eyes spot it and tell the central eyes where to look. And here's the truly bizarre part: If the secondary eyes are covered, the spider cannot track moving objects.

I find this almost impossible to imagine. As I write these words, I am focusing the sharpest parts of my eyes on the letters appearing on my screen. Meanwhile, in my peripheral vision, I can see the black shape of Typo, my corgi puppy, as he prowls around my living room in search of trouble. These tasks—sharp vision and motion detection—feel inseparable. And yet jumping spiders have separated them so thoroughly that they exist within *different sets of eyes*. The central ones recognize patterns and shapes and see in color. The secondary ones track movements and redirect attention. Different eyes for different tasks, and each set has its own distinct connections to the spider's brain.* Jumping spiders remind us that we share a visual reality with other sighted creatures, but we experience it in utterly different ways. "We don't have to look to aliens from other planets," Jakob tells me. "We have animals that have a completely different interpretation of what the world is right next to us."

Humans have two eyes. They're on our heads. They're equally sized. They face forward. None of these traits is the norm, and a cursory glance at the rest of the animal kingdom reveals that eyes can be as varied as the creatures that own them. Eyes can come in eights or hundreds. The eyes of the giant squid are as big as soccer balls; those of fairy wasps are the size of an amoeba's nucleus. Squid, jumping spiders, and humans have all independently evolved camera-like eyes, in which a single lens focuses light onto a single retina. Insects and crustaceans

---

* What about the other two pairs of eyes? One seems to detect motion behind the spider. The other is very reduced, and its purpose is unclear.

have compound eyes, which consist of many separate light-gathering units (or ommatidia). Animal eyes can be bifocal or asymmetric. They can have lenses made of protein or rock. They can appear on mouths, arms, and armor. They can accomplish all the tasks our eyes can perform, or just a few of them.

This smorgasbord of eyes brings with it a dizzying medley of visual Umwelten. Animals might see crisp detail at a distance, or nothing more than blurry blotches of light and shade. They might see perfectly well in what we'd call darkness, or go instantly blind in what we'd call brightness. They might see in what we'd deem slow motion or time-lapse. They might see in two directions at once, or in every direction at once. Their vision might get more or less sensitive over the span of a single day. Their Umwelt might change as they get older. Jakob's colleague Nate Morehouse has shown that jumping spiders are born with their lifetime's supply of light-detecting cells, which get bigger and more sensitive with age. "Things would get brighter and brighter," Morehouse tells me. For a jumping spider, getting older "is like watching the sun rising."

SONKE JOHNSEN OPENS HIS BOOK *The Optics of Life* by noting that vision "is about light, so perhaps we should start with what light is." And then, with admirable candor: "I have no idea." Though it surrounds us almost constantly, light's true nature is not intuitive. Physicists contend that it exists both as an electromagnetic wave and as particles of energy known as photons. The specifics of this dual nature needn't concern us. What matters is that neither guise is something living things should obviously be able to detect. From a biological perspective, perhaps the most wondrous thing about light is that we can sense it at all.

Look inside the eyes of a jumping spider, a human, or any other animal, and you'll find light-detecting cells called photoreceptors. These cells might vary dramatically from one species to another, but they share a universal feature: They contain proteins called opsins. Every animal that sees does so with opsins, which work by tightly embracing a partner molecule called a chromophore, usually derived from vitamin A. The chromophore can absorb the energy from a single photon of light. When it does, it instantly snaps into a different shape, and its contortions force its opsin partner to reshape itself, too. The opsin's transformation then sets off a chemical chain reaction that ends with an electrical signal traveling down a neuron. This is how light is sensed. Think of the chromophore as a car key and the opsin as an ignition switch. The two fit together, light turns the key, and the engine of vision whirs into life.

There are thousands of different animal opsins, but they are all related.* Their unity creates a paradox. If all vision relies on the same proteins, and if those proteins all detect light, then why are eyes so diverse? The answer lies in light's distinct properties. Since most light on Earth comes from the sun, its presence can hint at temperature, time of day, or depth of water. It reflects off objects, revealing enemies, mates, and shelter. It travels in straight lines and is blocked by solid obstacles, creating telltale features like shadows and silhouettes. It covers Earth-scale distances almost instantaneously, offering a fast and far-ranging source of information. Vision is diverse because light is informative in a multitude of ways, and animals sense it for myriad reasons.

The biologist Dan-Eric Nilsson says that eyes evolve through four stages of increasing complexity. The first just involves photoreceptors—cells that do little more than detect the presence of light. The hydra, a relative of jellyfish, uses photoreceptors to ensure that its stings fire more readily in dim light; perhaps it does this to save those stings for nighttime hours, when its prey is more common, or to deploy them when it senses the shadow of a passing target. Olive sea snakes have photoreceptors at the tips of their tails, which they will pull away from sources of light. Octopuses, cuttlefish, and other cephalopods have

---

* In 2012, evolutionary biologist Megan Porter compared almost 900 opsins from different species, and confirmed that they share a single ancestor. That original opsin arose in one of the earliest animals and was so efficient at capturing light that evolution never conjured up a better alternative. Instead, the ancestral protein diversified into a wide family tree of opsins, which now underlie all vision. Porter draws that tree as a circle, with branches radiating outward from a single point. It looks like a giant eye.

photoreceptors dotted throughout their skin, which might help to control their amazing color-changing abilities.*

In the second stage, photoreceptors gain shade—a dark pigment or some other barrier that blocks the light coming in from certain angles. Shaded photoreceptors can not only detect light's presence but also infer its direction. These structures are still so simple that many scientists don't even regard them as genuine eyes, but they are useful to their owners nonetheless. They can also show up anywhere. The Japanese yellow swallowtail butterfly has photoreceptors on its genitals. A male uses these cells to guide his penis over a female's vagina, and a female uses them to position her egg-laying tube over the surface of a plant.

In the third of Nilsson's stages, shaded photoreceptors cluster into groups. Their owners can now knit together information about light from different directions to produce images of the world around them. For many scientists, this is the point when light detection becomes actual vision, when simple photoreceptors become bona fide eyes, and when animals can truly be said to see.† At first, their vision is blurry and grainy, suitable only for crude tasks like finding shelter or spotting looming shapes. But with the addition of focusing elements like lenses, their view sharpens, and their Umwelt fills with rich visual detail. High-resolution vision is the fourth of Nilsson's stages. When it first appeared, it would have intensified the interactions between animals. Conflicts and courtships could play out over distances longer than touch or taste would allow and at speeds too fast for smell. Predators could now spot their prey from afar, and vice versa. Chases ensued. Animals became bigger, faster, and more mobile. Defensive armor, spines, and shells evolved. The rise of high-resolution vision might explain why, around 541 million years ago, the animal kingdom dramatically diversified, giving rise to the major groups that exist today. This

---

* There's always at least one person who writes in with a pompous and incorrect corrective, so let's get this out of the way: The word *octopus* is derived from Greek and not Latin, so the correct plural is not *octopi*. Technically, the formal plural would be *octopodes* (pronounced ock-toe-poe-dees) but *octopuses* will do.

† This distinction isn't universally agreed upon, and some researchers would argue that a stage-two eye—a photoreceptor plus a shading pigment—also counts as an eye.

flurry of evolutionary innovation is called the Cambrian explosion, and stage-four eyes might have been one of the sparks that ignited it.

Nilsson's four-stage model addresses a concern of Charles Darwin, who was unsure how complex modern eyes could have evolved. "To suppose that the eye, with all its inimitable contrivances . . . could have been formed by natural selection, seems, I freely confess, absurd in the highest possible degree," he wrote in *The Origin of Species*. "Yet reason tells me, that if numerous gradations from a perfect and complex eye to one very imperfect and simple, each grade being useful to its possessor, can be shown to exist . . . then the difficulty of believing that a perfect and complex eye could be formed by natural selection, though insuperable by our imagination, can hardly be considered real." The gradations Darwin imagined do indeed exist: Animals have every conceivable intermediate from simple photoreceptors to sharp eyes. And different animal groups have repeatedly and independently evolved diverse eyes using the same opsin building blocks. The jellyfish alone have evolved stage-two eyes at least nine times, and stage-three eyes at least twice. Eyes, far from being a blow to evolutionary theory, have proved to be one of its finest exemplars.*

Darwin was wrong, though, in calling complex eyes perfect and simpler ones imperfect. Stage-four eyes are not some Platonic ideal that evolution was striving toward. The simpler eyes that preceded them are all still around and are well suited to the needs of their owners. "Eyes didn't evolve from poor to perfect," Nilsson emphasizes. "They evolved from performing a few simple tasks perfectly to performing many complex tasks excellently." As we saw in the introduction, a starfish has eyes on the tips of its five arms. These eyes can't see color, detail, or fast movements, but they don't have to. They only have to detect large objects, so that the starfish can slowly amble back

---

* In 1994, Nilsson and Susanne Pelger simulated the evolution of a sharp stage-four eye from a simple stage-three one. The simulation began with a small, flat patch of photoreceptors. With every generation, the patch slowly thickens and curves into a cup. It gains a crude lens, which gradually improves. Assuming pessimistically that the eye improves by just 0.005 percent every generation, and that each generation lasts for a year, it would take just 364,000 years for the blurry stage-three eye to become something like ours. As far as evolution goes, that's a blink of an eye.

toward the safety of a coral reef. A starfish has no need for an eagle's acute eye, or even a jumping spider's. It sees what it needs to.* The first step to understanding another animal's Umwelt is to understand what it uses its senses *for*.

Primates, for example, probably evolved big, sharp eyes to capture tree-dwelling insects sitting on branches. We humans have inherited that acute vision, which sighted people now use to guide their dexterous fingers, to read symbols that they imbue with meaning, and to assess the cues hidden in subtle facial expressions. Our eyes suit our needs. They also give us a singular Umwelt that most other animals do not share.

IN 2012, WHEN AMANDA MELIN, a scientist who studies animal vision, met Tim Caro, a scientist who studies animal patterns, their conversation naturally turned to zebras.

Caro had become the latest in a long line of biologists to wonder why zebras have such conspicuous black-and-white patterns. One of the earliest and most prominent hypotheses, he told Melin, was that the stripes counterintuitively act as camouflage. They mess with the eyes of predators like lions and hyenas by breaking up the zebra's outline, or by helping it to blend in among the vertical trunks of trees, or by causing a confusing blur when it runs. Melin was dubious. "I had a look on my face," she recalls. "I said, 'I think most of the carnivores are hunting at night, and their visual acuity is going to be so much worse than humans'. They probably can't see the stripes.' And Tim went, '*What?*'"

Humans outshine almost every other animal at resolving detail.

---

* It's not the case, either, that advanced eyes always exist in advanced creatures and simple eyes always in simple ones. There are some microbes that consist entirely of single cells and which also double as surprisingly complex eyes. Consider the freshwater bacterium *Synechocystis*. Light that hits one side of its spherical cell becomes focused on the opposite side. The bacterium can sense where that light is coming from, and move in that direction. It is effectively a living lens, and its entire boundary is a retina. The warnowiids, a group of single-celled algae, also seem to be living eyes, and each cell has components that resemble a lens, an iris, a cornea, and a retina. What they see, and whether they see at all, are open questions.

Our exceptionally sharp vision, Melin realized, gives us a rarefied view of a zebra's stripes. She and Caro calculated that on a bright day, people with excellent eyesight can distinguish the black-and-white bands from 200 yards away. Lions can only do so at 90 yards and hyenas at 50 yards. And those distances roughly halve at dawn and dusk, when these predators are more likely to hunt. Melin was right: The stripes can't possibly act as camouflage because predators can only make them out at close range, by which point they can almost certainly hear and smell the zebra. At most distances, the stripes would just fuse together into a uniform gray. To a hunting lion, a zebra mostly looks like a donkey.*

An animal's visual acuity is measured in cycles per degree—a concept that, by happy coincidence, you can think of in terms of zebra stripes. Stretch out your arm and give a thumbs-up. Your nail represents roughly 1 degree of visual space, out of the 360 degrees that surround you. You should be able to paint 60 to 70 pairs of thin black-and-white stripes on that nail and still be able to tell them apart. A human's visual acuity, then, is somewhere between 60 and 70 cycles per degree, or cpd. The current record, at 138 cycles per degree, belongs to the wedge-tailed eagle of Australia.† Its photoreceptors are some of the narrowest in the animal kingdom, which allows them to be densely packed within the eagle's retinas. With these svelte cells, the eagle effectively sees the world on a screen with over twice as many pixels as ours. It can spot a rat from a mile away.

But eagles and other birds of prey are the only animals whose vision is substantially sharper than ours. Sensory biologist Eleanor Caves has been collating visual acuity measurements for hundreds of species, almost all of which are surpassed by humans. Aside from raptors, only

---

* So why are zebras striped? Caro has a definitive answer: to ward off bloodsucking flies. African horseflies and tsetse flies carry a number of diseases that are fatal to horses, and zebras are especially vulnerable because their coats are short. But stripes, for some reason, confuse the biting pests. By filming actual zebras, as well as normal horses dressed in zebra-striped coats, Caro showed that flies would approach the animals and then fumble their landings. It's not yet clear why this happens.

† One oft-quoted study from the 1970s suggested that the American kestrel has an acuity of 160 cpd, but other studies of the same bird have found much lower values on a par with humans.

other primates come close to our standards. Octopuses (46 cpd), giraffes (27 cpd), horses (25 cpd), and cheetahs (23 cpd) do reasonably well. A lion's acuity is only 13 cpd, just above the 10 cpd threshold at which humans are considered legally blind. Most animals fall below that threshold, including half of all birds (and surprising ones like hummingbirds and barn owls), most fish, and all insects. A honeybee's acuity is just 1 cycle per degree. Your outstretched thumbnail represents roughly one pixel of a bee's visual world, and all the detail within that nail would collapse into a uniform smudge. Around 98 percent of insects have vision that's even coarser. "Humans are weird," Caves tells me. "We're not the pinnacle of any sensory modality, but we're rocking it with visual acuity." And paradoxically, our sharp vision muddies our appreciation of other Umwelten, because "we assume that if we can see it, they can, and that if it's eye-catching to us, it's grabbing their attention," says Caves. "That's not the case."

Caves fell prey to this perceptual bias herself. She studies cleaner shrimps, which helpfully exfoliate fish of parasites and dead skin. "They're cleaning colorful coral reef fish, and they're colorful themselves, so I thought they'd have reasonable vision," Caves tells me. They do not. Their fish clients can see the vibrant blue spots on their bodies, and the bright white antennae that they wave about, but they themselves cannot. A cleaner shrimp's beautiful patterns are not part of a cleaner shrimp's Umwelt, even at very close range. "They probably can't even see their own antennae," Caves says.

Many butterflies also have intricate patterns on their wings, which might warn predators that these insects are toxic. Some scientists have suggested that the butterflies might recognize each other from these patterns, but that's unlikely when their vision isn't sharp enough. A blackbird can see the black spots that freckle the orange wings of a map butterfly, but another map butterfly probably just sees an orange blur. We've always looked at butterflies, cleaner shrimps, and zebras through the wrong eyes—ours.

Why, then, since animals are so frequently adorned with elaborate patterns, aren't sharp eyes more common? In some cases it's because eyes are constrained by their past. The curse of low resolution is baked

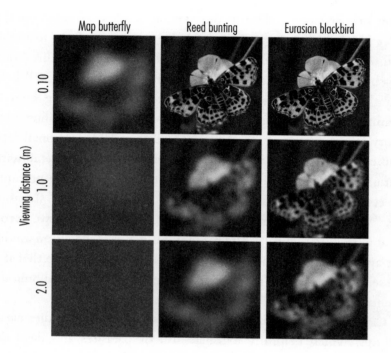

*A map butterfly as viewed through the eyes of
different species from varying distances*

into the structure of a compound eye, and having started off with eyes
of this kind, insects and crustaceans are now stuck. Robber flies manage 3.7 cycles per degree, but that's about the limit. For a fly's eye to be
as sharp as a human's, it would have to be a meter wide.

Acute eyes also come with a hefty drawback. As the wedge-tailed
eagle demonstrates, animals can achieve sharper vision by having
smaller and more densely packed photoreceptors. But each receptor
now collects light over a smaller area and is thus less sensitive. These
qualities—sensitivity and resolution—seesaw against each other. No
eye can excel at both. An eagle might be able to spot a far-off rabbit in
broad daylight, but its acuity plummets as the sun sets. (There are no
nocturnal eagles.) Conversely, lions and hyenas might not be able to
resolve a zebra's stripes at a distance, but their vision is sensitive enough
to hunt one at night. They, and many other animals, have prioritized

sensitivity over acuity. As ever, eyes evolve to suit the needs of their owners. Some animals simply don't need to see crisp images. And some animals don't need to see images at all.

DANIEL SPEISER NEVER THOUGHT he would spend his career trying to empathize with scallops. When he started graduate school in 2004, he thought about them the same way most people do—"as lumps of meat on a plate," he tells me. But those appetizing pan-seared lumps are merely the muscles that scallops use to close their shells. Look at a full, living scallop, and you'll see a very different animal. And that animal will see you, too. Each half of a scallop's fan-shaped shell has eyes arrayed along its inner edge—dozens in some species, and up to 200 in others. In the bay scallop, the eyes look like neon blueberries. Speiser finds them "funny and horrifying and charming," all at once.

It is strange enough that scallops have eyes when most other bivalves like mussels and oysters do not. It's even stranger that those eyes, as Mike Land showed in the 1960s, are complex. Each one sits at the end of a mobile tentacle. Each has a little pupil: "It's wild and creepy to see all of them opening and closing at the same time," Speiser says. Light passes through the pupil and hits the back of the scallop's eye, where it is reflected by a curved mirror. The mirror is a precisely tiled array of square crystals that collectively focus light onto the scallop's retinas. That's retinas, plural. There are two per eye, and they are about as different as two animal retinas could be.* Between them, they have thousands of photoreceptors, which gives them enough spatial resolution to detect small objects. "Their optics are really good," Speiser says.†

---

* There are two major groups of animal photoreceptors, known as ciliary and rhabdomeric. Both use opsins, but they function in very different ways. Scientists used to think that ciliary receptors were only found in vertebrates, and rhabdomeric ones were only in invertebrates. But that's not true: Both kinds of receptors are found in both groups. And both are found in the scallop, which has one retina full of ciliary photoreceptors and one full of rhabdomeric ones. Why? It's unclear, although one retina appears to be used to detect moving objects and the other is used for selecting habitats.

† It's not that scallop eyes are perfect. When light enters the eye, it must first get through the retina before the mirror can reflect and focus it. The retina gets two shots at absorbing that

But *why*? When scallops are threatened, they can swim away, opening and closing their shells like panicked castanets. Beyond these rare moments of action, though, they mostly sit on the seafloor, sieving edible particles from the water. They're "glorified clams," according to Sonke Johnsen. Why do they need such a complicated eye, let alone dozens or hundreds of them? What does a scallop use its vision for? To find out, Speiser ran an experiment that he called Scallop TV. He strapped their shells to small seats, placed them in front of a monitor, and showed them computer-generated movies of small, drifting particles. It was such a ridiculous setup that no one seriously thought that it would work. But it did: If the particles were large enough and moving slowly enough, the scallops opened their shells, as if ready to feed. "It was the craziest thing I've ever seen," Johnsen tells me.

At the time, Speiser thought that scallops must use their eyes to spot potential food. Now he thinks something else is happening. Interspersed between their eyes are tentacles that scallops use to smell molecules in the water. Speiser thinks they use smell to recognize predators like starfish and vision to detect things that are simply worth an investigative sniff. When they opened their shells in response to Scallop TV, they weren't trying to feed but were seeking to explore. "My guess is that we were seeing scallops being curious," Speiser says.

Speiser suspects that scallop vision works in a very different way than ours. Our brains combine the overlapping information from our two eyes into a single scene. A scallop *could* do the same across a hundred eyes, but that seems unlikely given how crude its brain is. Instead, each eye might simply tell the brain whether it has detected something moving or not. Think of the scallop's brain as a security guard watching a bank of a hundred monitors, each connected to a motion-sensing camera. If the cameras detect something, the guard sends sniffer dogs to investigate. Here's the catch: The cameras may be state-of-the-art, but the images they capture *are not sent to the guard*. All the guard sees on the monitors is a warning light for every camera that has spotted some-

light—once on its unfocused initial pass, and again in its more focused form. This means that the eye sees a focused image against a background of blurry haze.

thing. If Speiser is right about this bizarre setup, it means that even though each individual scallop eye has good spatial resolution, the animal itself might not have spatial *vision*. It knows when eyes in a certain region of its body have detected something, but it has no visual image of that object. It doesn't experience a movie in its head the same way we do. It sees without scenes.

This kind of vision is probably closer to our sense of touch than anything we experience with our eyes. We don't create a tactile scene of the world, even though we can feel with every part of our skin. Indeed, we largely ignore those sensations until something pokes us (or vice versa). And when we feel something unexpected, our most common reaction is to turn and look at it. Perhaps for a scallop, smell (not vision) is the fine-grained exploration sense and vision (not touch) is the crude, whole-body detection sense.*

But if that's the case, why does each individual eye have such good resolution? Why do sophisticated components like the mirrors and double retinas exist? Why are there so many eyes when just a few could cover the entire space around the scallop's shell? Why have such good eyes evolved in an animal whose brain can barely handle the information they convey?† No one knows. "Sometimes I feel like I can almost get my mind around it, and extend my empathy into scallops," Speiser tells me. "But a lot of the time I feel lost again."‡

---

* This idea is especially compelling because the eyes are actually modified chemosensory tentacles. It's a visual system jury-rigged from one originally used for smell and touch.

† In 1964, Mike Land, who was still a graduate student, looked into a scallop's eye and saw an upside-down image of himself. That's how he discovered that each eye contains a focusing mirror. He later showed that the mirror consists of layered crystals, and suggested (correctly) that the crystals are made of guanine—one of the building blocks of DNA. Guanine crystals don't naturally form squares, so the scallop must somehow control their growth. It's unclear how it manages this, or how it gets every crystal to the same exacting measurement—74 billionths of a meter thick.

‡ Scallops aren't the only animals with perplexing distributed vision. Chitons are mollusks that look like the disembodied forehead of a Klingon from *Star Trek*; their bodies are covered in armored plates, and those plates are dotted with hundreds of small eyes. Fan worms look like colorful feather dusters, extending from rocky tubes; those plumes are tentacles, which teem with eyes. Giant clams look like . . . well, very big clams; their meter-wide mantles contain several hundred eyes. Dan-Eric Nilsson likens all of these eyes to burglar alarms. They detect nearby movement and encroaching shadows, so their owners know when to take defen-

Some animals might have the scallop's distributed vision without possessing eyes at all. The brittle star *Ophiomastix wendtii* looks like a skinny, spiny starfish, or perhaps like five centipedes wriggling out of a hockey puck. It doesn't have any obvious eyes, but it clearly sees. It will scuttle away from light, crawl toward shady crevices, and even change color after sunset. In 2018, Lauren Sumner-Rooney showed that the brittle star has thousands of photoreceptors over the full lengths of its sinuous arms. It's as if the entire animal acts as a compound eye.* Weirder still, it's only an eye *during the day*.

When the sun is out, the brittle star expands sacs of pigment in its skin, which give it the deep red of a blood clot. At night, it shrinks these sacs, and becomes pale gray and striped. When expanded, the pigment sacs block light from reaching the photoreceptors at certain angles. This gives each receptor the directionality of a stage-two eye, and it gives the entire animal the spatial vision of a stage-three eye. But when the pigment sacs contract at night, the photoreceptors are fully exposed. Unable to tell the direction of incoming light, their spatial vision no longer works. "It knows when it's exposed to light, but doesn't know how to get away from it," Sumner-Rooney says.

It's anyone's guess what the brittle star itself makes of this change. Unlike a scallop, it doesn't even have a brain—just a decentralized ring of nerves surrounding its central disc. This ring coordinates the five arms but doesn't command them; they mostly act on their own. It's as if the brittle star has the same weird system of cameras as the scallop, but without the security guard. The cameras are just signaling each other. Do they do so across the entire animal? Is each separate arm its own eye? Is each arm a swarm of semi-autonomous eyes that happen to be linked? "It could be something so out there that we haven't even thought of it yet," Sumner-Rooney tells me. "Everything we know

---

sive measures. The chitons clamp down onto rocks, the fan worms pull their fans back into their tubes, and the giant clams close their shells. It's likely that, like the scallops, none of these animals sees scenes.

* Like brittle stars, sea urchins also seem to use their entire bodies as a crude eyeball. Each urchin is a spiky ball that crawls around on hundreds of tube feet. Its photoreceptors are on those feet, and they are shaded either by the animal's spines or by its hard exoskeleton. Its vision may not be especially sharp, but it can certainly amble toward dark shapes.

about animal vision to date relies on having an eye. We're basing everything on a century's worth of research on contiguous retinas, with photoreceptors that are close together and grouped. [The brittle star] violates a lot of those assumptions."

With multitudes of eyes, no heads, and sometimes no brains, brittle stars and scallops all reveal how strange vision can be. "An animal doesn't have to see a picture to be able to use vision," Sumner-Rooney says. "But humans are such visually driven creatures that trying to conceive of these completely alien systems is very hard." It is easier to imagine the visual worlds of more familiar creatures with heads and two eyes. But even then, we might miss what is right in front of us.

RISING HIGH ON COLUMNS of warm air, griffon vultures soar over rolling landscapes in search of food. Since they can spot carcasses on the ground, they should easily be able to see large obstacles ahead of them. And yet vultures, eagles, and other large raptors often fatally crash into wind turbines. In one Spanish province alone, 342 griffon vultures collided with wind turbines over a 10-year period. How could birds that fly by day and have some of the planet's sharpest eyes fail to avoid structures so large and conspicuous? Graham Martin, who studies bird vision, answered this question by addressing another: Where exactly do vultures look?

In 2012, Martin and his colleagues measured the griffon vulture's visual field—the space around its head that its eyes can cover. They got each bird to rest its beak on a specially fitted holder, and then looked into its eyes from all directions with a visual perimeter. "It's the same device that an optician would use when you get an eye test," Martin told me at the time. "It's just a question of sitting the bird down for half an hour. One tried to grab at me and I did lose a bit of my thumb."

The perimeter revealed that a vulture's visual field covers the space on either side of its head but has large blind spots above and below. When it flies, it tilts its head downward, so its blind spot is now directly ahead of it. This is why vultures crash into wind turbines: While soaring, they aren't looking at what is right in front of them. For most of their history, they never had to. "Vultures would never have encountered an object so high and large in their flight path," Martin says. It might work to turn off the turbines if the birds are near, or to lure the vultures away using ground-based markers. But visual cues on the blades themselves won't work.* (In North America, bald eagles also crash into wind turbines for the same reasons.)

When I think about Martin's study, I'm suddenly and acutely aware of the large space behind my head that I cannot see and that I seldom think about. Humans and other primates are rather odd in having two eyes that point straight ahead. The left eye gets a very similar view to the right, and their visual fields overlap a lot. This arrangement gives us excellent depth perception. It also means we can barely see things to our sides, and we can't see what's behind us without turning our heads. For us, seeing is synonymous with facing, and exploration is achieved through gazing and turning. But most birds (except for owls) tend to have side-facing eyes and don't need to point their heads at something to look at it.

A soaring vulture that's scanning the ground can also see other vultures flying next to it, without having to turn. A heron's visual field covers 180 degrees in the vertical; even when standing upright with its beak pointing straight ahead, it can see fish swimming near its feet. A mallard duck's visual field is completely panoramic, with no blind spot either above or behind it. When sitting on the surface of a lake, a mallard can see the entire sky without moving. When flying, it sees the world simultaneously moving toward it and away from it. We use the phrase "bird's-eye view" to mean any vista seen from on high. But a bird's view is not just an elevated version of a human one. "The human visual world is in front and humans move into it," Martin once wrote. But "the avian world is around and birds move through it."†

---

* Why don't vultures just have wider visual fields that allow them to look ahead while flying? Martin thinks it's because their large, sharp eyes are vulnerable to dazzling glare from the sun. In general, he says, birds with large eyes tend to have larger blind spots. Birds with panoramic vision, like ducks, tend to have smaller and less acute eyes that can better tolerate the presence of the sun.

† Chickens and many other birds rely on frontal vision only at close range, when they want to accurately grab something with their beaks or feet.

Birds also differ from humans in where their vision is sharpest. Many animals have an area in their retinas where their photoreceptors (and the attendant neurons) are densely packed, increasing the resolution of their vision. This region goes by many names. In invertebrates, it's called an *acute zone*. In vertebrates, it's an *area centralis*. If that area is also inwardly dimpled, as it is in our eyes, it's a *fovea*. For all our sakes (except the vision scientists, to whom I apologize), I'm just going to stick with *acute zone*. In humans, it's a bullseye—a round spot in the center of our visual field. It's what you are training upon these letters as you read them. Most birds also have circular acute zones, but theirs point outward, not forward. If they want to examine objects in detail, they have to look sideways, with just one eye at a time. When a chicken investigates something new, it will swing its head from side to side to look upon it with the acute zone of each eye in turn. "When chickens look at you, you never know what the other eye is doing," says Almut Kelber, a zoologist who studies bird vision. "They must have at least two centers of attention, which is very hard to imagine."

Many birds of prey, like eagles, falcons, and vultures, actually have two acute zones *in each eye*—one that looks forward, and another that looks out at a 45-degree angle. The side-facing one is sharper, and it's the one that many raptors use when hunting. When a peregrine falcon dives after a pigeon, it doesn't plunge straight at its prey. Instead, it flies along a descending spiral. That's the only way it can keep the pigeon within its murderous side-eye, while also pointing its head down and maintaining a streamlined shape.*

The peregrine prefers to use its right eye to track prey. Such preferences are common to birds; when eyes see distinct views, those eyes can be used for distinct tasks. The left half of a chick's brain is specialized for focused attention and categorizing objects; the bird can spot food grains among a bed of pebbles if it uses its right eye (directed by its left brain), but not its left eye. The right half of the brain deals with the unexpected; many birds use their left eyes (directed by their right

---

* Turning the eyes is out of the question because birds of prey can barely move their eyes without turning their heads. Indeed, their eyes are so big that they almost touch each other inside the skull.

brains) to scan for predators, and are quicker to detect a threat when it approaches from the left.

An animal's visual field determines where it can see. Its acute zones determine where it sees *well*. Without considering both traits, we can seriously misinterpret an animal's actions. In a video that went viral on TikTok, a male argus pheasant displayed his dazzling plumage to a female, who seemed to look off to the side. Viewers laughed at her apparent disinterest, not knowing that she was looking right at him with her side-facing visual field. A seal's visual field is more similar to ours but with excellent coverage above its head and poor coverage below, presumably to spot fish silhouetted against the sky. A seal that swims upside down might look relaxed to a human observer, but is actually scanning the seafloor for food.

Cows and other livestock also have a somnolent air because their gaze is so fixed. They rarely turn to look at you in the way another human (or a jumping spider) might. But they also don't need to. Their visual fields wrap almost all the way around their heads and their acute zones are horizontal stripes, giving them a view of the entire horizon at once. The same is true for other animals that live in flat habitats, including rabbits (fields), fiddler crabs (beaches), red kangaroos (deserts), and water striders (the surface of ponds). Except for the occasional aerial predators, *up* and *down* are largely immaterial to them. There is only *across,* in every possible direction. A cow can simultaneously see a farmer approaching it from the front, a collie walking up from behind, and the herdmates at its side. *Looking around,* which is inextricable from our experience of vision, is actually an unusual activity, which animals do only when they have restricted visual fields and narrow acute zones.

Elephants, hippos, rhinos, whales, and dolphins have two or three acute zones per eye, possibly because they can't quickly turn their heads.* Chameleons don't have to turn because their turret-like eyes can move independently; they can look in front and behind at the same

---

* A whale's pupil doesn't constrict by shrinking into a pinhole, like ours does. Instead, it pinches in the middle, creating what looks like an awkwardly smiling mouth with two small openings at either end. Each of these openings is effectively its own mini-pupil, and admits light onto a separate acute zone.

time, or track two targets moving in opposite directions. Other animals are steadier in their gaze. Many male flies focus upward: The large facets at the top of their compound eyes are called love spots, and allow them to detect the silhouettes of females flying overhead. Male mayflies have gone even further: The female-spotting parts of their eyes are so enormous that each eye looks like it is wearing a chef's hat. The fish *Anableps anableps,* which lives at the surface of South American rivers, also partitions its eyes. The top half sticks out of the water and is adapted for air vision, and the bottom half stays below the surface and is adapted for aquatic vision. It's also known as the four-eyed fish.

In the three-dimensional world of the deep ocean, above and below matter as much as in front and behind. Many deep-sea fish like barreleyes and hatchetfish have tubular eyes that point upward, allowing them to see the outlines of other animals silhouetted against the faint downwelling sunlight. The brownsnout spookfish, a kind of barreleye, has amended the upward eye of its kin with a downward-pointing chamber that has its own retina; with these two-part eyes, it can look up and down at the same time. So can the cock-eyed squid, whose left eye is twice the size of its right. It hangs in the water column with the small eye pointing downward to spot bioluminescent flashes and the big eye pointing up to spot silhouettes. Meanwhile, the deep-sea crustacean *Streetsia challengeri* has fused its eyes into a single horizontal cylinder, which looks like a corn dog. It can see in almost every direction circumferentially—above, below, and to the sides—but not ahead or behind.

It is almost impossible to imagine what it would be like to see like *Streetsia,* or a chameleon, or even a cow. The reverse-facing camera of my smartphone can show me what's going on over my shoulder, but that image still appears in my relentlessly forward-facing visual field. Again, as with the scallops, it helps to think about touch. I can simultaneously feel the sensations on the skin of my scalp, soles, chest, and back. If I concentrate, I can just about imagine what it might be like to fuse the omnidirectional nature of that sensation with the long range of sight. Vision can extend in any direction and every direction. It can envelop and surround. And it can vary in time as well as space. It can fill not just the empty voids around us but also the fleeting gaps between moments.

THE MEDITERRANEAN IS HOME to a small, unassuming fly called *Coenosia attenuata*. Just a few millimeters long, with a pale gray body and large red eyes, "it looks like a standard housefly," Paloma Gonzalez-Bellido tells me. In fact, it is a killer. From its perch on a leaf, it will take off in pursuit of fruit flies, fungus gnats, whiteflies, and even other killer flies—"anything that's small enough for them to subdue," Gonzalez-Bellido says. During the chase, it stretches out its legs. As soon as one touches the target, all six clamp shut, forming a cage. Often, it will fly the victim back to its original perch. If you can coax a killer fly to crawl onto your finger, it will repeatedly launch itself from your digit and return with prey, like a (very tiny) falcon to its falconer. This experience can be unexpectedly magical for a human. It's less so for the prey. While a typical housefly has a proboscis that resembles a sponge on a stick, used for dabbing and sucking at liquids, a killer fly's proboscis is part dagger and part rasp, used for stabbing and scraping flesh. The fly shoves it into its victim and hollows it out while it is still alive. Gonzalez-Bellido has a video in which you can see a killer fly's mouthparts scraping away a fruit fly's eye from the inside, leaving nothing behind but a grid of transparent lenses. Farmers and gardeners frequently introduce this insect into greenhouses to take care of pests, and it has now spread all around the world.

For killer flies, speed is everything. "Their prey can come from anywhere, and the Mediterranean is so dry that it's rare for them to have prey," Gonzalez-Bellido says. They immediately take off after anything that could conceivably be a meal and, once airborne, catch their prey as quickly as possible so that they themselves aren't cannibalized by others of their kind. Their chases are near impossible for even well-trained human eyes to follow. By filming these pursuits with high-speed cameras, Gonzalez-Bellido showed that they typically take a quarter of a second. They might even be over in half that time. A killer fly can capture its target in the space of a human blink.

Their ultrafast hunts are guided by ultrafast vision. It may seem strange to talk about animals seeing at different speeds, because light is the fastest thing in the universe, and vision seems instantaneous to us. But eyes don't work at light speed. It takes time for photoreceptors to react to incoming photons, and for the electrical signals they generate to travel to the brain. In killer flies, evolution has pushed these steps to their limits. When Gonzalez-Bellido shows these insects an image, it takes just 6 to 9 milliseconds for their photoreceptors to send electrical signals, for those signals to reach their brains, *and* for their brains to send commands to their muscles.* By contrast, it takes between 30 and 60 milliseconds for human photoreceptors to accomplish just the first of those steps. If you looked at an image at the same moment as a killer fly, the insect would be airborne well before a signal had even left your retina. "We don't know of a faster photoreceptor than the ones from these flies," Gonzalez-Bellido tells me. She says it with something approaching pride.†

The fly's vision also updates more quickly. Imagine looking at a light that flickers on and off. As the flickering gets faster, there will come a point when the flashes merge into a steady glow. This is called the critical flicker-fusion frequency, or CFF. It's a measure of how quickly a brain can process visual information. Think of it as the frame rate of the movie playing inside an animal's head—the point at which static images blend into the illusion of continuous motion. For humans, in good light, the CFF is around 60 frames per second (or hertz, Hz). For most flies, it's up to 350. For killer flies, it's probably higher

---

* The photoreceptors in a killer fly's eye fire quickly and reset quickly. Both traits demand a lot of energy. Compared to the photoreceptors of a fruit fly, those of a killer fly have three times more mitochondria—the bean-shaped batteries that supply animal cells with power.

† Other predatory insects, like dragonflies and robber flies, have large, high-resolution eyes with distinctive acute zones. As they pursue their targets, they turn their heads to keep the prey within the sharpest part of their visual field. Killer flies "have to pay attention in all directions," Gonzalez-Bellido says, so they don't have an acute zone, and their visual resolution isn't especially high. Despite that, they seem to have a more demanding hunting strategy. Dragonflies hunt against the sky, spotting the silhouettes of prey that fly above them. But killer flies somehow "do the impossible thing of hunting against the ground," Gonzalez-Bellido says. They'll pick out prey moving in front of complex backgrounds, and then chase those targets through leaves and other cluttered environments.

still. To its eyes, a human movie would look like a slideshow. The fastest of our actions would seem languid. An open palm, moving with lethal intent, would be easily dodged. Boxing would look like tai chi.

In general, animals tend to have higher CFFs if they're smaller and faster. Compared to human vision, cats are slightly slower (48 Hz) and dogs slightly faster (75 Hz). The eyes of a scallop are positively glacial (1 to 5 Hz), and those of nocturnal toads are slower still (0.25 to 0.5 Hz). Those of leatherback turtles (15 Hz) and harp seals (23 Hz) are faster but still sluggish. Those of swordfish aren't much better under normal conditions (5 Hz), but these fish can heat up their eyes and brains with a special muscle, boosting the speed of their vision by eight times. Many birds have naturally fast vision; with a maximum CFF of 146 Hz, the pied flycatcher—a small songbird—has the fastest vision of any vertebrate that's been tested, perhaps because its survival depends on tracking and catching flying insects.* And those insects have eyes that are faster still. Honeybees, dragonflies, and flies have CFFs between 200 and 350 Hz.

It's possible that each of these visual speeds comes with a different sense of time's passage. Through a leatherback turtle's eyes, the world might seem to move in time-lapse, with humans bustling about at a fly's frenetic pace. Through a fly's eyes, the world might seem to move in slow motion. The imperceptibly fast movements of other flies would slow to a perceptible crawl, while slow animals might not seem like they were moving at all. "Everyone asks us how we catch the killer flies," Gonzalez-Bellido says. "You just move toward them slowly with a vial. If you're slow enough, you're just part of the background."

FAST VISION REQUIRES A lot of light, so killer flies can only be active during the day. Other animals are not so limited.

After the sun's golden fingers withdraw from the Panamanian rainforest and the understory's shade thickens into an even deeper dark-

---

* Traditional fluorescent lights flicker at 100 Hz—that is, 100 times a second. That's too fast for humans to see, but not for many birds like starlings, for whom the lights must be stressful and irritating.

ness, a small bee emerges from a hollow stick. This is *Megalopta genalis*, a sweat bee. Its legs and abdomen are golden yellow. Its head and torso are metallic green. None of those beautiful hues are usually visible to human observers because the bee only emerges when there's too little light for humans to see, let alone see in color. But despite the darkness, *Megalopta* slaloms through a labyrinth of lianas and tracks down its favorite flowers. Having collected its fill of pollen, it somehow then returns to the very same thumb-width stick in which it nests.

Eric Warrant, who grew up collecting insects and now studies their eyes, first encountered *Megalopta* in 1999 on a research trip to Panama. He quickly confirmed, to his astonishment, that it uses vision to guide its nighttime flights. By filming the insect with infrared cameras, Warrant saw that when it first emerges from its stick, it turns around and hovers slowly in front of the entrance, memorizing the appearance of the surrounding foliage. Later, when it has finished foraging, it uses this visual memory to find its way home. If Warrant set up his own landmarks, like white squares, and moved them to another stick while the bee was away, it would return to the wrong place. The bee's feat would be hard enough in bright daylight: Rainforests are neither easy to navigate nor short of sticks. But *Megalopta* somehow finds its home "in the dimmest imaginable light," Warrant says. He has filmed the bee finding its nest on nights so dark that he couldn't even see his own hand in front of his face. He had to use night-vision goggles to see what the bee could with its own eyes. "They're no clumsier in the dark than a honeybee is in bright sunlight," Warrant tells me. "They come flying in quite rapidly, they don't hesitate, and they land incredibly quickly. It's one of the most amazing things I've ever seen."

Warrant suspects that *Megalopta*'s ancestors veered toward a nocturnal schedule to escape intense competition from daylight pollinators, including other bees. But life at night isn't easy for animals that rely on vision, for two major reasons. The first is obvious: There's much less light. Even the light of a full moon is a million times dimmer than full daylight. A moonless night that's illuminated by stars alone is a hundred times dimmer still. A night where starlight is obscured by clouds or tree cover is a hundred times dimmer again. These are the kinds of conditions in which *Megalopta* can still navigate—starless darkness that offers barely enough light for an eye to collect. The second challenge is less intuitive: Photoreceptors can accidentally go off on their own, and at night, these false alarms can easily outnumber the real signals from actual photons. So nocturnal animals must not only detect the little light that's there but also ignore the phantom lights that aren't. They must overcome both the limits of physics and the messiness of biology.

Some animals have simply dropped out of the struggle. Like all sensory systems, eyes are expensive to build and maintain. It takes a lot of energy to even prep photoreceptors and their associated neurons for the arrival of light, so that they can react when needed. Even when animals aren't seeing anything, the mere possibility of sight drains their resources. This drain is significant enough that if eyes stop being useful or effective, they tend to diminish or disappear. Sometimes animals invest in other senses that aren't yoked to light. (We'll meet these later; many exceptional senses were discovered because scientists noticed animals doing amazing things in total darkness.) Others unsubscribe from vision entirely. In underground realms, in caves, and in other dark corners of Earth where vision cannot earn its worth, eyes are often lost.*

Other animals, instead of ceding their vision to the dark, have evolved ways of seeing in the dimmest of conditions. Some use neural tricks, including the sweat bee that Warrant studied. It pools the responses from several different photoreceptors, turning lots of smaller pixels into a few large megapixels. Its photoreceptors might also collect photons for more time before firing, like a camera whose shutter is left open for a longer exposure. These two strategies group the photons reaching the bee's eye in both space and time, increasing the ratio of signal to noise. Its vision is grainy and slow as a result but remains bright when brightness seems impossible. And "seeing a coarser,

---

* There are many ways to break an eye, and evolution has explored them all. Lenses have degenerated. Visual pigments have disappeared. Eyeballs have sunk beneath the skin or been covered by it. One species alone, the Mexican cavefish, has lost its eyes several times over, as different sighted populations moved from bright rivers to dark caves and independently abandoned vision. As Eric Warrant tells me, "Why Gollum in *The Hobbit* had extra-big eyes makes no scientific sense."

slower, brighter world is better than seeing nothing at all," Warrant says.[*]

Animals can also see in the dark by grabbing every last photon they can. Some species, including cats, deer, and many other mammals, have a reflective layer called a tapetum, which sits behind their retinas and sends back any light that gets past their photoreceptors; those cells then get a second chance to collect the photons they initially missed.[†] Other animals have evolved exceptionally large eyes and wide pupils. The tawny owl's eyes are so big that they bulge out of its head. The tarsiers—small primates from Southeast Asia that look like gremlins—have eyes that are each larger than their brains. And the biggest eyes of all evolved in one of the darkest environments in the planet—the deep ocean.

TO DIVE INTO THE OCEAN is to enter the largest habitat on the planet—a realm with over 160 times more living space than all the ecosystems on the surface combined. Most of that space is dark.

At 10 meters down, 70 percent of the light from the surface has been absorbed. If you were descending in a submersible, anything red, orange, or yellow on your person would now look black, brown, or gray. By 50 meters, greens and violets have largely vanished, too. By 100 meters, there is only blue, at just 1 percent of its surface intensity, if that. By 200 meters, the start of the mesopelagic or twilight zone, that intensity has fallen by another 50 times. The blue is now almost laser-like—eerily pure and all-encompassing. Through it, silvery fish dart about. Gelatinous jellyfish and siphonophores slowly snake past. At 300 meters, it's as dark as a moonlit night, and getting darker.

---

[*] This doesn't fully account for *Megalopta genalis*'s night vision, though. "I can't explain how they do it," Warrant tells me. "I've got clues about some of the mechanisms they use to enhance vision in dim light, but I can't see the whole picture."

[†] Reflections from the tapetum are responsible for the eyeshine of dogs, cats, deer, and other animals illuminated by car headlights or camera flashes. The structure of a reindeer's tapetum changes in the dark winter to reflect even more light. Coincidentally, this also changes the tapetum's color, and thus the color of reindeer eyes, from golden yellow in the summer to a rich blue in the winter.

Gradually, the fish get blacker, the invertebrates redder. Increasingly, they produce their own light, and their bioluminescent flashes paint the outline of your descending submersible. At 850 meters, the residual sunlight is so faint that your eyes can no longer function. At 1,000 meters, no animal eyes can. This is the beginning of the bathypelagic or midnight zone. The complex visual scenes of the surface are long gone and have been replaced by a living star-field of bioluminescence, twinkling in the otherwise total darkness. Depending on where you are in the world, there might be another 10,000 meters of ocean left to go.

The deep ocean's consummate darkness creates a problem for the scientists who want to study its denizens. Researchers can't see what's around them unless they turn on their submersible's lights, but doing so is devastating for creatures that have adapted to a lightless life. Even moonlight can blind a deep-sea shrimp in a few seconds. A submersible's headlights will do much worse. Some deep-sea animals end up doing kamikaze runs at subs. Startled swordfish ram them with their swords. Other creatures freeze or flee. "The way to think about ocean exploration is that we probably create a sphere a hundred yards wide that keeps away anything that can get away," says Sonke Johnsen. "Most of the time, we're seeing terror and blindness. We see how animals behave when they think they're being killed by some glowing god."

To be more respectful of deep-sea Umwelten, Johnsen's mentor Edith Widder created a stealth camera called Medusa. It films deep-sea animals with red light that most of them can't see, and attracts them with a ring of blue LEDs that resemble a bioluminescent jellyfish. "The only real innovation is that we turned off the lights," he says. "Once we do that, really big stuff shows up."

In June 2019, Widder and Johnsen took Medusa on a 15-day research cruise through the Gulf of Mexico. Under what seemed to be the only storm in the Gulf, they would manually lower the 300-pound camera to the end of its 2,000-meter line, and then haul it up again the next night. "Have you ever pulled up a fridge-sized object for a mile?" Johnsen asks me. "It took three hours every night." After every deployment, Nathan Robinson would pore over Medusa's videos. And over the course of the first four, "we saw a shrimp making a little bio-luminescence," Johnsen says. "Yay?"

Then, on June 19, "I'm on the bridge, and all of a sudden, Edie's at the bottom of the stairs with a smile on her face that's practically cracking her ears off, and I thought: This can only be one thing." On its fifth outing, Medusa had filmed a giant squid.

The footage was unmistakable. At a depth of 759 meters, a long cylinder appears and snakes toward the camera before unfurling into a mass of writhing, suckered arms. It briefly grabs the camera with two long tentacles before losing interest and withdrawing back into the dark. The crew estimated that it was a 10-foot-long juvenile, which was nowhere close to the species' maximum size of 43 feet. Still, it was a *giant squid*—an almost mythic animal, and one with the largest and most sensitive eyes on the planet.

As I noted at the start of this chapter, the eyes of a giant squid (and the equally long but much heavier colossal squid) can grow as big as soccer balls, with diameters up to 10.6 inches. These proportions are perplexing. Yes, bigger eyes are more sensitive, and it makes sense for an animal in the dark ocean to have them. But no other creature, including those that live in the deep sea, has eyes that are even in the same ballpark as a giant or colossal squid's. The next-largest eyes, which belong to the blue whale, are less than half the size. A swordfish's eye, which is the largest of any fish at 3.5 inches, could fit inside a giant squid's pupil. The squid's eyes are not just big; they are absurdly and excessively bigger than those of any other animal. What does it need to see that it can't see with a swordfish-sized eye?

Sonke Johnsen, Eric Warrant, and Dan-Eric Nilsson think they know the answer. They calculated that in the deep ocean, eyes suffer from diminishing returns. As they get bigger, they cost more energy to run but offer little extra visual power. Once they get past 3.5 inches—that is, swordfish-sized—there's little point in enlarging them further. But the team found that extra-large eyes *are* better at one task, and one task alone: spotting large, glowing objects in water deeper than 500 meters. There's an animal that fits those criteria, and it is one that giant squid really need to see: the sperm whale.

The largest toothed predators in the world, sperm whales are the giant squid's main nemeses. Their stomachs have been found full of the squid's parrot-like beaks, and their heads often bear circular scars inflicted by the serrated rims of the squid's suckers. They do not produce their own light, but just like a descending submersible, they trigger flashes of bioluminescence when they bump against small jellyfish, crustaceans, and other plankton. With its disproportionately large eyes, the giant squid can see these telltale shimmers from 130 yards away, giving it enough time to flee. It is the only creature with eyes large enough to see these bioluminescent clouds at a distance, and also the only one that *needs* to do so. "No other animals are looking for things that are really large at depth," Johnsen says. Sperm whales and other toothed whales use sonar rather than vision to find their food. Large sharks tend to go after smaller prey. Blue whales subsist on tiny shrimp-like krill. Krill might benefit from seeing the bioluminescent cloud of a blue whale, but their compound eyes are too limited in resolution, and their bodies are too slow to do anything with that information. Giant (and colossal) squid are unique in being massive animals that need to see massive predators, and their singular need has led to a singular Umwelt. With the largest and most sensitive eyes that exist, they scan one of the darkest environments on Earth for the faint sparkling outlines of charging whales.*

TURN OFF THE LIGHTS, and our world becomes monochromatic. This shift occurs because our eyes contain two types of photoreceptors—

---

* The giant squid seems to be a global species that lives in every ocean. But for the longest time, it was known only from carcasses that washed ashore. The first photographs of this creature in the wild were only taken in 2004. The first natural footage was captured in 2012, when Widder and her colleagues deployed the then-new Medusa camera off the coast of Japan. Seven years later, the stealth camera proved its worth yet again, just 100 miles southeast of New Orleans. "That part of the Gulf is packed with oil rigs, and there are thousands of remotely operated vehicles there," Johnsen says. "Those pilots have never seen a giant squid, and we saw one on our fifth deployment. Either we are the luckiest people in the world, or it's that we turned our lights off." (They are pretty lucky. Half an hour after the crew saw the squid footage, lightning struck their ship, frying a lot of instruments but mercifully sparing Medusa's hard drive. Shortly after, the ship also dodged a waterspout.)

cones and rods. The cones allow us to see colors, but they only work in bright light. In the dark, the more sensitive rods take over, and a kaleidoscope of daytime hues is replaced by the blacks and grays of the night. Scientists used to think that all animals were similarly color-blind at night.

Then, in 2002, Eric Warrant and his colleague Almut Kelber did a pivotal experiment with the elephant hawkmoth. This beautiful European insect has a pink-and-olive body and a wingspan of almost 3 inches. It feeds entirely at night, hovering in front of flowers and drinking their nectar with a long, unfurled proboscis. Kelber trained hawkmoths to drink instead from feeders, which sat behind blue or yellow cards. Having learned to associate these colors with food, the moths could reliably distinguish them from equally bright shades of gray. And they kept on doing so as Kelber turned down the lights in her lab.

At light levels equivalent to a half-moon, Kelber's world turned black-and-white, but the moths were still going strong. At one point, "it took me 20 minutes sitting in my dark lab to be able to see the moth," she tells me. "I couldn't even see its proboscis," but it was still drinking from the right feeders. The lights then faded to the levels of dim starlight, and, though Kelber couldn't see at all, the elephant hawkmoth could still perceive the cards in all their glorious color. But those colors were probably very different from the ones we perceive.

3.

Rurple, Grurple, Yurple

Color

WHEN MAUREEN AND JAY NEITZ ADOPTED A TOY POODLE puppy, "like all good parents, we went out and read a book about how to raise a dog," Jay tells me. The book claimed that dog names should ideally have two syllables and hard consonants. The Neitzes brainstormed a few options, and Maureen, in joking reference to Jay's research on vision, suggested Retina. (I point out that Retina has three syllables. "Yes, but our version has two," Jay says. "Ret-na.") Black, fluffy, and very cute, Retina became a part of history. She was one of the dogs who first confirmed what colors dogs actually see.

In the 1980s, when the Neitzes were getting their PhDs, many people believed that dogs were color-blind. In *The Far Side,* cartoonist Gary Larson drew a dog praying at its bedside for "Mom, Dad, Rex, Ginger, Tucker, me, and all the rest of the family to see color." Scientists bought into this myth, too: One textbook claimed that "on the whole, mammals appear not to have color vision except for the primates." And yet, very few species had actually been carefully tested—including dogs, despite their popularity. "People would always ask me what their dogs see, and we had really no idea," Jay says. "Or we had ideas, but no evidence."

To get that evidence, he took Retina and two Italian greyhounds to his lab. He trained them to sit in front of three lit panels, one of which was differently colored. If they touched the odd panel with their noses, they earned a cheesy treat. And they did, repeatedly. Dogs do see color. They just don't see the same range that most people see. Nor do most other animals. To appreciate their varied visual palettes, we must first understand what color really is, how animals see it, and *why* they evolved to see it at all. Color vision is complicated enough that even a simplified explanation, which I'm about to lay out, can feel abstract and confusing. But bear with me: The details are the key to truly understanding birds, butterflies, and blossoms. We need to spend some time in the weeds to appreciate the flowers.

Light comes in a range of wavelengths. Those we can see span from 400 nanometers, which we perceive as violet, to 700 nanometers, which we perceive as red. Our ability to detect these wavelengths, and the rainbow that lies between them, depends on our opsin proteins—the foundation of all animal vision. Opsins come in different varieties, and each is best at absorbing a particular wavelength of light. Normal human color vision depends on three of these opsins, each of which is deployed by a different type of cone cell in our retinas. Based on their preferred wavelengths, the opsins (and the cones that contain them) are called long, medium, and short. More familiarly, they're called red, green, and blue.* When light bounces off a ruby and enters our eyes, it

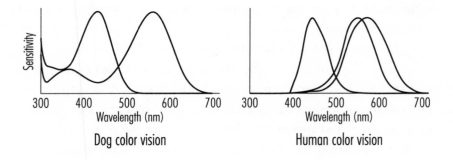

Each curve represents one class of cone cell; the peak of each curve shows the wavelength of light to which the cone is most sensitive. Note that dogs have two cone classes, while humans have three.

---

* Technically, based on the wavelengths of light that most excite them, the long and short cones should really be called yellow-green and violet instead of red and blue.

stimulates the long (red) cones strongly, the medium (green) ones moderately, and the short (blue) ones weakly. If that light bounces off a sapphire, the opposite happens—the short (blue) cones react most strongly, and the others less so.

But color vision involves more than merely *detecting* different wavelengths of light. It's about *comparing* them. The signals from the three types of cones are added and subtracted by a complex network of neurons. Some of these neurons are excited by inputs from the red cones but inhibited by inputs from the green ones; they allow us to discriminate reds from greens. Other neurons are excited by the blue cones but inhibited by the red *and* green ones; they allow us to distinguish blues and yellows. This simple neural arithmetic—R − G and B − (R + G)—is called opponency. It's how the raw signals from just three cones are transformed into the glorious rainbows that we perceive.

Opponency is the basis of (almost) all color vision. Without it, an animal doesn't really see colors in the way we imagine. *Daphnia* water fleas, for example, have four opsins that are sensitive to orange, green, violet, and ultraviolet wavelengths. But those wavelengths just trigger hardwired and almost reflexive responses. Ultraviolet means sun, so swim away. Green and yellow mean food, so swim toward. Water fleas can respond to four specific kinds of light that *we* see as colored. But being unable to compare the signals from their four opsins, they can't perceive a spectrum.

Color, then, is fundamentally subjective. There's nothing inherently "green" about a blade of grass, or the 550-nanometer light that it reflects. Our photoreceptors, neurons, and brains are what turn that physical property into the sensation of green. Color exists in the eye of the beholder—and also in their brain. Consider the story of the artist Jonathan I., as told by Oliver Sacks and Robert Wasserman in "The Case of the Colorblind Painter." After a life of seeing and painting in colors, he suffered a brain injury that turned his world monochrome. His retinas were healthy, his opsins were present, and his cones were working. But his brain could only conjure up a world of blacks, whites, and grays. Even when he closed his eyes, his imagined world was drained of color.

A small proportion of people, and entire species of animals, also see only in shades of gray, not because of brain damage but because their retinas aren't set up for color vision. They are called monochromats. Some, like sloths and armadillos, only have rod cells, which work well in dim light but aren't geared toward color. Others, like raccoons and sharks, only have one cone, and since color vision depends on opponency, having one cone is effectively like having none. Whales have just one cone, too: To paraphrase the vision scientist Leo Peichl, for a blue whale, the ocean is not blue. Cone cells are unique to vertebrates, but other animals have wavelength-specific photoreceptors that play a similar role. Surprisingly, the cephalopods—octopuses, squid, and cuttlefish—have just one class of these, which means they are also monochromats.* They can rapidly change the colors of their skin yet are unable to see their own shifting hues.

The existence of so many monochromats hints at one of the most counterintuitive things about color vision: It isn't necessary. Almost all the things that animals use their eyes for—navigating, foraging, communicating—can be done with shades of gray. What, then, is the point of seeing colors at all?

Physiologist Vadim Maximov suggested that the answer might lie around 500 million years in the past, during the Cambrian era, when the ancestors of modern animal groups arose. Many of those ancestral creatures lived in shallow seas, with rays of sunlight flickering around them. These rippling rays are beautiful to our modern eyes but would have been enormously confusing to ancient monochromatic ones. If the brightness of a given spot of water can change by a hundred times from one second to the next, it becomes much harder to spot relevant objects against a background. Is that dark shape that just appeared the looming shadow of a predator, or merely the shadow of a sunbeam that briefly strayed behind a cloud? Monochromatic eyes that only deal in brightness and darkness would struggle to tell. But eyes that see in color would fare much better. That's because different wavelengths of

* The firefly squid is an exception. It's the only cephalopod known to have three different classes of photoreceptors, and may well have color vision.

light tend to keep the same relative proportions, even when the total amount of light brightens or dims. A strawberry that looks red in bright sunlight still looks red in the shade, and its green leaves are still obviously green even under the reddish tint of a sunset. Color—and specifically color vision with opponency—offers *constancy*. If an animal can compare the outputs of photoreceptors that are tuned to different wavelengths, it can stabilize its view of a world where light dances and flickers. Even two classes will do the job. That's the basis of dichromacy, the simplest form of color vision. It's what Retina, other dogs, and most mammals have.

Dogs have two cones—one with a long, yellow-green opsin and another with a short, blue-violet one. They see mostly in shades of blue, yellow, and gray. When my corgi Typo looks at his red-and-violet toy, he probably sees the red as a dark, muddy yellow, and the violet as a deep blue. When he looks at the bright-green ring that he likes to chew, the green stimulates both his cones equally. Because of opponency, those signals cancel out, and Typo sees white.

Horses are dichromats, too, and their cones are sensitive to wavelengths very similar to those a dog's respond to. This means that horses struggle to make out the orange markers that are used to highlight obstacles at racecourses. These orange blazes stand out to trichromatic human vision, but Sarah Catherine Paul and Martin Stevens showed that they blend into the background to a horse's dichromatic eyes. If we designed racecourses for horse vision, we'd paint the markers fluorescent yellow, bright blue, or white.

Then again, if we designed racecourses for *inclusive* human vision, we'd probably do the same. Most "color-blind" people are also dichromats, because they're missing one of the three usual cones. They still see colors, albeit in a narrower range. There are many kinds of color-blindness, but deuteranopes, who lack the medium green cones, come closest to seeing like dogs and horses. Their world is painted in yellows, blues, and grays, while reds and greens are hard to distinguish. Color-blind people might be confused by traffic lights, electrical wiring, or paint swatches. They might struggle to read packaging or charts, to distinguish sports teams that are wearing ostensibly distinct colors, or to complete seemingly simple school assignments like drawing a rainbow. In some countries, they might be disqualified from flying planes, joining the military, or even driving. Color-blindness shouldn't be a disability, but it can be because humans have built cultures that are predicated on trichromacy. And what's so special about trichromacy, other than that most people have it? If dichromacy is good enough for most mammals, why are we and other primates different? Why do we see the colors we do?

THE FIRST PRIMATES WERE almost certainly dichromats. They had two cones, short and long. They saw in blues and yellows, like dogs. But sometime between 29 and 43 million years ago, an accident occurred that permanently changed the Umwelt of one specific lineage of primates: They gained an extra copy of the gene that builds their long opsin. Such duplications often happen when cells divide and DNA is copied. They're mistakes, but fortuitous ones, for they provide a redundant copy of a gene that evolution can tinker with without disrupting the work of the original. That's exactly what happened with the long-opsin gene. One of the two copies stayed roughly the same, absorbing light at 560 nanometers. The other gradually shifted to a shorter wavelength of 530 nanometers, becoming what we now call the medium (green) opsin. These two genes are 98 percent identical, but the 2 percent gulf between them is also the difference between seeing only in blues and yellows and adding reds and greens to the mix.* With the new medium opsins joining the earlier long and short ones, these primates had evolved trichromacy. And they passed their expanded vision to their descendants—the monkeys and apes of Africa, Asia, and Europe, a group that includes us.

This story explains *how* we came to see the colors we see, but not

---

* Both medium and long genes lie on the X chromosome. If someone with two X chromosomes inherits a faulty copy of either gene, they usually have a working backup. But if someone with an X and a Y chromosome inherits a faulty copy, they're stuck with it. This is why red-green color-blindness, which is typically caused by the loss of either the M or L cones, is much more common in men than in women.

why. Why exactly did the duplicated long-opsin gene shift toward a medium wavelength? The answer might seem obvious: to see more colors. A monochromat can make out roughly a hundred grades of gray between black and white. A dichromat adds around a hundred steps from yellow to blue, which multiplies with the grays to create tens of thousands of perceivable colors. A trichromat adds another hundred or so steps from red to green, which multiplies again with a dichromat's set to boost the color count into the millions. Each extra opsin increases the visual palette exponentially. But if dichromats can flourish with just tens of thousands of colors, why do trichromats benefit from millions?

Since the nineteenth century, scientists have suggested that trichromats would do better at spotting red, orange, and yellow fruit against green foliage.* More recently, some researchers have argued that their advantage lies more in finding the most nutritious rainforest leaves, which tend to flush red when they are young and rich in protein. These explanations aren't mutually exclusive: Most primates eat fruits, but at times when those aren't ripe or available, larger species can make do with young leaves. That's the "perfect setting for the evolution of trichromacy," says Amanda Melin, who studies primate vision (and occasionally, as we saw in the last chapter, zebra stripes). "It's useful for finding your main food and your fallback food."†

The monkeys of the Americas complicate this story. They also evolved trichromacy, but in a distinct way with very different consequences. In 1984, Gerald Jacobs noticed that some squirrel monkeys were sensitive to red light, but others were not. And with help from Jay Neitz, he worked out why. These monkeys never developed a sec-

---

* Kentaro Arikawa, who studies color vision, first realized that he has a red-green color deficiency when he was six; his mother asked him to pick strawberries from their garden for breakfast and he failed, disappointing her. In several lab experiments, trichromats do outperform dichromats at finding fruit.

† Primates also have unusually acute vision, which might explain why trichromacy didn't evolve in other fruit- or leaf-eating mammals. "You can give a mouse trichromacy, but what good would that be to a nocturnal mammal with poor acuity?" says Melin. By contrast, sharp-eyed primates can use trichromacy to spot fruit and young leaves from afar, and reach them before competitors realize they have appeared.

ond copy of the long-opsin gene.* Instead, their original gene now comes in several versions, some of which still produce long cones and some of which make medium cones. The gene also sits on the X chromosome, which means that the male monkeys (which are XY) can only ever inherit one version. Medium or long, it doesn't matter: They're destined to dichromacy. The female monkeys, however, are XX. Some of them inherit *both* the medium and long versions, one on each of their X chromosomes. That gives them trichromacy.† So when a group of these monkeys cavorts through the treetops in search of food, some will see red fruits against green leaves, while others will only see yellows and grays. Even brothers and sisters can perceive different colors.

It's easy to assume that the dichromats must be at a disadvantage. But after 15 years of studying white-faced capuchins in the forests of Costa Rica, Amanda Melin thinks differently. By following several groups of these monkeys, she learned to identify every individual on sight. And by collecting their poop and sequencing their DNA, she worked out which were trichromats and which were dichromats. Neither group, she found, is more likely to survive or reproduce than the other. The trichromats are indeed better at finding brightly colored fruit, but the dichromats surpass them at finding insects disguised as leaves and sticks. Without a riot of colors to confuse or distract them, they're better at detecting borders and shapes, and seeing through camouflage. Melin has watched them nabbing insects that she, a trichromat, didn't even know were there. Seeing extra colors has both drawbacks and benefits. More isn't necessarily better, which is why some females are still dichromats and all males are.

Or, I should say, *almost* all males. In 2007, the Neitzes added the human long-opsin gene to the eyes of two adult male squirrel mon-

---

* Howler monkeys are the exception. They live in the Americas, but unlike the other monkeys they share a continent with, they are all trichromatic, males and females. That's because they evolved trichromacy in the same way as their cousins in Africa and Eurasia—by duplicating the long-opsin gene. And they did so independently.

† It's even more complicated than this, because many of these American monkeys have three possible versions of the same gene. Females might inherit two of the three versions or a pair of the same ones, which means that these animals have six different forms of color vision—three dichromacies and three trichromacies.

keys, giving them three cones instead of two, and turning them into trichomats. The two monkeys—Dalton and Sam—suddenly performed differently on the same vision tests that they had been doing every day for two years, and could distinguish new colors that were previously invisible to them. Dalton died from diabetes shortly after the experiment. But as of April 2019, when I last spoke to Jay, Sam was still alive and in his 12th year of trichromacy. I wondered what his life was now like. Does he behave any differently? Does he react to fruit in new ways? "I tried to talk to him," Jay said, laughing. "How cool *is it*? That's the interesting thing, right? But he's very nonchalant."

To me, Sam's silence speaks volumes. He reminds us that seeing more colors isn't advantageous in and of itself. Colors are not inherently magical. They become magical when *and if* animals derive meaning from them. Some are special to us because, having inherited the ability to see them from our trichromatic ancestors, we imbued them with social significance. Conversely, there are colors that don't matter to us at all. There are colors we cannot even see.

IN THE 1880S, JOHN LUBBOCK—banker, archeologist, polymath—split a beam of light with a prism and shone the resulting rainbow onto ants. The ants scurried away from the light. But Lubbock noticed that they also fled from a region just beyond the rainbow's violet end, which looked dark to his eyes. This area wasn't dark to the ants, though. It was bathed in ultraviolet—literally "beyond violet" in Latin. Ultraviolet (or UV) light has wavelengths ranging from 10 to 400 nanometers.* It is largely invisible to humans, but must be "apparent to the ants as a distinct and separate colour (of which we can form no idea)," Lubbock

---

* Visible light is just a small part of the vast electromagnetic spectrum, and there are reasons it's the only slice that our eyes can detect. Electromagnetic waves with very short wavelengths, like gamma rays and X-rays, are largely absorbed by the atmosphere. Those with very long wavelengths, like microwaves and radio waves, don't have enough energy to reliably excite opsins. For these reasons, no animal can see microwaves or X-rays. There's only a narrow Goldilocks zone of wavelengths that are useful for vision, and they range from 300 to 750 nanometers. Our eyes, which work from 400 to 700 nanometers, already cover much of that available visual space. But in the margins, a lot can happen.

presciently wrote. "It would appear that the colours of objects and the general aspect of nature must present to them a very different appearance from what it does to us."

At the time, some scientists believed that animals either are colorblind or see the same spectrum that we do. Lubbock showed that ants are exceptional. Half a century later, bees and minnows turned out to see ultraviolet, too. The narrative shifted: *Some* animals can see colors we can't, but the skill must be very rare. But after another half century, in the 1980s, researchers showed that many birds, reptiles, fish, and insects have UV-sensitive photoreceptors. The narrative changed again: UV vision exists in many groups of animals, but not in mammals. Still wrong: In 1991, Gerald Jacobs and Jay Neitz showed that mice, rats, and gerbils have a short cone that is tuned to UV. Okay, *fine,* mammals *can* have UV vision, but only small ones like rodents and bats. Not so: In the 2010s, Glen Jeffery found that reindeer, dogs, cats, pigs, cows, ferrets, and many other mammals can detect UV with their short blue cones. They probably perceive UV as a deep shade of blue rather than a separate color, but they can sense it nonetheless. So can some humans.

Our lenses typically block out UV, but people who have lost their lenses to surgeries or accidents can perceive UV as whitish blue. This happened to the painter Claude Monet, who lost his left lens at the age of 82. He began seeing the UV light that reflects off water lilies, and started painting them as whitish blue instead of white. Monet aside, most people can't see UV, which probably explains why scientists were so eager to believe that the ability was rare. In fact, the opposite is true. Most animals that can see color can see UV. It's the norm, and we are the weirdos.*

Ultraviolet vision is so ubiquitous that much of nature must look

---

* Why don't most humans see UV? It might be the cost of having sharp eyesight. When light passes through our lenses, shorter wavelengths are bent at sharper angles. Even if the lens admitted UV, it would focus these wavelengths at a point well in front of the others, blurring the image on the retina. This is called chromatic aberration. It's less of an issue for small eyes, or for those that don't need to be very acute. But for big-eyed animals with sharp vision, it's a problem. This may be why primates don't see UV, and why raptors see much less of it than other birds.

different to most other animals.* Water scatters UV light, creating an ambient ultraviolet fog, against which fish can more easily see tiny UV-absorbing plankton. Rodents can easily see the dark silhouettes of birds against the UV-rich sky. Reindeer can quickly make out mosses and lichens, which reflect little UV, on a hillside blanketed by UV-reflective snow. I could go on.

I'm going to go on. Flowers use dramatic UV patterns to advertise their wares to pollinators. Sunflowers, marigolds, and black-eyed Susans all look uniformly colored to human eyes, but bees can see the UV patches at the bases of their petals, which form vivid bullseyes. Usually, these shapes are guides that indicate the position of nectar. Occasionally, they are traps. Crab spiders lurk on flowers to ambush pollinators. To us, these spiders seem to match the colors of their chosen blooms, and they've long been treated as masters of camouflage. But they reflect so much UV that they are highly conspicuous to a bee, which makes the flowers they sit upon that much more alluring. Rather than blending in, some of them attract their UV-sensitive prey by standing out.

Many birds also have UV patterns in their feathers. In 1998, two independent teams realized that much of the "blue" plumage of blue tits actually reflects a lot of UV; as one of them wrote, "Blue tits are ultraviolet tits." To humans, these birds all look much the same. But thanks to their UV patterns, males and females look very different from each other. The same is true for more than 90 percent of songbirds whose sexes are indistinguishable to us, including barn swallows and mockingbirds.

It's not just humans who can't see UV patterns. Since UV light is heavily scattered by water, predatory fish that have to spot prey at a distance are often insensitive to it. Their prey, in turn, have exploited this weakness. The swordtail fish of Central American rivers look drab to us, but as Molly Cummings and Gil Rosenthal showed, males of some species have strong UV stripes along their flanks and tails. These markings are alluring to females, but they're invisible to the swordtails' main pred-

---

* Some scientists think that the first kind of color vision to evolve was dichromacy with a green photoreceptor and a UV one. If that's true, animals have been seeing UV for as long as they've been seeing color.

ators. And in places where those predators are more common, swordtails have more vivid UV markings. "They could get away with being super-flamboyant" without attracting danger, Cummings says. Similar secret codes exist in Australia's Great Barrier Reef, home to the ambon damselfish. To human eyes, it resembles a lemon with fins, and looks identical to other closely related species. But Ulrike Siebeck found that its head is actually streaked with UV stripes, as if invisible mascara had run all over its face. Predators can't see these markings, but the ambons themselves use them to distinguish their own kind from other damselfish.

For us, UV feels enigmatic and intoxicating. It's an invisible hue lying just on the edge of our vision—a perceptual void that our imaginations are keen to fill. Scientists have often attributed special or secret significance to it, treating it as a channel for covert communication. But aside from the ambon damselfish and swordtails, most such claims have foundered.* The reality is that UV vision and UV signals are extremely common. "My personal view is that it's just another color," Innes Cuthill, who studies color vision, tells me.

Imagine what a bee might say. They are trichromats, with opsins that are most sensitive to green, blue, and ultraviolet. If bees were scientists, they might marvel at the color we know as red, which they cannot see and which they might call "ultrayellow." They might assert at first that other creatures can't see ultrayellow, and then later wonder why so many do. They might ask if it is special. They might photograph roses through ultrayellow cameras and rhapsodize about how different they look. They might wonder whether the large bipedal animals that see this color exchange secret messages through their flushed cheeks. They might eventually realize that it is just another color, special mainly in its absence from their vision. And they might wonder what it would be like to add it to their Umwelt, bolstering their three dimensions of color with a fourth.

---

* Other claims about UV vision have also fallen apart. In 1995, a Finnish team suggested that kestrels can track voles by looking for UV reflecting off their urine. This claim has been frequently repeated in books and documentaries, but "it's wrong," says Almut Kelber. In 2013, she and her colleagues showed that vole urine doesn't actually reflect much UV and isn't distinguishable from water. Kestrels can't possibly see it from afar.

———

NESTLED 9,500 FEET UP in Colorado's Elk Mountains, the town of Gothic was once home to a thriving silver mine. When the value of silver crashed in the late nineteenth century, Gothic became a ghost town. But in 1928, it was reborn as, of all things, a research station. Today, the Rocky Mountain Biological Laboratory, affectionately known as Rumble, attracts scientists from around the world. Hundreds of them migrate there every summer to live and work among what looks like the set of a Western, to study the local soils and streams, ticks and marmots. When Mary Caswell "Cassie" Stoddard arrived there in 2016, she had hummingbirds on her mind.

"I grew up watching birds, but it wasn't until I got to college that I learned birds can perceive colors humans can't," Stoddard tells me. "I found that mind-blowing." Most birds have four types of cone cells, with opsins that are most sensitive to red, green, blue, and either violet or UV. That makes them *tetrachromats*. Theoretically, they should be able to distinguish a multitude of colors that are imperceptible to us. To confirm that they can, Stoddard and her team tested Rumble's resident broad-tailed hummingbirds—a beautiful species with iridescent green feathers and, in the males, bright magenta bibs.

Exploiting the hummingbirds' natural instinct to feed from colorful flowers, Stoddard attracted them to feeders placed near special lights, which had been customized to produce colors that a tetrachromat should be able to see. One light might illuminate a nectar-containing feeder with a mix of green and ultraviolet, while another might shine pure green onto a water-containing feeder. Stoddard couldn't tell the difference between these colors, but the hummingbirds could, with minimal experience. Over the course of a day, they would increasingly flock toward the nectar feeder, having "learned to distinguish between lights that look identical to us," she says. "That's what we always predicted, but seeing it with our own eyes was thrilling."*

---

* If Stoddard set both lights to produce the same colors, the hummingbirds could no longer reliably arrive at the nectar-baited feeder. This suggests that they're not just learning the position of the right feeder, or relying on other senses like smell.

Even with experiments like this, it is easy to underestimate what other birds can see. They don't just have human vision plus ultraviolet, or bee vision plus red. Tetrachromacy doesn't just widen the visible spectrum at its margins. It unlocks an entirely new *dimension* of colors. Remember that dichromats can make out roughly 1 percent of the colors that trichromats see—tens of thousands, compared to millions. If the same gulf exists between trichromats and tetrachromats, then we might be able to see just 1 percent of the *hundreds of millions* of colors that a bird can discriminate. Picture trichromatic human vision as a triangle, with the three corners representing our red, green, and blue cones. Every color we can see is a mix of those three, and can be plotted as a point within that triangular space. By comparison, a bird's color vision is a *pyramid,* with four corners representing each of its four cones. Our entire color space is just *one face* of that pyramid, whose spacious interior represents colors inaccessible to most of us.

If our red and blue cones are stimulated together, we see purple—a color that doesn't exist in the rainbow and that can't be represented by a single wavelength of light. These kinds of cocktail colors are called non-spectral. Hummingbirds, with their four cones, can see a *lot* more of them, including UV-red, UV-green, UV-yellow (which is red + green + UV), and probably UV-purple (which is red + blue + UV). At my wife's suggestion, and to Stoddard's delight, I'm going to call these rurple, grurple, yurple, and ultrapurple.* Stoddard found that these non-spectral colors and their various shades account for roughly a third of those found on plants and feathers. To a bird, meadows and forests pulse with grurples and yurples. To a broad-tailed hummingbird, the bright magenta feathers of the male's bib are actually ultrapurple.

Tetrachromats also have a different concept of white. White is what we perceive when all our cones are equally stimulated. But you'd need a different blend of wavelengths to excite a bird's quartet of cones than you would a human's trio. Paper is treated with dyes that happen to absorb UV, so it wouldn't look white to a bird. Many supposedly

---

* I am still hung up on whether UV-purple should be called ultrapurple or purpurple.

"white" bird feathers reflect UV and wouldn't necessarily look white to birds, either.

It's hard to know what birds make of rurples, grurples, and other non-spectral colors, Stoddard says. As a violinist, she knows that two simultaneously played notes can either sound separate or merge into completely new tones. By analogy, do hummingbirds perceive rurple as a blend of red and UV, or as a sublime new color in its own right? When they make choices about which flowers to visit, "do they group rurple with reds, or do they see it as an entirely different hue?" she asks. They can tell that it's different from pure red, "but I can't articulate what it looks like to them."

Birds aren't the only tetrachromats. Reptiles, insects, and freshwater fish, including the humble goldfish, have four cones as well. By looking at tetrachromats among modern animals and working backward, scientists can deduce that the first vertebrates were likely tetrachromats, too. Mammals, probably because they were all initially nocturnal, lost two of their ancestral cones and became dichromats. But they scurried beneath the feet of dinosaurs, which were almost certainly tetrachromats and "probably saw all kinds of cool non-spectral colors," Stoddard says. It's ironic that for the longest time, illustrators and filmmakers portrayed dinosaurs in dull shades of brown, gray, and green. Only recently have artists started painting these animals with bright colors, inspired by the revelation that they are the ancestors of birds. But even these vivid hues, applied with a trichromat's eye, capture just a tiny proportion of the colors that dinosaurs probably wore or saw.

It is much easier for most people to imagine a dog's sense of color than a bird's (or a dinosaur's). If you are a trichromat, you can simulate dichromatic vision by using apps that remove certain colors. You could even simulate what a different trichromat (like a bee) might see by mapping their blue, green, and UV system onto our red, green, and blue one. But there is no way of representing a tetrachromat's color vision for a trichromatic eye. "People often ask if we can engineer goggles to allow humans to see these non-spectral colors—and I wish!" Stoddard says. You could use a spectrophotometer to find the rurples and grurples on a bird's feathers, but you'd then have to recolor them with our more limited range of colors. Four into three just won't go. Frustrating though it might be, most of us simply cannot imagine what many animals actually look like to each other, or how varied their sense of color can be.

EVEN FOR A BUTTERFLY, the red postman has a peculiarly delicate style of flight. With fast wingbeats but surprisingly little forward motion, it seems to be trying very hard to be nowhere in particular. Its languid movements befit its defenses: Full of toxins, and clad in red, black, and yellow warning colors, it's in no rush to avoid predators. But there is nothing off-putting about them to human eyes. In a greenhouse in Irvine, California, I watch as two dozen of these butterflies flutter by my head, between the red and orange flowers of lantana plants. Between their bright colors and soothing movements, the world feels both richer and more tranquil. The technical name for these butterflies is *Heliconius erato,* and both parts feel fitting. In Greek mythology, Mount Helicon was the home of the Muses and a source of poetic inspiration; Erato was the Muse of love poetry.

One erato butterfly lands on the shoot of a lantana plant, curls her abdomen, and deposits a tiny golden egg. Five more sit sociably together on a nearby leaf, slowly opening and closing their wings. Another alights on the display of the greenhouse's climate control system, which reads 97 degrees Fahrenheit and 59 percent humidity. Jeans, I realize, were a mistake. Next to me, Adriana Briscoe, who is more sensibly dressed, is looking around and beaming broadly. This greenhouse is hers, and it is both a workplace and a retreat, somewhere she goes to feel happy and calm. "I love being here," she says wistfully. "You can see why many scientists have devoted their careers to studying these butterflies."

Throughout Central and South America, erato typically lives alongside a close relative—*Heliconius melpomene,* named after the Muse of tragedy. Both erato and melpomene are toxic, and they mimic each other so that any predator that learns to avoid one will also avoid the other. In any one place, these two species look almost identical. But across their range, they vary considerably. In Tarapoto, Peru, both erato and melpomene have red bands on their forewings and yellow bands on their hindwings. But in Yurimaguas, just 80 miles away, both species have yellow splotches and red bases on their forewings, and red stripes on their hindwings. You'd scarcely believe that eratos from the two sites were actually the same species, and you'd struggle to distinguish between eratos and melpomenes at any one site. Briscoe's greenhouse could have been full of both, and I would never have known. So how do the butterflies themselves tell the difference? When Briscoe started studying them in the late 1990s, it struck her as odd that no one knew. "For such visual animals that are also very popular, it seems like it would have been an obvious thing to do to look at their eyes," she says.

Most butterflies are trichromats. Like bees, they have three opsins that are most sensitive to UV, blue, and green, and can see colors ranging from red to UV. But in 2010, Briscoe discovered that *Heliconius* butterflies differ from their relatives in two important ways. First, they're tetrachromats. Alongside the usual blue and green opsins, they have *two* UV opsins that peak at different wavelengths. Second, while related butterflies pattern their wings with yellow pigments, *Heliconius* uses yurple—the non-spectral color that mixes UV and yellow. These two traits are related. With two UV opsins, the *Heliconius* species can carve up the UV part of the spectrum into finer gradations, and discriminate between subtly different shades of UV-based colors. And by painting their wings with those colors, they can better tell the difference between their own kind and their mimics. Even birds, with their single UV opsin, don't seem to discriminate between yellow and the shade of yurple the butterflies use.

The male erato butterflies can't, either. In 2016, Briscoe's student Kyle McCulloch found that only female eratos are tetrachromats. The males are trichromats. They have the gene for the second UV opsin, but for some reason they suppress it. Just like squirrel monkeys, the female eratos have an extra dimension to their color vision that males lack.* In Briscoe's greenhouse, we watch as two eratos start to have sex. Their abdomens join, but before they can separate, the female takes off with the male still stuck to her. They flutter off as one, briefly conjoined by their genitals but forever separated by their Umwelten.

These butterflies are not the only species with a sex difference in tetrachromacy. *Humans* share that trait. Somewhere in Newcastle, England, lives a woman known in the scientific literature as cDa29. She's a private person who doesn't do interviews, and her real name isn't publicly known. But according to psychologist Gabriele Jordan, who has worked with her extensively, cDa29 aces tests that only a tetrachromat could pass. Much like Stoddard's hummingbirds, she can pick out one shade of green among other extremely similar ones, "like a cherry from a tree," Jordan tells me. "For us, it's just green among greens. Other people look and look and look and then maybe have a guess. She can spot the odd one out within milliseconds."

Human tetrachromats are usually women, because the genes for the long and medium opsins both sit on the X chromosome. Since most women have two X chromosomes, they can inherit two slightly different versions of either gene. They would then end up with *four* different kinds of opsins that are tuned to different wavelengths—short, medium, long-a, and long-b, for example. Around one in eight women has this pattern . . . but most of them are not tetrachromatic. To possess that ability, a lot of other pieces need to fall in place. Normally, the red and green cones respond best to wavelengths that are just 30 nanometers apart. To produce a new and distinct dimension of color, the fourth cone has to sit almost exactly in the middle of that range, 12 nanometers away from the green. (That's what cDa29 has.) To build an

---

* There's another twist to this story, which readers of my first book, *I Contain Multitudes*, will be delighted by. Every now and then, Briscoe would find a female erato with male-like eyes that only had three opsins. This pattern confused her, until she realized that all of these females were infected by a bacterium called *Wolbachia*. *Wolbachia* is one of the most successful bacteria on the planet, and infects a huge proportion of insects and other arthropods. It only passes down the female line from mother to daughter, and has many tricks for doing away with useless males. Sometimes it kills males outright. Sometimes it transforms them into females. Sometimes it allows females to reproduce asexually without needing males at all. What it's doing in erato is a mystery, but one that Briscoe is now trying to solve.

opsin with that exact specification, "you almost have to split an atom genetically," Jordan says. Even if women can make the right kind of fourth cone, they need to have it in the right part of the retina—the central fovea, where our color vision is sharpest. And most important, they need the right neural wiring to perform opponency with the signals from these cones.

This combination of traits is rare enough that only a very small proportion of women with four cones are truly tetrachromatic. Jordan tells me that many people who say they are actually aren't. Artists, in particular, are often convinced that they can see more colors than others, but being more attentive to hues because of your work is not the same as seeing a whole other dimension of color. "I've tested many who turned out not to be tetrachromatic," Jordan says. "It's very attractive, the idea of superhuman vision.* But it isn't as common as people make out." The first confirmed tetrachromat was cDa29; Jordan estimates that there are around 48,600 others in the United Kingdom, but they are not easy to find.† They're not walking around with amazing technicolor clothes, just as dichromats aren't filling their lives with drab colors. Until cDa29 got tested, "she never thought there was anything special about her vision," Jordan says. "You're viewing the world with a given set of retinas and a given brain, and if you can't see with someone else's, it doesn't really cross your mind that you're better."

When Jordan first told me this, I confess to feeling a little disappointed, as I did when Jay Neitz told me that Sam the genetically engineered squirrel monkey was nonchalant about his newfound trichromacy. Colors matter to us. Color TVs, printers, and books are more prized than their black-and-white cousins. It's natural to expect that an extra dimension of color would be a spectacular thing to see.

---

* Note that cDa29 and other genuine tetrachromats can't see ultraviolet like birds can, so their vision would cover the same range of wavelengths as a normal trichromat's. They still see an extra dimension of color, and their color space can still be represented by a pyramid instead of a triangle. But it's a pyramid that fits inside the one that birds have.

† In 2019, Jordan developed a test that could quickly tell if women have a fourth cone with exactly the right 12-nanometer spacing to offer true tetrachromacy. "That would allow us to go around and very quickly find out how many of them there are," she says. "And then COVID-19 came."

To learn that it could be taken for granted threatens to drain color of its magic. But of course, *all of us*—monochromat, dichromat, trichromat, or tetrachromat—take the colors that we see for granted. Each of us is stuck in our own Umwelt. As I wrote in the introduction, this is a book not about superiority but about diversity. The real glory of colors isn't that some individuals see more of them, but that there's such a range of possible rainbows.

When thinking about human tetrachromats and erato butterflies, I'm struck by how absurd it is that people once thought all animals saw the same spectrum of colors as humans. Humans don't even see the same colors as each other.* We have varying forms of partial or complete color-blindness. Some of us are tetrachromats. Look across the rest of the animal kingdom and you'll find even greater variations. Color vision varies considerably within the 6,000 species of jumping spiders, the 18,000 species of butterflies, and the 33,000 species of fish.

At least three kinds of color vision exist just within the eye of a larval zebrafish. The part of the fish's retina that looks up at the sky sees in black and white, because color isn't necessary for spotting the silhouettes of aerial predators. The part that looks straight ahead is dominated by UV detectors, which help it to spot tasty plankton. And the part that scans the horizon and the space below the fish is tetrachromatic. From black-and-white vision to more colors than humans can see, the eyes of these baby fish have it all.

To appreciate the colors that another animal sees, you can't just add an Instagram filter over your own view. You can't assume that those colors stay the same across a scene or a season, or from one individual to another. And you can't just count the numbers of opsins or photoreceptors that an animal has and reconstruct its visual palette. Kentaro Arikawa has found that many butterflies have a frankly excessive number of photoreceptor classes. The cabbage white butterfly has eight, but one exists only in females and another only in males. The Japanese

---

* Amanda Melin tells me that human color vision is far more varied than what she and others have seen in chimps, baboons, and other primates. It's unclear why, but it might be that our survival is now less closely tied to the colors we see, allowing for variants that might once have been detrimental to remain.

yellow swallowtail has six but uses only four, for tetrachromatic vision; the other two are likely hardwired for specific tasks, like spotting objects of a specific color flying past. The champion among butterflies—the common bluebottle—has 15. But these insects are not pentadecachromats, with 15-dimensional color vision. Only three of the photoreceptors are found all over the eye, while four are confined to the top half, and eight to the bottom. Arikawa expects that he'd find even finer segregations if he looked for them. The bluebottle butterfly, he thinks, is probably a tetrachromat that uses its other 11 classes of photoreceptors to detect very specific things in narrow parts of its visual field.

Indeed, color vision doesn't ever need to be more sophisticated than tetrachromacy. Based on the colors that reflect off natural objects, animals can see everything they could possibly need to with just four classes of photoreceptors, evenly spaced across the spectrum. Birds have close to the ideal setup. Anything more would be a wasteful and inefficient extravagance. So when scientists find animals with a lot more than four kinds of photoreceptors, there's probably something strange afoot.

"IF YOU PUT YOUR FINGERS in there, it's going to hit you," Amy Streets tells me, gesturing at a small aquarium tank in Brisbane, Australia. "If you want to try it . . ."

I do want to try it, but the animal in the tank has a reputation, and I'm nervous about testing it.

"How hard is the hit?" I ask.

"It's enough to surprise you," Streets says. "Do it."

I stick my pinky into the water. Almost instantly, there's a flash of green as a two-inch-long animal darts out and attacks me. There's a loud click, and a sharp but tolerable pain in my finger. I feel strangely proud to have taken a punch from a purple spot mantis shrimp.

Mantis shrimps, also known as stomatopods (or, more affectionately, pods), are marine crustaceans. They're related to crabs and prawns but have evolved on their own for around 400 million years. The back half looks much like a small lobster. But the front half includes two folded arms that are slung underneath the animal's body like the mantises for which they are named. In the "spearer" species, these arms end in a row of fiendish spikes; in the "smashers," they end in a bludgeoning hammer. Both groups can unfurl these weapons at astonishing speeds, and need little excuse to do so. They punch their prey into submission. They punch anything that intrudes upon their burrows. They punch each other at first contact. Mantis shrimps throw punches like humans throw opinions—frequently, aggressively, and without provocation.

Their punches are the fastest and most powerful in the world. The clubs of a large smasher can accelerate like a high-caliber bullet and hit speeds of 50 miles per hour *in water*. These animals can punch their way into crab shells, out of aquaria, and through flesh and bone. For good reason, they've been nicknamed thumb-splitters, finger-poppers, and knuckle-busters. You can understand why I was nervous about letting one hit me. Even that individual, which was too small to do any damage, moved quickly enough to vaporize the water in front of its club. This created small bubbles, which made a popping sound as they collapsed—hence the click that I heard. "The different species sound slightly different in their smacking, which is kind of fun," Streets tells me.

She takes me to another tank that contains a peacock mantis shrimp, a gaudily colored smasher whose carapace is streaked with reds, blues, and greens. Of the 500 stomatopod species, this is the most famous. It is also one of the most powerful. "*Don't* get hit by these guys," Streets says, emphatically. I take her advice. Instead of testing the peacock pod's patience, I stare at its eyes. There are two of them, which look like pink muffins wrapped in blue foil. They sit at the top of the animal's head, at the ends of mobile stalks. The left one is staring at me. The right one is looking at Streets. They are arguably the strangest eyes on the planet, and they see color in a way that no other animal shares. Of all the creatures we have encountered so far, the mantis shrimp's Umwelt is the hardest to imagine. After more than three decades, Justin Marshall, who runs the lab where Streets works, still struggles to do so.

Marshall's mother was a natural history illustrator, and his father was a marine biologist and curator of fish at London's Natural History Museum. They filled his childhood with beaches and boats, and his mind with a love of colors and marine life. In 1986, when his PhD advisor, Mike Land (whose work we met in the last chapter), asked him to choose between studying spiders, butterflies, or stomatopods, the decision was obvious. "I pretty rapidly chose mantis shrimps," Marshall tells me, "because they lived in the tropics."

He began his study by dissecting the eye of a peacock pod. Like other crustaceans, these animals have compound eyes, which consist of many separate light-gathering units. But, uniquely, each eye is split into three sections. There are two hemispheres with a distinct midband running between them, like the tropics wrapping around Earth. When Marshall looked at the midband under a microscope, he found a beautiful surprise—a kaleidoscopic array of colored blobs that were red, yellow, orange, purple, pink, and blue. At the time, crustaceans were thought to be color-blind. This animal clearly wasn't. "I remember exactly what Mike said when I showed him the slide, which was, 'Fuck! Fuck, fuck, fuck! Fuck!'" Marshall says. "I thought, oh, this must be good."

Marshall guessed that the mantis shrimp uses these colored blobs to filter the light that reaches a single class of photoreceptors. In this way, it could see colors with an eye that would normally be color-blind. To test this idea, he traveled from England to the United States to work with Tom Cronin, who had both the right equipment and a burgeoning interest in stomatopods. Over a few intense weeks, the duo worked their way through the eye, analyzing any photoreceptors they could find. And to their shock, they found not one class, but at least *11*. "It didn't make sense," Cronin tells me. "We found a new one every time we looked at a new part of the eye. That was the most miraculous period of my whole career, Justin and I working together and discovering this." The mantis shrimp "could have a color vision system that outperforms anything previously described," the duo wrote in 1989. Or as Marshall puts it, "There were even more *fucks*."

The midband consists of six rows of light-gathering units. Forget the bottom two for now; only the top four are used for color vision. Each row has three unique photoreceptors that are arranged in tiers. Row 1 has violet and blue receptors, row 2 has yellow and orange, row 3 has orange-red and red, row 4 has cyan and green, and each row has its own unique UV photoreceptor sitting on top of the others.[*] That makes 12 photoreceptor classes, including four that are devoted to ultraviolet.[†] Mantis shrimps have more classes of photoreceptors covering the ultraviolet spectrum than we have *in total*. What could they possibly be doing with so many? Could they be dodecachromats, with 12-dimensional color vision? Or are they performing four kinds of trichromacy in each of the midband rows? Either way, they must surely be connoisseurs of color, able to tell even the subtlest differences between nigh-indistinguishable hues. A coral reef looks stunning enough to us; what must it look like to a stomatopod? Speculations have run amok. Imaginations have run wild. *The Oatmeal,* an online comic strip, suggested that "where we see a rainbow, a mantis shrimp sees a thermonuclear bomb of light and beauty."

It does not. In 2014, Marshall's student Hanne Thoen did a decisive experiment that upended the mantis shrimp's growing reputation. She trained them to attack one of two colored lights in exchange for a rewarding snack. She then altered the colors until they were similar enough that the animals could no longer tell them apart. Humans can distinguish colors whose wavelengths differ by between 1 and 4 nanometers. But the mantis shrimps failed with colors that were between 12 and 25 nanometers apart, which is roughly the gap between pure yellow and orange. For all their optical extravagance, they turned out to be abysmally bad at discriminating colors. Humans, bees, butterflies, and goldfish can all outperform them.

---

* The colored blobs that Marshall first noticed are found in rows 2 and 3. As he suspected, they do act as filters, but their job is to sharpen the sensitivity of the underlying photoreceptors.

† You might have read that they have 16 photoreceptor classes. Aside from the 12 in the first four rows of the midband, there are two in the last two rows, and two more in the hemispheres. As far as anyone knows, these other four are not involved in color vision. Also, not all mantis shrimps have 12 classes. While most species live in colorful shallows, some inhabit deeper waters and have lost all but one or two of their photoreceptor classes.

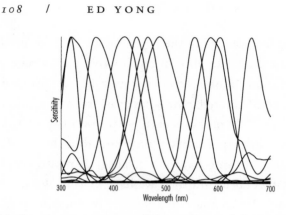

*Each curve represents one of the 12 classes of photoreceptor cells
in a mantis shrimp's eye. The peak shows the wavelength of light
to which that class is most sensitive.*

Marshall now thinks that the mantis shrimp sees colors in a unique
way. Rather than discriminating between millions of subtle shades, its
eye actually does the opposite, collapsing all the varied hues of the
spectrum into just 12 colors, like a child's coloring book. Every kind of
red stimulates the bottom photoreceptor of row 3. All shades of violet
stimulate the top receptor of row 1. And rather than comparing the
outputs of these 12 receptors through opponency, the retina just sends
its raw signals directly to the brain. The brain then uses these patterns
to recognize specific colors, as if the visible spectrum were a barcode
and its midband were a supermarket scanner. You could imagine that if
receptors 1, 6, 7, and 11 go off, the brain recognizes these signals as
prey, and the mantis shrimp attacks. If receptors 3, 4, 8, and 9 go off,
that might be a mate, and since it's a mantis shrimp, "some very careful
wooing ensues," Marshall says. The animal might not even have any
conception of color at all.

All of this remains a highly educated guess. None of the stomatopod researchers I spoke to claim to really know what these animals see.
It's possible that they might use different kinds of color vision for different tasks. For recognizing food, as in Thoen's experiment, a 12-
color look-up table might suffice. But when recognizing each other,
they might use a more conventional system that can discriminate between similar colors. After all, many of them are vividly colored them-

selves, and display their markings to each other when they meet. "For a mate, maybe the subtleties matter," Cronin says. "But that's a very difficult experiment to do."

Studying animal behavior is always challenging. But studying the behavior of mantis shrimps borders on masochism. In Marshall's lab, as part of a new experiment, Streets has been trying to train the peacock pods to attack zip ties with particular colors on them. But when she demonstrates this for me, the animals consistently make the wrong choice. At one point, one of them punches the wall of the aquarium. Another just air-punches (water-punches?) nothing in particular. I ask Streets if they're hard to train. "Oh my god," she says, shaking her head slightly. They're not motivated by food, because they don't need to eat very often. They seem to lose interest very easily, so she can only test them once a day. "I swear to god they know what the task is but they're just spiteful," Streets says.

"Do you love or hate working with them?" I ask.

"It's mixed," she says, resignedly. "At first, it's super-cool. I'm working with *mantis shrimps*! Everyone who likes this sort of thing has heard of them. But then you start working with them, and you just sit there and wonder why you're doing this."

WE, LIKE STREETS, are going to stick with mantis shrimps for a little while longer, because there's even more to their eyes than meets the . . . well, you know. Indeed, their eyes have proved to be so unusual, so complicated, and so hard to understand that many scientists around the world now study them. Nicholas Roberts and Martin How do so in Bristol, England. They take me to a room where they also have peacock mantis shrimps—eight individuals, which live in separate aquaria for each other's safety. Their tanks are at eye level, which makes it easier to see how inquisitive they are. As we approach, several of them notice and start looking at us. I press a finger against one of the tanks, and a pod named Nigel swims up. I move the finger, and he follows. It feels like I'm dragging him around.

Nigel's eyes are constantly moving, in every conceivable direction.

They move up and down, and side to side. They rotate clockwise and anticlockwise.* They rarely move together, or in the same direction. Roberts sometimes does experiments where he films mantis shrimps from above as they look at a screen. "Quite often, they'll have one eye doing the task on its own, and one eye pointed up at the camera," he tells me. As I noted in the previous chapter, we interpret active eyes as a sign of an active mind. But mantis shrimps actually have small, weak brains. The hypermobile nature of their eyes is not a sign of a probing intelligence. But it *is* the key to understanding how and what they see.

Our retinas have cone-rich foveae, where our vision is sharpest and most colorful. We train this zone onto different parts of the world by flicking our eyes from place to place. And when we spot something interesting in our peripheral vision, we redirect our gaze at it to analyze it in detailed color. Mantis shrimps do something similar. The midband sees color, but its view is confined to a thin strip of space. The hemispheres probably only see in black-and-white, but their view is panoramic. As the mantis shrimp moves its eyes around, it looks for movements and objects of interest with the hemispheres. When it spots something, it flicks its eyes across and scans the midbands over the area, as if waving two supermarket scanners along a shelf. Does the mantis shrimp start with a monochrome view, which it gradually paints with colors? "I don't think so," Marshall tells me. He suspects that "they never construct a solid two-dimensional representation of color" in their brains. Instead, as they scan with their midbands, they simply wait for anything that excites the right combination of photoreceptors.

Imagine that you're a mantis shrimp. It is a truth universally acknowledged that you are in want of something to punch. Your eyes are in constant, uncoordinated motion, the right one perusing one part of the reef, the left glancing somewhere else. Your view is monochrome because what you're after is not color but movement. You spot it to

---

* While we can perceive depth by comparing the images from our two eyes, a mantis shrimp can do the same with the three zones of a single eye. Each eye has trinocular vision and can gauge distance independently of its twin. This is a handy skill for a pugnacious animal that often loses one of its eyes in combat.

your right, and flick both your eyes across. They're scanning together now, sweeping their midbands over the mystery object. Suddenly, photoreceptors 3, 6, 10, and 11 fire. Your brain recognizes a fish. Your arms lash out and hit their mark.

This style of vision is highly efficient, and means less work for the mantis shrimp's small brain.[*] But it comes with a catch. It's very hard to detect movement with an eye that's also moving. When we walk along a street or stare out a vehicle window, our eyes actually fix on specific points ahead of us, rapidly flicking from one to the next. These flicks, or saccades, are some of the fastest movements we make, which is just as well, because as they're happening, our visual system shuts down. Our brains fill the millisecond-long gaps to create a sense of continuous vision, but that's an illusion. The same thing happens to mantis shrimps when they do their slow midband scans. "It could be that in that time, they have to turn off their motion vision," How tells me. "Their eye is moving, the world is blurring, and it's probably harder to see a predator coming in." But when the eye *isn't* scanning, most of the mantis shrimp's view is black-and-white. The jumping spiders we met in the previous chapter split different visual tasks—motion and colorful detail—among separate eyes. The mantis shrimps do the same among different portions of the same eye, and *among different periods of time*. To see movement, they have to give up color. To see color, they give up movement. "It's a time-sharing system," Cronin says. "It's not really one you'd build, but they discovered it and it has worked for them."

By this point, dear reader, you might reasonably be feeling overwhelmed by talk of photoreceptors and midbands and hemispheres and all the other absurd complications that mantis shrimps have packed into their eyes. Or maybe, after all of that, you're feeling a touch of clarity, as if you're on the cusp of imagining the stomatopod Umwelt. In either case, I have bad news for you. *There is more.*

---

[*] Imagine that you're trying to build a robot that can sneak into a local diner and find a hamburger for you. You could equip that robot with two state-of-the art cameras and an algorithm that can learn to analyze and classify the images from those cameras. But "surely it's better to just build a hamburger-detector," Marshall says. "And the best way to do that is to build a line-scan device. It's much more efficient."

Remember that light is a wave. As it moves, it oscillates. Those oscillations can usually occur in any direction perpendicular to the line of travel, but they're sometimes confined to just one plane—imagine attaching a rope to a wall and then shaking it up and down, or side to side. This kind of light is said to be *polarized,* and it is common in nature. It is formed when light is scattered by water or air, or when it reflects off smooth surfaces like glass, waxy leaves, or bodies of water. Humans are largely oblivious to polarization, but most insects, crustaceans, and cephalopods can see it in much the same way that they see color. Their eyes typically have two classes of photoreceptors that are stimulated by horizontally or vertically polarized light. By comparing their two receptors, they can distinguish between light that's polarized to different extents, or at different angles. You could call these animals *dipolats.*[*]

Mantis shrimps have this arrangement in the top hemisphere of their eyes. But in the bottom hemisphere, their polarization receptors are rotated by 45 degrees. And in rows 5 and 6 of the midband, they have something unique. Polarized light usually oscillates in a single fixed plane, but that plane can sometimes rotate, so the light travels along a twisting helix. This is called *circular polarization.* And as Marshall's postdoc Tsyr-Huei Chiou found in 2008, mantis shrimps are the only animals that can see it. The bottom rows of their midbands have photoreceptors that are tuned to circularly polarized light, spiraling either clockwise or anticlockwise. So mantis shrimps have *six* classes of polarization receptors—vertical and horizontal, two diagonals, clockwise and anticlockwise. Ever the exceptions, they are *hexapolats.*[†]

I have explained polarization and color separately, and these topics often occupy separate chapters in textbooks. But there's no reason to think that mantis shrimps treat them differently. They might well treat

---

[*] Cephalopods are more sensitive to polarization than any other animals. Shelby Temple and his colleagues found that the mourning cuttlefish can spot the difference between two kinds of polarized light whose planes of vibration differ by just one degree. These animals are color-blind, but they might use polarization as a replacement, to add rich detail to their visual world.

[†] They can also rotate their eyes to enhance the polarization contrast between an object and its background, making them the first known animals with dynamic polarization vision.

the six kinds of polarization signals as yet more colors—more channels of information that they use to recognize objects around them. But why do they need six more, when they already have 12? Why is their vision so inordinately complicated? "There are animals with much simpler visual systems that are very effective on the reef," Tom Cronin tells me. So, with mantis shrimps, "there remains the question: *What's it all for?* And no one knows."

WAIT A MINUTE. BACK UP a bit. *Why exactly can mantis shrimps see circularly polarized light?*

Unlike linearly polarized light, circularly polarized light is very rare, which is probably why no other animal has evolved the ability to see it. Indeed, the only things in the mantis shrimps' environment that reliably give off circularly polarized light . . . are the mantis shrimps themselves. One species reflects it from the large keel on its tail, which males use during courtship. Another reflects it from body parts that it displays to rivals during combat. Perhaps, then, mantis shrimps com-

Linearly polarized light

Circularly polarized light

municate using a form of light so secretive that only they can see it. There's something unsatisfyingly circuitous about this explanation, though. Circularly polarized signals would be useless if the mantis shrimps didn't already have eyes that could see them. But why would those eyes have evolved that ability if there wasn't anything for them to see? Which came first, the eye or the signal?

Tom Cronin thinks it was the eye. In the bottom two rows of the midband, the photoreceptors are arranged in a way that just happens to untwist circularly polarized light so that it becomes linearly polarized instead. That's how mantis shrimps can sense it. This arrangement might have been an anatomical fluke—a quirk of their compound eye that gave them the ability to see circularly polarized light, even when there was little of that light around to see. The ancestral mantis shrimps effectively had an accidental sense. They exploited it by slowly developing structures on their shells that reflect circularly polarized light, evolving signals that suited their eyes. This happens a lot. Signals are meant to be seen, and so the colors that adorn the fur, scales, feathers, and exoskeletons of animals are shaped by the colors that the animals' eyes can perceive. In viewing nature's paintings, eyes define its palette.

Primates, for example, evolved trichromacy to better spot young leaves and ripe fruits. And once they added red to their Umwelt, they began evolving patches of bare skin that could convey messages by flushing with blood. The red faces of rhesus macaques, the red rumps of mandrills, and the comically red and bald heads of uakaris are all sexual signals made possible by trichromatic vision.

Most of the fish in coral reefs are also trichromats. But since red light is strongly absorbed by water, their sensitivities are shifted toward the blue end of the spectrum. This explains why so many reef fish, like the blue tang that stars in Pixar's *Finding Dory*, are blue and yellow. To their version of trichromacy, yellow disappears against corals, and blue blends in with the water. Their colors look incredibly conspicuous to snorkeling humans, because our particular trio of cones excels at discriminating blues and yellows. But the fish themselves are beautifully camouflaged to each other, and to their predators.

The color vision of predators diversified the patterns of Central

America's strawberry poison frog—a single species that comes in 15 incredibly different forms. One is lime green with cyan stockings. Another is orange with black spots. These colors are so varied as to seem almost random, but there's method to the visual madness. These frogs are poisonous, and the most toxic ones are also the most conspicuous. But as Molly Cummings and Martine Maan discovered, they are conspicuous only to birds and not to other predators like snakes. It is likely that tetrachromatic avian eyes drove the evolution of the outlandish amphibian skins. This makes sense: The colors are intended as warnings, and across the generations, frogs whose hues were best suited to the vision of their predators were more likely to go unattacked. And Cummings and Maan showed that you can work out who those predators are—in this case, birds—by studying the colors of their prey. Since eyes define nature's palette, an animal's palette tells you whose eyes it is trying to catch.

You can apply the same logic to flowers. In 1992, Lars Chittka and Randolf Menzel analyzed 180 flowers and worked out what kind of eye would be best at discriminating their colors. The answer—an eye with green, blue, and UV trichromacy—is exactly what bees and many other insects have. You might think that these pollinators evolved eyes that see flowers well, but that's not what happened. Their style of trichromacy evolved hundreds of millions of years before the first flowers appeared, so the latter must have evolved to suit the former. Flowers evolved colors that ideally tickle insect eyes.

I find these connections profound, in a way that makes me think differently about the act of sensing itself. Sensing can feel passive, as if eyes and other sense organs were intake valves through which animals absorb and receive the stimuli around them. But over time, the simple act of seeing recolors the world. Guided by evolution, eyes are living paintbrushes. Flowers, frogs, fish, feathers, and fruit all show that sight affects what is seen, and that much of what we find beautiful in nature has been shaped by the vision of our fellow animals. Beauty is not only in the eye of the beholder. It arises because of that eye.

———

IT'S A SUNNY AFTERNOON in March 2021, and I'm taking Typo, my corgi, for a walk. As we approach a neighbor who is rinsing his car with a hose, Typo stops, sits, and stares. As I wait with him, I notice a rainbow in the water arcing from the hose. To Typo's eyes, it goes from yellow to white to blue. To mine, it goes from red to violet, with orange, yellow, green, and blue in the middle. To the sparrows and starlings perched in a tree behind us, it goes from red to ultraviolet, with perhaps even more gradations in between.

I noted at the start of this chapter that color is fundamentally subjective. The photoreceptors in our retinas detect different wavelengths of light, while our brains use those signals to construct the sensation of color. The former process is easy to study; the latter is extremely difficult. This tension between reception and sensation, between what animals can detect and what they actually experience, exists for most of the senses. We can dissect a mantis shrimp's eye and work out what every component does, but still never really know how it actually sees. We can work out the exact shape of the taste receptors on a fly's feet without ever understanding what it experiences when it lands on an apple. We can chart how an animal reacts to what it senses, but it's much harder to know how it *feels*. And that distinction becomes especially difficult—and important—when thinking about pain.

The side-facing slits of a dog's nostrils allow its exhalations to waft more odors into its nose.

Clonal raider ants have been marked with paint so they can be easily tracked.

*Organs of smell come in
varied forms, including the
trunks of elephants, the
beaks of albatrosses, and the
forked tongues of snakes.*

*With receptors on their feet, butterflies and
other insects can taste things by landing on them.*

*Catfish are swimming tongues, with taste buds
dotted all over their skins.*

*A jumping spider's central eyes offer sharp vision,
while the pair on the side tracks movements.*

*The killer fly's ultrafast vision allows it to capture
quick-flying insects in the span of a human blink.*

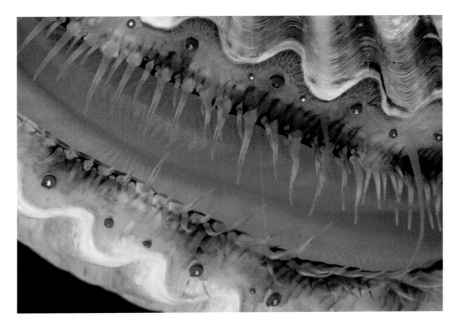

The bay scallop has dozens of bright blue eyes along the rim of its shell.

The brittle star's entire body is an eye, but only during the daytime.

The huge top part of a male mayfly's eye allows it to spot passing females.

A chameleon can look forward and behind simultaneously with its independent eyes.

E. A. Lazo-Wasem
Yale Peabody Museum

With two eyes fused into a single cylinder, Streetsia challengeri can see above, below, and to the sides, but not in front.

*In darkness so intense that you couldn't see your own hand,
this nocturnal sweat bee can still spot its small jungle nest.*

*The elephant hawkmoth can see the colors
of flowers, even under dim starlight.*

Typo the corgi, a very good boy, is modeling the difference between the trichromatic color vision of (most) humans and the dichromatic vision of dogs.

Many natural patterns, including the markings on
flowers and the facial stripes of the ambon damselfish, are visible only to eyes that can see ultraviolet.

The bib of the broad-tailed hummingbird and the
wing bars of the Heliconius erato *butterflies reflect
ultraviolet colors that humans can't perceive.*

The peacock mantis shrimp sees color in a completely different way than other animals do, using the midband of its three-part eyes.

The naked mole-rat is insensitive to the pain of acids and capsaicin, the chemical that gives chilies their kick.

The thirteen-lined ground squirrel can hibernate through the winter because it is insensitive to cold temperatures that we'd find painful.

These animals can all sense the infrared radiation emanating from warm objects. Fire-chaser beetles do so to find burning forests, while vampire bats and rattlesnakes track down warm-blooded prey.

Sea otters use their sensitive paws to quickly feel
for prey they can't see, while red knots do the same
by probing into sand with their bills.

Tactile organs come in many forms—
the nose of the star-nosed mole, the sting
of the emerald jewel wasp, the facial
feathers of the crested auklet, and the
whiskers of a mouse.

Manatees manipulate objects and greet each other with their exquisitely touch-sensitive lips.

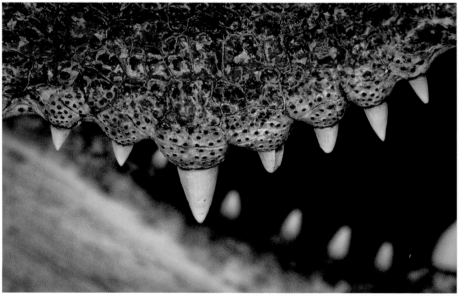

The bumps on a crocodile's snout can detect gentle ripples made by its prey.

4.

The Unwanted Sense

Pain

IN A WARM ROOM THAT SMELLS SWEETLY OF CORN, I'M HOLDING a small rodent in my gloved hand. Pink and mostly hairless, it seems less like a rat or guinea pig and more like a finger that's been soaking in a bath too long. It almost looks embryonic, even though it is a fully grown adult. Its eyes are black pinpricks. Its long incisors stick out in front of its lips. Its loose skin feels tough but is so translucent that I can make out its internal organs, including the dark outline of its liver. It is a naked mole-rat. Its appearance is the least strange thing about it.*

Naked mole-rats are exceptionally long-lived for rodents, with life spans of up to 33 years. Their lower incisors can splay apart and come together to grasp objects. Their sperm are misshapen and sluggish. They can survive for up to 18 minutes without oxygen, a hardship that no mouse can endure for more than a minute. They live in cooperative colonies like those of ants and termites, with one or more breeding queens and dozens of sterile workers. A single naked mole-rat, like the one I am holding, is an unusual sight. So is a naked mole-rat in the open. They normally live within labyrinthine underground tunnels, which they constantly expand, remodel, and patrol in their quest for

---

* Naked mole-rats are so weird that their bizarre traits have often been mythologized, and many of the claims that surround them are untrue. I highly recommend the paper "Surprisingly Long Survival of Premature Conclusions About Naked Mole-Rat Biology," which is an important corrective to some of these myths.

nutritious tubers. Thomas Park replicates this network in his Chicago-based lab with interconnected plastic cages filled with toilet paper rolls and wood chips. Some of the mole-rats are instinctively chewing the walls of these containers in a bid to expand their artificial tunnels, and kicking their legs back as if removing loosened dirt. Others are resting in the nesting chamber, a pile of wrinkled bodies curled around their queen. She is much larger than they are, and her belly bulges with unborn pups. "For naked mole-rat people, that's a beautiful sight," Park tells me. I take his word for it.

In their wild burrows, naked mole-rats also sleep in large huddled piles to keep warm. Those at the bottom rapidly run out of oxygen, which is probably why they have evolved to withstand the gas's absence. They've also been forced to tolerate carbon dioxide, which builds up in the nesting chambers with every exhalation. Carbon dioxide normally makes up 0.03 percent of the air in an average room. If levels shot up to 3 percent, you'd hyperventilate and panic. Meanwhile, the gas would dissolve in the wet surfaces of your mucous membranes, acidifying them. Your eyes would sting. Your nose would burn. You'd wince in distress. You'd try to get away. But a naked mole-rat wouldn't flee or flinch.

Park demonstrated this with an arena that's infused with carbon dioxide at one end and regular air at the other. A mouse would scurry to the latter region. But naked mole-rats were comfortable in the thick carbon dioxide, moving away only when levels reached a preposterous 10 percent. They simply don't find acids painful. They'll sniff strong vinegary fumes with no sign of discomfort. They don't register drops of acid beneath their skin—the equivalent of squirting lemon juice into a cut in your hand. And they're similarly unperturbed by capsaicin, the chemical that gives chili peppers and pepper spray their burn. While capsaicin inflames our skin, leaving it hypersensitive to heat, it has no such effect on naked mole-rats. It's not that these animals can't feel pain, as is commonly said. They dislike pinches and burns, and they'll recoil from the chemical responsible for mustard's sting. But they're oblivious to several noxious substances that we find painful.

Our experience of pain depends on a class of neurons called nociceptors. (The word is pronounced with a soft *c,* and comes from the Latin word *nocere,* meaning "to harm.") The naked tips of these neurons pervade our skin and other organs. They are loaded with sensors that detect harmful stimuli—intense heat or cold, crushing pressures, acids, toxins, and chemicals released by injuries and inflammation.* Nociceptors vary in their size, how excitable they are, and how quickly they transmit information—qualities that collectively sculpt a landscape of pricks, stabs, burns, throbs, cramps, and aches that we are unfortunate enough to experience.

Almost all animals have nociceptors, and naked mole-rats are no exception. But theirs are fewer in number and have been disabled in several ways. Those that would normally be activated by acids are instead blocked by them. Those that detect capsaicin still do so, but don't produce the neurotransmitters that normally convey their signals to the brain. Some of these changes seem easy to explain: If naked mole-rats could still feel acidic pain, the carbon dioxide in their nesting chambers would probably lead to an agonized sleep. "But we don't know why they can't respond to capsaicin," Park tells me. Perhaps they eat an especially spicy tuber to which they've become resistant? Or perhaps it was the opposite: After millions of years in a relatively safe environment, they simply lost sensory abilities that they no longer needed. Either way, their imperviousness tells us that there's nothing inherently painful about either capsaicin or acids.

Several hibernating mammals that, like the naked mole-rat, must deal with building levels of carbon dioxide are also insensitive to acids. Birds that carry the seeds of pepper plants don't feel the burn of capsaicin. Humans are insensitive to nepetalactone, a chemical produced by the catnip plant that is intensely irritating to mosquitoes. The grasshopper mouse, a surprisingly ferocious predator of scorpions, can shrug off stings that feel to humans like cigarettes being stubbed out on our skin. The mouse's nociceptors have evolved to *stop* firing when

---

* Unlike vision, smell, and hearing, which detect specific stimuli—light, molecules, sound— nociception detects a class of very different stimuli that are united by their potential to cause harm. It's a mishmash sense, combining elements of smell, which we've already explored, and others, like touch, that we are about to.

they recognize a scorpion's toxins, turning venom that would normally be excruciating into a painkiller.

People often assume that pain feels the same across the entire animal kingdom, but that is not true. Much like color, it is inherently subjective and surprisingly variable. Just as wavelengths of light aren't universally red or blue, and odors aren't universally fragrant or pungent, nothing is universally painful, not even chemicals in scorpion venom that specifically evolved to inflict pain. Pain, in warning animals of injury and danger, is crucial to their survival. And while all animals have things to be wary of, they differ in what they must avoid and what they must tolerate. That makes it notoriously tricky to tell what an animal might find painful, whether an animal is experiencing pain, or whether it even can.

IN THE EARLY 1900S, the neurophysiologist Charles Scott Sherrington noted that the skin has "a set of nerve-endings whose specific office it is to be amenable to stimuli that do the skin injury." Those nerves would "evoke skin pain" if connected to the brain, but they could still trigger defensive reflexes "devoid of psychical feature" if said connections were cut. A dog, for example, would still pull its paw away from a hard squeeze even after a spinal injury. Sherrington wanted a separate term to describe the act of sensing harmful stimuli as distinct from the painful feelings they produce—a term that would have "the advantage of greater objectivity." He came up with *nociception*.

Over a century later, scientists and philosophers still make the distinction between nociception and pain. Nociception is the sensory process by which we detect damage. Pain is the suffering that ensues. Last week, when I accidentally touched a hot pan, the nociceptors in my skin sensed the scalding temperatures. That's nociception, which triggered a reflex that forced my arm to withdraw *before I realized what was happening*. Shortly after, signals from those nociceptors reached my brain, which produced feelings of discomfort and distress. That's pain. The two are intimately linked but also distinct. Nociception occurred in my hand (and spinal cord); the pain was produced by my brain. They are the sensory and emotional halves of a process that, to most of us, feel inseparable.

But they *can* be separated. Amputees who feel the phantom remnants of their old limbs can experience pain without nociception. Other people are congenitally indifferent to pain—from birth, they're aware of sensations that others would find painful, but aren't distressed by them.* Some painkillers duplicate this effect by acting on the central nervous system to dull pain without affecting nociception. "I took Vicodin after having surgery on my jaw," Robyn Crook, a neuroscientist who studies pain, tells me. "I would still be fully aware that the sensation was there, but I felt very serene about it." People can also learn to ignore or even enjoy things that trigger nociceptors, like mustard, chilies, or intense heat.†

To be clear, the separation between nociception and pain does not make the latter any less real. People (and especially women) with chronically painful disorders have long been disbelieved and neglected by the medical establishment. They've been wrongly told that their suffering is just in their heads, or the result of mental health problems like anxiety. Pain is easy to dismiss in this way because it is subjective. And thanks to the unfortunate persistence of dualism—the outdated belief that the mind and body are separate—people often equate *subjective* with *woolly,* and *psychological* with *imagined.* This is harmfully wrong. It's not the case that nociception is a physical process of the body, while pain is a psychological process of the mind. Both arise from the firing of neurons. It's just that in humans, nociception can be confined to the peripheral nervous system, while with pain, the brain is

---

* This condition can be dangerous. Children and babies who have it don't learn that injuries are dangerous, and often bite their own fingers, bang their heads against objects, or scald themselves. Those who survive are sometimes exploited. The first documented case of congenital indifference to pain was a man who made a living at a circus, as a human pincushion. One Pakistani boy who had the condition would perform on the streets by stabbing knives through his arms. He died on his 14th birthday after jumping off a roof.

† I highly recommend Leigh Cowart's *Hurts So Good*—an exploration of masochists, ultramarathoners, icy ocean bathers, and other people who engage with pain on purpose.

always involved. Pain requires some degree of conscious awareness. Nociception can exist without it.

Nociception is an ancient sense. It is so widespread and consistent across the animal kingdom that the same chemicals, opioids, can quell the nociceptors of humans, chickens, trout, sea slugs, and fruit flies—creatures separated by around 800 million years of evolution. But since pain is subjective, it is difficult to tell which creatures have it. Humans can barely do that with each other. "You can tell me you have a screaming headache and I'd have no idea what that means for you," Crook tells me, "and we're the same species, with brains that are basically the same." Scientists who study human pain still largely rely on people's own accounts, and animals obviously can't talk about their feelings.* Our only recourse is to read the tea leaves of their behavior.

Pinch the foot of a mouse (or naked mole-rat) and it will pull its limb away, and probably lick and groom it. Offer painkillers and it will accept. These actions resemble what a hurt human might do, and since a rodent's brain is similar enough to ours, we can reasonably guess that its nociceptive reflex is accompanied by pain. But such arguments by analogy are always fraught, especially when it comes to animals with very different bodies and nervous systems. A leech will writhe when pinched, but are those movements analogous to human suffering, or to an arm unconsciously pulling away from a hot pan? Other animals may hide their pain. Social creatures can call for help by whining when they're injured, but an anguished antelope would likely keep quiet lest its distress calls convey weakness to a lion. The signs of pain vary from one species to another. How, then, do you tell if an animal is experiencing it?

For many historical thinkers, who believed animals incapable of emotions or conscious experiences, the question was irrelevant. The seventeenth-century dualist René Descartes thought of them as automata. Paraphrasing his views, the philosopher and priest Nicolas Malebranche wrote that "animals eat without pleasure, cry without

---

* Brain scanners aren't helpful: It's unclear what patterns of brain activity would indicate a conscious mind, let alone a conscious mind in pain, let alone a conscious, non-human mind in pain.

pain, grow without knowing it: they desire nothing, fear nothing, know nothing." Such views have changed in recent decades, and most scientists would now agree that mammals can feel pain. But fierce debates are still raging around other animal groups, including fish, insects, and crustaceans.* At the core of these lingering controversies is the distinction between nociception and pain. That distinction "is a relic of attempts to emphasize differences between humans and other animals or between 'higher' and 'lower' animals," wrote Donald Broom, a biologist who specializes in animal welfare. After all, in other senses, the actions of sensory receptors and the subjective experiences produced by the brain don't get different names. Scientists who study eyes don't get into arguments about whether humans have vision and fish merely have photoreception.

But as we saw in the earlier chapters, there *is* a difference between what the cells in a retina detect and the conscious experience of seeing. Vision scientists *do*, in fact, make distinctions between simple photoreception and spatial vision—remember the four stages of Dan-Eric Nilsson's model of eye evolution. They suspect that some creatures, like scallops, might stretch our concept of vision by seeing without scenes. They recognize that some aspects of our visual world, like colors, are constructs of the brain, and that some animals that can sense different wavelengths of light, like mantis shrimps, might not perceive colors at all.

In the chemical senses—smell and taste—it's also possible to sense and react to a stimulus without being aware of it. You're doing it right now. Humans have taste receptors throughout our bodies—not over our skin or feet, but in our internal organs. Sweet receptors in our gut control the release of appetite-controlling hormones. Bitter receptors in our lungs recognize the presence of allergens and trigger an immune response. All of this happens without us knowing. Similarly, the taste receptors on a mosquito's foot could trigger a reflex that makes it withdraw from DEET without ever passing information to the insect's

---

* Until the 1980s, there were still debates about whether premature or newborn human babies could perceive pain or would benefit from painkillers.

brain. The taste receptors on a fly's wing can initiate a grooming reflex if they detect microbes, without the fly needing to know what a microbe or a wing is. To an observer, those behaviors look remarkably like disgust, but we have no idea if such emotions are playing out in the insect's brain.

Broom is right that we rarely distinguish between the raw act of sensing and the subjective experiences that ensue. But that's not because such distinctions don't exist. It's because they usually don't matter. Questions about what a scallop sees, or whether birds and humans see the same red, are philosophically interesting. But the distinction between pain and nociception is a morally, legally, and economically vital matter, which affects our cultural norms around catching, killing, eating, or experimenting on animals. Pain (or nociception, if you prefer) is the unwanted sense. It is the only one whose *absence* (in naked mole-rats or grasshopper mice) feels like a superpower. It is the only one that we try to avoid, that we dull with medication, and that we try to avoid inflicting upon others.

SCIENTISTS WHO WORK ON vision or hearing can play images and sounds at the animals they're studying. But those who study pain have to harm the creatures they work with in the pursuit of knowledge that might improve the welfare of those same creatures. They want to use as few animals as possible, but have to use enough that their results are statistically sound. Their work is morally challenging and often frustrating. "People either feel that animals absolutely feel pain like we do so it's a stupid question to research, or that they don't feel pain like we do so it's a stupid question to research," Robyn Crook tells me. "There's not a lot of middle ground where people are agnostic."

Fish exemplify the fraught nature of pain research. In the early 2000s, Lynne Sneddon, Mike Gentle, and Victoria Braithwaite injected trout in the lips with bee venom or acetic acid, the substance that gives vinegar its kick. Unlike fish that had been injected with saline, these unfortunate individuals began breathing heavily. They stopped eating for several hours. They lay on the gravel bottoms of their tanks and rocked from side to side. Some of them rubbed their lips against the gravel or the walls of their tanks. They no longer kept their distance from unfamiliar objects, as if something was distracting them—an effect that vanished when they were injected with morphine. Sneddon and her colleagues couldn't see how these actions, which persisted well after the injections, could be attributed to mere nociception. They saw animals in pain.

These studies, published in 2003, were groundbreaking. Scientific texts, angling magazines, and Nirvana lyrics had all promulgated the belief that fish don't feel pain. The struggles of a hooked fish were meant to be simple reflexes, rather than signs of suffering. No one even knew if fish had nociceptors, until Sneddon's team confirmed that they do. She tells me that when she began her work, she would ask veterinary students or angling groups if fish experience pain. "A few people would say yes," she says. Now, after 17 years of mounting evidence, "pretty much everyone puts their hand up."

When fish nociceptors fire, the signals travel to parts of the brain that deal with learning and other behaviors more complex than simple reflexes. Sure enough, when the animals are pinched, shocked, or injected with toxins, they'll behave differently for hours or days—or until they get painkillers. They'll make sacrifices to get those drugs, or to avoid further discomfort. In one experiment, Sneddon showed that zebrafish prefer to swim in an aquarium full of plants and gravel than in one that's empty. But if she injected the fish with acetic acid and dissolved a painkiller in the water of the barren aquarium, they abandoned their normal preferences and chose the boring but soothing environment instead. In another study, Sarah Millsopp and Peter Laming trained goldfish to feed in a specific part of an aquarium, and then gave them an electric shock. The fish fled and stayed away for days, forgoing food in the process. They eventually returned, but did so more quickly if they were hungry or if the shock had been mild. Their initial escape might have been reflexive, but they then weighed up the pros and cons of avoiding further harm. As Braithwaite wrote in her book, *Do Fish Feel Pain?,* "There is as much evidence that fish feel pain and suffer as there is for birds and mammals."

But a group of vocal critics remains unconvinced.* They accuse Sneddon and others of anthropomorphism, and seeing the fish in their studies through human eyes. More likely, they argue, those fish were behaving unconsciously. After all, their brains are capable of little else. Our brains are topped by a thick mushroom cap of neural tissue called the neocortex. It's organized like an orchestra, with many specialized sections that act together to produce the music of consciousness and the lament of pain. But fish brains lack a neocortex, much less a highly organized one. "Fishes are neurologically equipped for unconscious nociception and emotional responses, but not conscious pain and feelings," seven skeptics wrote in 2014, in a paper entitled "Can Fish Really Feel Pain?"

Ironically, this argument is itself grossly anthropomorphic. It blithely assumes that the neocortex must be necessary for pain in all animals, since that's the case in humans. But if that's true, then birds can't feel pain, either, since they also lack a neocortex. And by the same faulty logic, fish must lack all the other mental skills that are rooted in the neocortex, like attention, learning, and many of the other abilities that they plainly possess. Animals often evolve different solutions to the same problems, and different structures for the same tasks. To argue that fish can't feel pain because they lack a human-like neocortex is like saying that flies cannot see because they lack camera eyes.

The critics do have a point, though: We cannot assume that all animals are capable of pain or other conscious experiences. Consciousness isn't an inherent property of all life. It arises from nervous systems, and while those systems might not need a neocortex, they do need enough processing power. For perspective, crabs and lobsters use a cluster of about 30 neurons to control the rhythmic movements of their stomachs. Meanwhile, the nematode worm *C. elegans* has 302 neurons in total. Can the worm produce subjective experiences with just 10 times as many neurons as a crab needs to churn its stomach? That doesn't

---

* For a sense of the debate, compare reviews written by Sneddon and by a group of authors led by James Rose. You can also read Brian Key's paper entitled "Why Fish Do Not Feel Pain," and the dozens of replies that largely contradict him.

seem likely. "At some point the nervous system is just too small," says Robyn Crook. "But how much brainpower is enough?" Is it the 86 billion neurons of humans, the 2 billion of a dog, the 70 million of a mouse, the 4 million of a guppy, or the 100,000 of a fruit fly? Crook doubts that the 10,000 neurons of a sea slug are enough, but "it's not like someone can say you need 10,057 neurons," she tells me.

What matters is not just the total tally of neurons but the connections between them. In human brains, hundreds of thousands of neurons connect the different sections of our cortical orchestra. These links allow us to play the full symphony of a painful experience, melding sensory cues with negative emotions, bad memories, and more. But such links are much sparser in the brains of insects. A fruit fly's nociceptors connect to a part of the brain called the mushroom body, which is critical for learning. But the mushroom body only has 21 output neurons that lead to other brain regions. The fly may well learn to avoid a nociceptive stimulus, but do those lessons come with the bad feelings that are so inherent to human suffering? Insects might not even have a brain region that processes emotions, like the amygdala does in humans. "That makes it difficult to understand what the subjective experience of pain would be like in an insect," Shelley Adamo, a physiologist who studies insect behavior, tells me.

Then again, Adamo adds, how would you know what an insect's emotional center looks like? Given how little we know about how *human* brains work, let alone how those of other animals are wired, it feels premature to make definitive proclamations about whether any neurological feature is necessary for experiencing pain. And some animals seem to overperform the limits of their simple brains.

IN 2003, AT A PUB in Killyleagh, Northern Ireland, the biologist Robert Elwood bumped into celebrity chef Rick Stein. "We've got a mutual interest in crustaceans: I study their behavior and you cook them," Elwood remembers saying. And Stein immediately asked, "Do they feel pain?" Elwood didn't think they could, but he didn't really know. Afterward, the question gnawed at him, and he started trying to answer it. "I thought it would be a quick project, and we could move on," he tells me. "It didn't turn out that way."

Elwood studied the common hermit crab, which frequents European beaches and tucks its soft abdomen into empty seashells. These shells are valuable property, and the crabs are vulnerable without them. But Elwood and his colleague Mirjam Appel found that they will nonetheless evacuate if given a small electric shock. These flights looked reflexive, but the crabs didn't always flee. It took a stronger shock to force them out of their favored periwinkle shells than it did to evict them from the less desirable flat-top shells. And they were half as likely to abandon their shells if they could smell the scent of predators in the water. "That told me that this isn't a reflex," Elwood says. Instead, evacuation is a decision the crabs make after weighing up several sources of information.

The crabs also behaved differently long after the shocks. After fleeing, they wouldn't return to their shells, despite being dangerously exposed. They groomed the part of their abdomens that got shocked. And even when they didn't relinquish their shells, they were quicker to accept a new one without the usual careful investigations. These data, Elwood says, are consistent with the idea of pain, but it's impossible to know what crustaceans are really feeling. "I'm often asked if crabs and lobsters feel pain," he tells me, "and after 15 years of research, the answer is maybe."

Crustaceans are the evolutionary cousins of insects and have similarly simple nervous systems. And yet, Elwood's crabs behaved in apparently complex ways. How do we reconcile that inconsistency? If an animal's actions don't match what its brain is theoretically capable of, are we overinterpreting its behavior or underestimating its nervous system? Sneddon and Elwood argue that it's the latter. Adamo would say it's the former. And it really isn't clear who is right, or if they all are.*

---

* Debates around animal pain can be extremely acrimonious. But notably, Adamo, Sneddon, and Elwood all jointly published a review on defining animal pain, and all speak good-naturedly about each other's views, even though they disagree.

"Fussing about the size of the brain may be a red herring," Adamo tells me. Instead, she prefers to think about the evolutionary benefits and costs of pain. By costs, she means energy, not agony. Evolution has pushed the nervous systems of insects toward minimalism and efficiency, cramming as much processing power as possible into small heads and bodies. Any extra mental ability—say, consciousness—requires more neurons, which would sap their already tight energy budget. They should pay that cost only if they reaped an important benefit. And what would they gain from pain?

The evolutionary benefit of nociception is abundantly clear. It's an alarm system that allows animals to detect things that might harm or kill them, and take steps to protect themselves. But the origin of pain, on top of that, is less obvious. What is the adaptive value of suffering? Why should nociception *suck*? Some scientists suggest that unpleasant emotions might have intensified and calcified the effect of nociceptive sensations, so that animals not only avoid what is currently hurting them but also learn to avoid it in the future. Nociception says, "Get away." Pain says, ". . . and don't go back." But Adamo and others argue that animals can learn to avoid dangers perfectly well without needing subjective experiences. After all, look at what robots can do.

Engineers have designed robots that can behave as if they're in pain, learn from negative experiences, or avoid artificial discomfort. These behaviors, when performed by animals, have been interpreted as indicators of pain. But robots can perform them without subjective experiences. This is not to claim, as Descartes did, that animals are unthinking, unfeeling automata; as Adamo says, "No robot is as sophisticated as an insect." Her point is that insect nervous systems have evolved to pull off complex behaviors in the simplest possible ways, and robots show us how simple it is possible to be. If we can program them to accomplish all the adaptive actions that pain supposedly enables without also programming them with consciousness, then evolution—a far superior innovator that works over a much longer timeframe—would surely have pushed minimalist insect brains in the same direction. For that reason, Adamo thinks it's unlikely that insects (or crustaceans) feel pain. Or, at least, their experience of pain is likely to be very different from ours. The same goes for fish. "I would expect they have *something,* but what?" she says. "It's probably not the same."

This point is crucial. The controversies about animal pain often assume that they either feel exactly what we feel or nothing at all, as if they're either little people or sophisticated robots. This dichotomy is false, but it persists because it's difficult to imagine an intermediate state. We know that some people have different *thresholds* of pain than others, just as we know that some have blurrier vision. But a qualitatively different version of pain is as conceptually challenging as a scallop's scene-less vision. Could pain exist without consciousness? If you strip the emotion out of pain, are you just left with nociception, or a gray area that our imaginations struggle to fill? Perhaps more than for other senses, it is easy to forget that pain can *vary,* and hard to conceive of how it might.

IN SEPTEMBER 2010, the European Union extended its regulations on animal research to cephalopods—the group that includes octopuses, squid, and cuttlefish. Being invertebrates, cephalopods aren't usually covered by laws that protect the welfare of backboned lab animals like mice or monkeys. But they also have much larger nervous systems than most invertebrates—500 million neurons in an octopus, compared to 100,000 in a fruit fly. They show intelligent and flexible behaviors that surpass those of some vertebrates like reptiles and amphibians. And, as the EU noted in its directive, "there is scientific evidence of their ability to experience pain, suffering, distress and lasting harm." That statement came as a surprise to Robyn Crook, who had worked with cephalopods and knew of no such evidence. The EU seemed to have assumed that apparently intelligent animals must be capable of suffering. But at the time, no one even knew if they had nociceptors, let alone if they experienced pain. "There was a huge disconnect between what science knew at that point and what legislators presumed science knew," Crook tells me.

She began to bridge that gap, starting with the longfin squid— a foot-long species that is commonly fished in the North Atlantic. This animal frequently loses the tips of its arms, either to aggressive rival squid or to the pincers of crabs. Crook mimicked these wounds with a scalpel. As expected, the squid jetted away while releasing clouds of distracting ink, and changed colors to blend in with their surroundings. A few days later, they were still quicker to flee and hide. But surprisingly, they never touched, groomed, or cradled their wounds, the way humans, rats, and even hermit crabs do. They could easily reach their stump with any of their other seven arms, but they didn't try.

Even more surprisingly, Crook found that injured squid behave as if their entire bodies were sore. When humans and other mammals get cut or bruised, the damaged area is painful but the rest of the body isn't. If I singe my hand, it hurts when I prod the burn but not when I poke my foot. But when Crook damaged one of the squid's fins, the nociceptors on the opposite fin were just as excitable as those on the wounded side. Imagine if your entire body became delicate to the touch whenever you stubbed your toe: That's a squid's reality. "When they're injured, their whole body becomes hypersensitive," Crook tells me. "They go from being normal to this potential world of pain." This might explain why they don't groom their wounds. They can sense that they've been hurt, but they might not be able to tell *where*.

For mammals, the localized nature of pain allows us to protect and clean vulnerable body parts, while getting on with the rest of our lives. Why should squid lack such a useful source of information? One possibility, Crook says, "is that everything in the ocean will eat a squid." Injured squid are especially attractive to predatory fish, either because they are more conspicuous or because they look (or smell) like easier prey. By setting their entire bodies on high alert, they might be better at evading attacks that could come from any direction.* Body-wide sensitivity also makes sense for animals that cannot physically reach most of their bodies. What good would it do them to know that a fin has been injured when they can't do anything about it?

Octopuses are different. Unlike squid, they *can* touch every part of

---

* Crook confirmed this in an experiment. She showed that sea bass will specifically target injured squid, which take evasive maneuvers earlier than uninjured animals. If she treated them with anesthetic, she also slowed their escapes, and reduced their odds of survival.

their bodies. They can even reach inside themselves to groom their own gills—the equivalent of a human putting a hand down their throat to scratch their lungs. And unlike squid, which are stuck in open-water groups and can't take a day off, octopuses can hole up in solitary dens until they feel better. Since they have the time and dexterity to tend to their injuries, it would make sense for them to know where their wounds are. And as Crook showed, they do. Octopuses will sometimes break off an arm if its tip is injured. When that happens, the stump will be more sensitive than the arms around it, and octopuses will cradle that stump in their beaks. In her latest study, published in 2021, Crook found that octopuses will avoid places where they've been injected with acetic acid, but gravitate to places where they receive painkillers. And once they're injected with local anesthetic, they stop grooming their injured arms. In her latest paper, Crook is unambiguous: "Octopuses are capable of experiencing pain."

Even before that study was published, Crook told me that she runs her lab on the assumption that cephalopods feel pain. She does studies that could improve cephalopod welfare, like checking if anesthetics work on them. She uses as few animals as possible (while still being statistically robust) and ensures that their injuries are minimal. Thinking through the ethics of animal research, especially when that research is about pain, is not easy, "but I think it *should* be hard," she says. "You should be distressed by what you're doing to an animal in an experiment, even if it's not painful. Animals don't sign up for this. Even if my broader goal is to alleviate animal suffering, the animal sitting in the tank doesn't know that."

Many scientists who study pain feel the same. They argue that whether cephalopods, fish, or crustaceans feel what humans do or experience something radically different, there is enough evidence to invoke the precautionary principle. "It's highly possible that these animals can suffer," Elwood says, "and we should consider ways of avoiding that suffering."

THE MANY DEBATES ABOUT pain in animals often revolve around a simple question: *Do they feel it?* And lurking behind that question are several implicit ones. *Is it okay to boil a lobster? Should I stop eating octopus? Can I go fishing?** When we ask if animals can feel pain, we're asking less about the animals themselves, and more about *what we can do to them*. That attitude limits our understanding of what animals actually sense.

There is much more to pain than its presence or absence. Shelley Adamo is right that we need to understand more about its benefits and costs. Pain does not exist for its own sake. There's no reason why anything should hurt. Things hurt so that animals can do something with that information. And without understanding their needs and their limitations, it's hard to interpret their behavior correctly.

Insects, for example, often do alarming things that seem like they should be excruciating. Rather than limping, they'll carry on putting pressure on a crushed limb. Male praying mantises will continue mating with females that are devouring them. Caterpillars will continue munching on a leaf while parasitic wasp larvae eat them from the inside out. Cockroaches will cannibalize their own guts if given a chance. These behaviors "strongly suggest that if a pain sense is present, it is not having any adaptive influence on the behavior," wrote Craig Eisemann and colleagues in 1984. But maybe they simply show what insects are willing to endure? Maybe cockroaches and mantises prioritize protein and procreation over pain, tolerating it in the same way that athletes and soldiers can in the middle of competition or combat. Perhaps caterpillars don't feel the pain of being eaten alive because they can't alleviate that pain.

Consider also the squid and octopuses. Both are cephalopods, but they've been evolving separately for more than 300 million years, roughly the same amount of time that separates mammals and birds.

---

* The answers to these questions could fill a completely different book. Here, I will only note that subjective pain is just one thing to consider when thinking about animal welfare, and may not even be the most important. "We could simply accept that nociception itself is more than enough to affect an animal's welfare, and thus may require treatment," the veterinarian Frederic Chatigny wrote. "Pain, while defined by consciousness, is not necessary for an animal's wellbeing to be negatively affected."

Their bodies and lifestyles are utterly different, so it's no surprise that their nervous systems function very differently after injury. Rather than asking if cephalopods experience pain, we might ask *which ones* experience it, *and how*. The same goes for the 34,000 known species of fish, the 67,000 known species of crustaceans, and the who-knows-how-many-million species of insects. It's ridiculous to treat these groups as monolithic when we know, from other senses like vision and smell, that even closely related animals differ in how they perceive the world.

Instead of focusing on whether pain even exists, we might ask, as physiologist Catherine Williams told me, "In which conditions and from which stimuli is it an advantage to have it, experience it, and display it?" And we would find that pain manifests differently in a burrowing mole-rat than in a scorpion-hunting mouse, or in a long-armed octopus than in a short-armed squid. We might possibly find different forms of pain in sociable animals that can call for help or solitary ones that must fend for themselves, or in short-lived animals that have few chances to repeat their mistakes versus long-lived ones that have many chances to. And we would certainly learn that pain can vary in animals that must tolerate extremes of temperature, from baking heat to freezing cold.

5.

So Cool

Heat

I'M COLD. OUTSIDE, THE AUTUMNAL AIR IS A WARM 24°C (75°F), but I'm inside what's essentially a large walk-in fridge that's been chilled to just 4°C (39°F). It's an artificial hibernaculum—a room designed to mimic the dark, frigid conditions in which hibernating animals spend the winter. Since I am apparently incapable of packing suitable clothing for a reporting trip, I've turned up in a light T-shirt. As heat bleeds from my bare-skinned arms, I instinctively rub them. Meanwhile, Maddy Junkins, who is more sensibly dressed, reaches into a box of shredded paper and pulls out a small furry sphere. It's a thirteen-lined ground squirrel. Roughly the size and weight of a grapefruit, it's curled into a ball, with its tail brushing its nose. It looks like a big, fancy chipmunk, with 13 black stripes streaking down its back and light spots within those stripes. I can see those patterns because my eyes can detect the red light that illuminates the room. The squirrel's eyes cannot, and, regardless, they're tightly shut. It's mid-September, and the long hibernation season has begun.

Hibernation isn't sleep but a more intense state of inactivity that allows the ground squirrel to survive the harsh winters of northern America. During this time, its metabolism almost totally shuts down.*

---

* The two processes are so different that hibernating ground squirrels actually incur sleep debt and must periodically rouse from inactivity and raise their body temperatures so they can get some actual sleep.

When Junkins delicately places it in my latex-gloved hand, I'm instantly struck by how still it is. There's none of the manic, twitchy energy that rodents possess. Its flanks, which ought to be vibrating with frenetic breaths, aren't moving. Its heart, which beats at least 5 times a second in the summer, now ekes out the same number of beats over a minute. "There's usually so much life in your hand, but this is not that at all," Junkins says. "It's an inactive, cold lump." Indeed, the squirrel soon becomes uncomfortably chilly to the touch. Its body has abandoned the summertime norm of 37°C (99°F) and instead hovers at 4°C (39°F), just like every inanimate object in the room. It, too, feels uncannily inanimate—devoid of warmth and so seemingly devoid of life. Only its paws confirm that it's actually living: They're still pink from blood, and, when gripped, they recoil, albeit in slow motion. If I held the squirrel for too long, the warmth of my hand would rouse it, so I put it back in its makeshift den before leaving the hibernaculum. Outside, Elena Gracheva, who runs the facility, is waiting.

"How was it?" she asks.

"So cool," I say.

Gracheva is a scholar of heat and the ways animals detect it. Having studied vampire bats and rattlesnakes (which we'll get to later), she recently turned her attention to the more adorable thirteen-lined ground squirrels and their remarkable ability to withstand low temperatures. "If you put me in a cold room, I'll start to feel pain, and then hypothermia," she tells me. "I probably couldn't survive for more than 24 hours." A thirteen-lined ground squirrel, however, can stay between 2°C (36°F) and 7°C (45°F) for half a year. The closely related Arctic ground squirrel can go even lower, withstanding subzero temperatures of −2.9°C (27°F). These feats of endurance depend on an essential ability that often goes unnoticed: The squirrel *doesn't mind the cold.*

Vanessa Matos-Cruz, who worked with Gracheva, demonstrated this by putting ground squirrels on a pair of heatable plates. If one of these was heated to 30°C (86°F) and the other to 20°C (68°F), where would the animal decide to stand? Rats, mice, and humans almost always go for 30°C since that produces a pleasant sensation of warmth—

think about how luxurious heated floors feel. But to thirteen-lined ground squirrels, 20°C is just as delightful as 30°C. They only start gravitating toward the 30°C plate when the alternative falls below 10°C (50°F)—a temperature that rats and mice will completely avoid because it's painfully cold. Even if the second plate falls to 0°C (32°F), the ground squirrels will still stand on it.

Without this tolerance for low temperatures, a ground squirrel wouldn't be able to hibernate. Instead, it would do what we do when we get too cold while asleep: start burning fat to produce heat, and, if that doesn't help, automatically wake up. For us, that's a lifesaver. For a ground squirrel in the dead of winter, it would be lethal. It *needs* to hibernate, and for that to happen, its senses have been adjusted accordingly. It's not that the ground squirrel ignores the cold. Instead, it has a different conception of what "cold" is—a different minimum temperature at which its body can no longer cope and its senses raise an alarm.

All living things are deeply affected by temperature. If conditions are too cold, chemical reactions slow to a useless crawl. If they are too hot, proteins and other molecules of life lose their shape and fall apart. These effects constrain most of life to a Goldilocks zone where the temperature is just right. The limits of that zone vary, but they always exist, which is why every animal with a nervous system has a way of sensing and responding to temperature.

Animals use a variety of temperature sensors, and the most thoroughly studied of these are a group of proteins called TRP channels. They are found throughout the body on the surface of sensory neurons, where they act as tiny gates that open when they reach the right temperature. When this happens, ions enter the neurons, electrical signals travel to the brain, and we feel the sensations of hot or cold. Some TRP channels are tuned to hot temperatures, and others to cold ones. (Cold isn't just the absence of hot; it's a different sense in its own right.)*

---

* In the 1880s, Magnus Blix used a pointed metal tube, connected to bottles of water at varying temperatures, to show that certain spots on his hand were sensitive to hot and others to cold. Two other scientists, Alfred Goldscheider and Henry Donaldson, independently made the same discovery at the same time.

TRP channels also respond to different severities of temperature: Some detect mild and innocuous ranges, while others fire at dangerous and painful extremes. Certain chemicals can trigger these channels as well, producing heating and cooling sensations. Chili peppers burn because the capsaicin within them triggers TRPV1—a TRP channel that detects painfully high temperatures.* Mint cools because it contains menthol, which activates the cold sensor called TRPM8.

These same sensors are found throughout the animal kingdom, but each species has its own subtly different versions that are calibrated to its body and lifestyle. Warm-blooded animals produce their own heat, and their versions of the cold sensor TRPM8 alert them if their body temperatures start dipping below a narrow comfortable range. In a rat, that set point occurs at around 24°C (75°F). In a chicken, whose body runs at a slightly hotter temperature, TRPM8 is tuned to 29°C (84°F) instead. Cold-blooded animals, by contrast, rely on their environment for warmth, and their body temperatures fluctuate over a wide range. Consequently, their versions of TRPM8 are typically set much lower—at 14°C (57°F) in frogs. Fish seem to lack TRPM8 altogether, and most of them can tolerate temperatures close to freezing. Even if they do feel pain, it seems that they have no idea what it's like to be agonizingly cold. Individual humans might feel comfortable at different temperatures, but that variation is even greater across the entire animal kingdom.

And what about the ground squirrels? Matos-Cruz found that their version of TRPM8 is very similar to that of other warm-blooded rodents but has a few mutations that make it much less sensitive. It still responds to menthol, but barely reacts to temperatures as low as 10°C (50°F). That partly explains why these squirrels can hibernate so comfortably in conditions we'd find intolerably frigid.†

---

* Contrary to popular belief, this isn't a matter of taste. As I can attest, having once taken a shower immediately after chopping habanero peppers, if you get enough capsaicin on your hands and other delicate body parts, you'll experience that burning sensation everywhere you touch.

† There's a version of the human TRPM8 that becomes increasingly common at higher latitudes and might reflect an adaptation to colder climates. It's still unclear whether people who carry this version perceive cold in a different way.

The TRPV1 sensor, which detects painful heat, is also tuned to the needs of its owners, and especially to their body temperatures. It activates at 45°C (113°F) in chickens, 42°C (108°F) in mice and humans, 38°C (100°F) in frogs, and 33°C (91°F) in zebrafish (which might have no use for a cold sensor but clearly benefit from a hot one). Each species has its own definition of hot. The temperature at which we live would be painful to a zebrafish. The temperature that would start to agonize a mouse wouldn't bother a chicken. And even chickens are overshadowed by two species that have the least sensitive versions of TRPV1 thus far tested, enabling them to shrug off heat that other creatures can't bear. For obvious reasons, one of these is the desert-dwelling Bactrian camel. Unexpectedly, the second is—drumroll, please—the thirteen-lined ground squirrel! The unassuming rodent that I held not only can cope with temperatures that are close to freezing but *also* can abide extreme heat. In Gracheva's hotplate tests, the squirrels will scurry over to a colder plate only if the one they're standing on reaches a scorching 55°C (131°F). No wonder they thrive throughout the United States, from Minnesota in the north to Texas in the south. Their temperature sensors influence their geographical range, the seasons in which they're active, and much else besides. By defining the temperatures that animals can sense and tolerate, and by tweaking their personal limits of "hot" and "cold," these sensors define where, when, and how they live.

Those lives can be extreme. The Saharan silver ant forages under the midday heat of Earth's greatest desert, over sands that can reach 53°C (127°F), while the Pompeii worm, which lives near volcanic undersea vents, can also resist brief spells at similar temperatures. Snow flies are active at −6°C (21°F), while ice worms spend their entire lives on glacial ice; both animals will die if you hold them. When scientists study these so-called extremophiles, they tend to focus on adaptations like heat-reflecting hairs in their bodies or self-made antifreeze in their blood. But such adaptations would be useless if an animal's sensory system were constantly screaming at it, triggering feelings of pain (or nociception). If you want to live in the Sahara, or at the bottom of the ocean, or on a glacier, you'd better tweak your senses to *like it*.

This concept is intuitive, and yet when we watch extremophiles, from emperor penguins braving the Antarctic chill to camels trekking over scorching sands, it's easy to think that they are suffering throughout their lives. We admire them not just for their physiological resilience but also for their *psychological* fortitude. We project our senses onto theirs and assume that they'd be in discomfort because we'd be in discomfort. But their senses are tuned to the temperatures in which they live. A camel likely isn't distressed by the baking sun, and penguins probably don't mind huddling through an Antarctic storm. Let the storm rage on. The cold doesn't bother them, anyway.

THE THERMOSTAT IN MY HOUSE currently reads 21°C (70°F). But the entire house isn't at the same temperature. I'm working in the south-facing living room, which is considerably warmer than the other parts. And as I type these words, my head is being warmed by a sunbeam while my feet are chilling in the shade beneath my desk. Such variations exist at small scales, too: The air 5 millimeters above my skin might be 10°C (18°F) cooler, so a fly that landed on my arm might experience very different temperatures on its legs than on its wings. Being small, the fly would quickly take on the temperature of its environment. If it landed on my head, the sun would heat its body to harmful temperatures within a few seconds. That's unlikely to happen, though, thanks to temperature sensors at the tips of its antennae.

Neuroscientist Marco Gallio demonstrated how good those sensors are by putting fruit flies in chambers whose quadrants were heated to varying degrees—essentially the same experiment that Matos-Cruz did with ground squirrels and hotplates. Gallio showed that flies could easily stay within air spaces that are kept to 25°C (77°F), which they love, while avoiding neighboring zones of 30°C (86°F), which they dislike, or 40°C (104°F), which kills them. They could also make these decisions at incredible speed. Whenever they'd hit the edge of a hot zone, they'd immediately execute a sharp midair U-turn, as if they'd run into an invisible wall.

Such maneuvers are possible because the chitin that makes up a fly's antennae is very good at conducting heat and because the antennae themselves are tiny. They can so quickly equilibrate with their surroundings that a fly can instantly tell if it has blundered into air that's too hot or cold. Gallio found that it can even use its antennae as stereo thermometers to track gradients of heat, much as a dog uses its paired nostrils for odors. The fly can tell if one antenna is just 0.1°C hotter than the other, and uses those comparisons to steer toward the more comfortable temperature. When Gallio tells me about these results, I suddenly reconsider the movements of every fly I've ever seen. Their paths, which always seemed so random and chaotic, now take on an air of purpose, as if the insect is threading its way through an obstacle course of hot and cold that I can't perceive, don't care about, and oafishly wade through.

The fly's ability is called thermotaxis, and it's common in the animal kingdom.* Creatures big and small use their thermosensors to tell if their surroundings have become intolerable, and to gauge how the temperature around them changes as they move. Like children who are told if they're getting warmer or colder as they approach a hidden object, most animals use changes in ambient temperature to follow the gradients of heat that are created by sunbeams and shadows, breezes and currents. But some have transformed this common ability into something rarer. They can tell if point B is hotter than point A without having to move there. They can actively seek out sources of heat from a distance.

AT 11:20 A.M. ON August 10, 1925, a bolt of lightning struck an oil depot near the town of Coalinga, California. The strike ignited a lake of fire that burned for three days. Flames rose so high that, at night,

---

* Fish, from tiny larvae to 30-foot whale sharks, will control their temperatures by ascending to warmer shallows or diving to colder depths. Sulfide worms that live in hydrothermal vents, where scalding volcanic fluids bubble out of the seafloor, can find pockets of cooler water amid the roiling plumes. Butterflies that are warming their flight muscles in a sunbeam will stop basking when temperature sensors in their wings tell them that they're overheating. Turtle embryos can even pull off thermotaxis within the confines of their eggs, shuffling over to bask on the warmest side before they've even hatched.

people could read by that light from nine miles away. And while they read, they might have also noticed tiny black specks that flew against the billowing curtains of smoke and traveled toward the inferno. These specks were fire-chaser beetles. They were living up to their name.

Moths are famously drawn to flames, but it's the light that attracts them.* The fire-chasing *Melanophila* beetles, however, are drawn to heat. These black, half-inch-long insects have been found in what entomologist Earle Gorton Linsley described as "unbelievable numbers" in smelting plants, the kilns of cement factories, and the vats of hot syrup in sugar refineries. One summer, Linsley saw them swarming an outdoor barbecue where "large quantities of deer meat were being prepared." In the 1940s, the insects would regularly bother football fans in Berkeley's California Memorial Stadium "by alighting on the clothing or even biting the neck or hands," Linsley wrote. It's possible that "the beetles are attracted by the smoke from some twenty thousand (more or less) cigarettes which on still days sometimes hangs like a haze over the stadium." These incidents are unfortunate for both species, because industrial plants, barbecues, and football stadiums are unhelpful distractions that waylay the beetles from their true targets: forest fires.

Arriving at a fire, the beetles have perhaps the most dramatic sex in the animal kingdom, mating as a forest burns around them. Later, the females lay their eggs on charred, cooled bark. When the wood-eating grubs hatch, they find an Eden. The trees they devour are too injured to defend against insect larvae feeding within them. The predators that might eat them are put off by the smoke and heat emitted from the embers and ashes. In peace, they thrive, mature, and eventually fly off in search of their own blazes. But forest fires are rare and unpredictable, and the beetles must have some means of detecting them from afar. Being active during the day, the beetles can't spot distant flames in

---

* Naomi Pierce, who showed that butterflies have temperature sensors in their wings, isn't fully convinced that moths are only drawn to the light of candle flames. She and her colleague Nanfang Yu have spent years investigating the possibility that moth antennae can act as infrared detectors.

the way that nocturnal insects easily could. They can't rely on seeing plumes of smoke since their eyes probably aren't sharp enough to distinguish such plumes from clouds. And though their antennae can certainly detect the smell of scorched wood, such clues are heavily influenced by the direction of the wind. For them, the most reliable cue is heat.

The atoms and molecules in all objects are constantly jiggling about, and this motion produces electromagnetic radiation. As an object gets hotter, its molecules move faster, and it emits more radiation at higher frequencies. That radiation includes some visible light—think about the glow of heated metal—but most of it lies in the infrared spectrum.* We can't see infrared, but we might be able to feel it. When you stand near a fireplace, infrared light radiates from the burning wood. When it reaches you, its energy is absorbed and heats up the closest parts of your skin, triggering the temperature sensors within it. You feel the heat. You can also work out where it's coming from because the parts of your body under infrared illumination are getting hotter while those that lie in the infrared shadow are not. But this trick only works at close range. Infrared light spreads from the fireplace in all directions, and some of it gets absorbed on its travels. As you walk farther from the logs, less and less of that light reaches you, until its imparted energy no longer warms your body to a noticeable degree. To detect the infrared light from a distant source, either the source must be extremely intense (like the sun) or you need specialized equipment. *Melanophila* beetles have the latter.

Below their wings and just behind their middle legs, these insects have a pair of pits. Each one contains a cluster of around 70 spheres that together look like a malformed raspberry. When zoologist Helmut

---

* Infrared light covers such a huge range of wavelengths that if you represent it by the length of your arm, the visible spectrum would be no wider than a hair. The shortest of these wavelengths, also known as near-infrared, can be seen by certain animals like the migrating salmon we met in Chapter 1 or humans wearing night-vision goggles. Mid-wavelength infrared lies beyond the scope of such sensors; these are the wavelengths that heat-seeking missiles seek, that forest fires emit, and that fire-chaser beetles chase. Far-infrared is what warm bodies give off. It's what thermal imaging cameras and rattlesnakes detect.

Schmitz examined these spheres under a microscope, he saw that each is filled with fluid and encloses the tip of a pressure-sensitive neuron. When infrared radiation hits the spheres, the fluid inside them heats up and expands. It can't bulge outward because the spheres have hard exteriors, so instead it squeezes the nerves, causing them to fire. This is a different kind of heat sensing than what we've seen earlier in this chapter. Unlike hibernating ground squirrels or zipping fruit flies, the beetles aren't just measuring the ambient temperature of their surroundings. Instead, much as we do when we bask by a fireplace, they're sensing radiant heat that travels from hot sources in the form of infrared light.

The beetles' spherical sensors must be extraordinarily sensitive, since the insects frequently travel to burning forests and other hot places from dozens of miles away. The Coalinga oil depot that was struck by lightning in 1925 lies in the middle of an arid, treeless region, and most of the beetles that arrived there likely came from forests that lay 80 miles to the east. Based on this distance, and simulations of the 1925 blaze, Schmitz calculated that the beetles' pits are more sensitive than most commercial infrared detectors and on a par with state-of-the-art quantum detectors that must first be cooled with liquid nitrogen. Schmitz thinks that the pits couldn't possibly be this sensitive on their own. The beetles must have ways of making them more responsive.

During flight, their beating wings create vibrations that travel into the nearby pits, shake the spherical sensors, and push the sensory neurons within to the edge of firing. It now takes much less infrared radiation to fully push them over that edge. Think of this another way: Imagine a brick that's lying flat on its side. If a fly crashed into it, it wouldn't budge. But if it was instead balanced on its edge, even a fly would be enough to topple it. In that state, the brick would be primed to react to a tiny amount of energy. Schmitz argues that a fire-chaser beetle's beating wings prime its heat sensors in a similar way, setting them up to detect sources of infrared that would normally be too weak. A beetle that's sitting on a tree would be relatively insensitive. But as soon as it takes off in search of fires, its body automatically widens its search area and transforms even faint traces of distant heat into blazing beacons.*

The beetles' bodies are relevant in another way. As with all insects, their outer surface is very good at absorbing the kinds of infrared radiation that fires emit. The beetles were effectively pre-adapted for chasing fires. Their ancestors merely needed to develop a sensor that could make sense of the infrared light that their bodies naturally absorb. The 11 species of *Melanophila* did this and became so successful that they spread across five continents. They never reached Australia, though. There, three other types of insects independently evolved infrared sensors that allow them to exploit the tranquil paradise of a charred forest. Fire-chasing is a trick so useful that it has evolved at least four times over. And fires are not the only sources of heat that an animal might want to track. Some species search for the warmth of bodies.

"YOU'RE DEFINITELY NOT ALLOWED to come in here," Astra Bryant tells me. I dutifully oblige, hovering outside while Bryant rummages through a fridge. After a few minutes, she emerges with a pipette that holds 5 microliters of clear liquid in its tip. It's a volume so small I can barely see it. I certainly can't see the thousands of nematode worms that are swimming inside.

Nematodes are one of the most diverse and numerous groups of animals, including tens of thousands of species that are mostly harmless to humans. The exceptions include the species that Bryant is carrying—*Strongyloides stercoralis,* the threadworm. Its larvae abound in soil and water that are contaminated by feces. If an unlucky person stands or walks through such places, the worms swim toward them and penetrate their skin. Threadworms, along with hookworms and

---

* For now, this idea is speculative, and very hard to test. Schmitz would have to take electrical recordings from a beetle's neurons, and do so in a way that doesn't leach any heat from the pits. And if his theory about the flapping wings is right, he'd have to do this in a flying insect. "That's very hard," he says, with Germanic understatement.

other skin-penetrating nematodes, infect around 800 million people around the world, from Vietnam to Alabama. They cause gastrointestinal problems, stunted development, and sometimes death. They're also very hard to treat. Bryant and her mentor Elissa Hallem are trying to find out how the worms find their hosts in the first place, in order to create new ways of preventing infections. Odors are certainly part of the equation. So is heat.

Bryant carries her pipette of monstrosities to a steel chamber with a biohazard sign on its door. Inside, there's a slab of translucent gel that has been asymmetrically heated so its right side is at room temperature, while its left is the temperature of a human body. Bryant squeezes the worms out onto the middle of the slab, and they appear on a nearby monitor as a ring of white dots. With horrific immediacy, the dots start to move. The ring quickly stretches into a cloud that drifts leftward toward the heat. Drifts? More like zooms. Each worm is just a millimeter or two long, but quickly covers a distance several hundred times greater. I'm starting to understand why hundreds of millions of people are infected. Within three minutes, they're all huddled on the leftmost edge, searching for the source of the heat they can sense but not find. "This was *shocking* the first time I saw it," says Bryant, who expected the worms to spend hours traveling distances that in fact they covered in minutes. "I show this in my talks, and I generally get groans out of the audience."

Parasitism may be grisly, but it's one of the most common lifestyles in nature. It's likely that the majority of animal species are parasites, which survive by exploiting the bodies of other creatures. Many of these freeloaders are fastidious about their choice of hosts and need some way of finding the right targets. Smells provide good cues. But hundreds of millions of years ago, another possibility emerged.

The ancestors of birds and mammals independently evolved the ability to produce and control their own body heat, divorcing their temperatures from the temperatures of their surroundings. This ability, known technically as endothermy and colloquially as warm-bloodedness, endowed birds and mammals with speed and stamina, durability and possibility. It allowed them to survive in extreme environments and stay active over long durations and distances. It also made them very easy to track. Their unwavering body heat made them perpetually blaring beacons, which parasites could use to find hosts, and especially blood vessels. Blood, after all, is a superb source of food—rich in nutrients, well balanced, and usually sterile. It's no surprise that at least 14,000 animal species have evolved to feed on it, or that many of these—bedbugs, mosquitoes, tsetse flies, and assassin bugs—are attuned to heat.

Among mammals, only three species of vampire bats feed exclusively on blood. Two mostly drink from birds, but the common vampire targets mammals, especially large ones like cows or pigs. It's a small animal that measures 3 inches from nose to tail and has a flattened, pug-like face. On the ground, its wings fold back, and it adopts a sprawling, four-legged stance. It approaches targets like this, either landing directly on their backs or alighting nearby and crawling over in a most un-bat-like way. Once near, it painlessly inflicts a small cut with its blade-like incisors and laps up the blood that flows out. A compound in its saliva, aptly known as draculin, stops the blood from clotting, allowing the bat to feed for up to an hour. It can drink its own body weight in blood and must do so once a night to survive. Other senses help it to track a target from afar, but once it gets at least 6 inches away, it uses a thermal sense to pick a good bite site.

The vampire's heat sensors lie in its nose, which consists of a heart-shaped flap lying over a semicircular pad. Sandwiched between these layers are a trio of millimeter-wide pits, each one riddled with heat-sensing neurons. Among infrared-sensing animals, vampire bats have a unique problem because they are themselves warm-blooded. The neurons in their pits ought to be bamboozled by their own body heat, but a dense network of tissue insulates them and keeps them 9°C (16°F) cooler than the rest of the bat's face.

Elena Gracheva studied those neurons in the days before she started working with those adorable ground squirrels. Her colleagues in Venezuela rode to caves where the bats roost, lured them out using their own horses as bait, dissected out their pit neurons, and shipped the tissue samples to Gracheva in the United States. By analyzing those samples, she showed that the neurons are loaded with a special version of TRPV1—the same temperature sensor that we met earlier in this chapter, which usually detects painful heat and the sting of chilies. TRPV1 is calibrated to different temperatures depending on what respective animals would find painfully hot—33°C (91°F) in a cold-blooded zebrafish, and 42°C (108°F) in a warm-blooded mouse or human. In the vampire bat, TRPV1 is set at a typical mammalian level, except in the pit neurons, where it instead goes off at a much lower temperature, 31°C (88°F). The bat has retuned this sensor from one that detects extreme heat into one that detects body heat.

Ticks also suck blood, but their heat sensors are found on the tips of their first pair of legs. When they wave these legs around—a behavior known as questing—it looks like they're waiting to grab something. They are, but they're also sensing. Jakob von Uexküll, coiner of the Umwelt concept, wrote that ticks track their hosts through scent and use temperature only to check if they've landed on bare skin. But this isn't true. Ann Carr and Vincent Salgado recently found that ticks can detect body heat from up to 13 feet away. More surprisingly, the duo showed that common repellents like DEET and citronella don't disrupt a tick's sense of smell but *do* stop them from tracking heat. This discovery might lead to new ways of preventing tick bites, and it might force scientists to reevaluate a lot of previous tick studies. How many past experiments have been misinterpreted because researchers have had an inaccurate picture of a tick's Umwelt?

In hindsight, the tick's thermal sense should have been clear. The organs at the tips of their questing legs were mostly thought to be odor detectors. But these structures also include tiny spherical pits with neurons at their bases, much like those on a vampire bat's face. Tellingly, these pits are covered with a thin sheet that has a small hole in it. That's a terrible design for a nose, because the sheet would block most odorants from reaching the underlying neurons. It is, however, an excellent design for an *infrared* sensor. Infrared radiation, emanating from the blood of a distant host, would be mostly blocked by the sheet, but some would pass through the hole to partly illuminate the pit below. By analyzing which bits were lit up, the tick could work out the direction of the radiation, and the whereabouts of its source. This idea still needs to be confirmed, but it makes sense. After all, it's how the most sophisticated heat sensors in nature work. To find them, you need a little courage, some shin guards, and a long pole.

WE CAN'T FIND JULIA. We know she's right in front of us, lurking within a rat nest that's inside a prickly pear bush, but we can't see her. We can hear a loud telltale beep as our antenna picks up a radio signal from the transmitter inside her, but she herself is silent. She's not even rattling. We leave her be and head off in search of another snake.

My wife, Liz Neeley, and I have come looking for rattlesnakes in a fenced tract of California scrubland that's owned by the U.S. Marine Corps. Leading us are Rulon Clark—who spent his childhood running around after snakes and lizards and never really stopped—and his student Nate Redetzke. Redetzke regularly has to relocate snakes that show up in nearby homes and has implanted several of them with radio trackers. Having parked on a dirt lane aptly called Rattlesnake Canyon Road, we donned Kevlar shin guards and tromped off through the sagebrush, breathing in the fennel-scented air, dodging poison oak, and clambering over boulders.

"Working with reptiles makes you very sensitive to temperature and weather," says Clark. He started our expedition in the early morning, hoping to find rattlesnakes that were openly basking in what forecasts said would be an unseasonably warm October day. But the forecasts were wrong. It's actually cold and overcast, so although we are out, the snakes are not. Powers was hidden deep within a cactus. Truman was somewhere inside a pile of boulders. Julia was out of sight. (Redetzke has named them all after former presidents and first ladies.) We're about to give up when he hears a loud beep, perks up, and bounds off around the hillside. Moments later, he shouts that he's found Margaret. He prizes apart the branches of a bush, reaches in with a pair of tongs, and pulls out a red diamond rattlesnake—rust-colored and 3 feet long. Red diamonds are supposedly docile, but even they have their limits. As Redetzke lowers Margaret into a bag, she strikes it, leaving globs of yellow venom on the cloth. Once inside, she rattles, but she's cold and the sound is dull.

Later, Redetzke nudges Margaret into a plastic tube that's just wider than her body. Gently gripping her tail at one end, I stare down the other into her face. The pupils are vertical slits. The mouth curves upward in what looks like a grimace. The lidless eyes are overhung by large horizontal scales that create what I call resting viper face—a look of perpetual anger. It's a visage that normally instills fear. But I find her beautiful. Who knows what she makes of me, but at this distance, she can certainly see me, and not just with her eyes. With a pair of small pits nestled just behind her nostrils, she can detect the infrared radiation that's flooding from my warm face and, to a lesser extent, from my clothed body. Against the cool morning sky, I must be shining.

Heat-sensitive pits have evolved independently among three groups of snakes. Two of these, pythons and boas, are non-venomous constrictors that kill with suffocating coils.* The third are the highly venomous and aptly named pit vipers—cottonmouths, copperheads, moccasins, and rattlesnakes.† Rattlesnakes will strike at warm objects, preferring freshly killed mice over long-deceased ones, and they'll hit their targets in complete darkness. Even a congenitally blind rattlesnake that was born without eyes could kill mice as effectively as a sighted individual. Thanks to its pits, its aim was good enough not only to hit the rodents but to specifically strike them in the head.

A pit viper's thermal sensitivity comes from the structure of its pits (which are similar to those on a tick's legs). To get an idea of their shape, imagine placing a miniature trampoline on the bottom of a

---

* In some ways, the pit organs of boas and pythons are very different from those of pit vipers. Their membranes aren't suspended, and are likely less sensitive. They have several pairs of pits running up the sides of their heads, instead of a single pair at the front of their heads—a pattern that George Bakken compares to the compound eyes of insects. And yet, Elena Gracheva found that all three groups rely on the same heat sensor—TRPA1.

† The first Western scientist to describe their pits, back in 1683, correctly guessed that they were sense organs, but wrongly supposed that they were ears. Just as wrongly, others suggested that they were nostrils, tear ducts, or sensors of smells, sounds, or vibrations. No one hit on the right answer until 1935, when Margarete Ros—no relation to Margaret the snake—noticed that she could stop her pet python from slithering toward warm objects by covering its pits with Vaseline. She deduced that the snake uses its pits to sense the body heat of its prey.

goldfish bowl and turning the whole thing on its side. There's a narrow opening, leading into a wider air-filled chamber, across which a thin membrane is stretched. When infrared radiation passes through the opening, it falls upon the membrane and heats it up. This happens readily because the membrane is exposed to the elements, is suspended in midair, and is a sixth as thick as a page of this book. It is also riddled with some 7,000 nerve endings that detect the slightest rise in temperature. Those nerves, as Elena Gracheva discovered, are packed with the heat sensor TRPA1, carrying 400 times as much of it as neurons elsewhere in the snake's body. They'll respond if the membrane rises in temperature by as little as 0.001°C. This astonishing sensitivity means that a pit viper can detect the warmth of a rodent from up to a meter away. A blindfolded rattlesnake that's sitting on your head could sense the warmth of a mouse on the tip of your outstretched finger.*

The pits are structurally similar to eyes. The membrane, which detects infrared light, is like a retina. The opening, which allows that light to enter, is like a pupil. And just like a pupil, the opening is narrow, which means that some regions of the membrane are heated by incoming infrared while others lie in cool shadow. The snake can use these patterns of hot and cold to map a heat source in its vicinity just as it uses the light falling on its retina to construct an image of a scene. These similarities aren't just metaphorical. Some scientists think that the pits really are a second pair of eyes, tuned to the infrared wavelengths of light that are invisible to the main pair. Signals from the two organs are initially processed by different parts of the brain but eventually feed into a single region called the optic tectum. There, the two streams are combined, and information inputs from the visible and infrared spectrums are seemingly fused together by neurons that respond to both. It's possible that the snakes really are *seeing* infrared, treating it as just another color. "It is a fallacy to consider the pit organs as an independent sixth sense," neuroscientist Richard Goris once wrote. "What the pits do is improve vision for their owners." They might

---

* Just trust me on this one and don't try it at home.

provide more detail at night, reveal warm objects that are obscured by undergrowth, or direct the snake's attention to scurrying prey.*

But if the pits are eyes, they're very simple ones with blurry vision. They only have thousands of sensors compared to the millions in a typical retina, and they have no lens to focus the incoming infrared. Nature documentaries get this wrong when they try to show what rattlesnakes see by filming the world with thermal cameras. Those images, with white and red rodents moseying in front of blue and violet backgrounds, are always unrealistically detailed. *Predator,* the 1987 movie in which Arnold Schwarzenegger encounters a trophy-hunting alien, did a better job of depicting the blurriness of infrared vision. (This is perhaps the only time that anyone has accused *Predator* of being realistic.)

Recently, physicist George Bakken simulated what the pits would pick up when a mouse runs across a log. He got grainy images of small warm blobs moving over large cool blobs. A mouse on your finger might be detectable to a blindfolded rattlesnake on your head, but it would be shapeless unless it ran onto your biceps. Pit vipers compensate for this shortcoming by carefully choosing their ambush sites. Sidewinders tend to point toward thermal edges where the environment rapidly flips between hot and cold and a moving warm-blooded animal might be easier to spot. And on China's Shedao Island, the local pit vipers choose ambush sites that face into open sky, allowing them to more easily detect the migrating birds they gorge upon in spring.

How do the snakes actually perceive heat? Chinese herpetologist Yezhong Tang found a hint by working with short-tailed pit vipers. If he blocked one eye and one pit on the same side, the snakes bit their victims 86 percent of the time. If he blocked either both eyes or both

---

* Some researchers have claimed that ground squirrels can fool a rattlesnake's infrared sense. When confronted, they raise their tails and heat them up by pumping warm blood into them. This would increase the size of their thermal silhouette and make them seem more intimidating to a heat-sensing predator. Tellingly, the squirrels only do this to rattlesnakes, and not to harmless gopher snakes that can't sense infrared. This has been billed as the first known example of infrared communication between two species. But Clark and others aren't convinced. The squirrels might just be raising their tails and pumping blood into them because they're alarmed. And they might be doing this to rattlesnakes instead of gopher snakes because the former are more alarming!

pits, their accuracy fell slightly to 75 percent. But if he blocked one eye and one pit *on opposite sides,* they landed just 50 percent of their strikes. That unexpected result suggests that the snakes are combining visual and infrared information. But how do they manage when those senses operate at such different resolutions? Bakken wonders if the brain could learn to better interpret the coarse information it gets from the pits using the much sharper information from their eyes. After all, humans can program artificial intelligences to classify pictures or spot hidden patterns by training them on a large enough set of images. Maybe a snake's eyes provide the training set that its brain needs to interpret the blurry information from the pits.

Whatever advantage the pits provide, it must be significant. The nerves in their membranes are loaded with tiny batteries called mitochondria, far more than exist in typical sense organs. This suggests that the infrared sense demands a lot of energy, so it must provide benefits that are worthy of that cost. It certainly seems to give pit vipers an edge over pit-less snakes.* But the more I ask Clark about the infrared sense, the more unanswered questions I'm left with.† Why did pit vipers evolve it when most of them also have excellent night vision? If the infrared sense bolsters vision, then why didn't it also evolve in other nocturnal vipers? Why did pythons and boas, which are separated from vipers by some 90 million years of evolution and hunt in very different

---

* Ecologist Burt Kotler, based in Israel, demonstrated this by setting pitted sidewinder rattlesnakes against horned vipers from the Middle East—very similar to the rattlers except that they lack an infrared sense. When Kotler placed both snakes in large outdoor enclosures, the pit-less horned vipers became less active on moonless nights, ceding the darkness to the sidewinders, which could still use heat to hunt. The Israeli rodents in those enclosures also came to treat the alien sidewinders as a threat greater than their own native vipers. Kotler describes the pits as a "constraint-breaking adaptation"—an innovation that shunts snakes to the next level of predatory effectiveness by allowing them to hunt in even the dimmest light.

† One of Clark's students, Hannes Schraft, found several confusing results when he tried to study pit vipers in the wild. At night, sidewinders lie in wait in bushes, which are slightly warmer than the surrounding sands and should act like glowing landmarks. But Schraft found that blindfolded sidewinders are appalling at finding bushes, and will wander erratically without success. He also wondered if the snakes use their infrared vision to gauge the temperature of their prey, since colder targets should be slower and easier to catch. They don't. Schraft presented them with lizard carcasses that had been warmed with a hot water bottle, and the snakes didn't care.

ways, evolve the same trick when more closely related snakes, like cobras and garter snakes, did not? And most puzzlingly, why do the pits seem to work better when they're cold?* "There's something that we're missing," Clark tells me. "Maybe the infrared sense is simply about targeting prey, but I think they're using it in ways that we don't understand."

To understand another animal's Umwelt, you have to watch its behavior. But a pit viper's behavior mostly consists of waiting. Since they don't generate their own body heat, they can go without eating for months and can sit in ambush until exactly the right moment. The few researchers brave enough to study them end up with animals that mostly sit around doing nothing, which makes them very hard to train—or comprehend. After all, even animals that we already understand and that we know how to train can sense heat in ways that are hard to explain.

WHEN ZOOLOGIST RONALD KRÖGER got a dog—a golden retriever named Kevin—he started wondering about its nose. Sleeping dogs tend to have warm noses. But shortly after they wake up, the tips become wet and cool. Kröger found that in a warm room, a dog will keep its nose around 5°C cooler than the ambient temperature, and between 9° and 17°C colder than the nose of a cow or pig in the same space. Why? Vampire bats and rattlesnakes both seem to cool their heat-sensitive pits. Could dogs be doing the same? Could their noses be infrared sensors as well as organs of smell?

---

* In 2013, Viviana Cadena found that rattlesnakes can control the way they exhale to actively cool their pits, keeping them a few degrees below their body temperatures. A few years later, Clark and Bakken kept rattlesnakes at various temperatures and measured their ability to spot a warm pendulum moving over a cooler background. To their surprise, the colder the snakes, the better they were at tracking the pendulum. "We were gobsmacked," says Bakken. This pattern doesn't make sense if the main heat sensor is TRPA1, which ought to work better at higher temperatures. It doesn't make sense since cold-blooded animals should be more effective as they get warmer. As a rattlesnake heats up, it becomes a faster and more active hunter . . . just as one of its main hunting senses becomes less sensitive? "It's backward and I don't know what to make of it yet," says Clark. In a refreshing act of academic straight talk, he and Bakken published their results under the title "Cooler Snakes Respond More Strongly to Infrared Stimuli, but We Have No Idea Why."

Kröger certainly thinks so. His team successfully trained three dogs—Kevin, Delfi, and Charlie—to tell the difference between two panels that looked and smelled the same but that differed in temperature by 11°C. In double-blind tests, when handlers didn't know the right answer and couldn't unconsciously influence the dogs, the three canines still picked the right panel between 68 and 80 percent of the time. The team suggests that wolves, the ancestors of domestic dogs, might have benefited from detecting the infrared radiation coming off their large prey. But since such radiation rapidly weakens with distance, how would it benefit animals that already have acute senses of hearing and smell? Surely a wolf would be able to sniff its meals well before its nose could detect telltale hints of warmth. And at close quarters, surely its eyes and ears would help it track a running target without any help from an infrared-sensing snout. "It's hard to imagine how this could be actually useful," says Anna Bálint, who worked on the study. "I guess we have to think outside of the box."

When thinking about another Umwelt, distance always matters. Under the right conditions, smell and vision operate over vast scales. Infrared senses work over shorter distances, unless they're honed to detect a blazing forest fire. And some senses are more intimate still, requiring the closeness of contact.

6.

A Rough Sense

Contact and Flow

At FIRST, EVERYONE THOUGHT THAT SELKA WAS SLEEPING. An adolescent sea otter, Selka was living in an enclosure at the Long Marine Laboratory in Santa Cruz, with a pool that had a fiberglass table resting just above the water's surface. She had taken to swimming under the table, sticking her nose into the narrow air space just below it, and having a nap—or so it seemed. It turned out that between snoozes, Selka had also been slowly unscrewing the nuts that held the table legs in place. One day, Sarah Strobel, a sensory biologist who had been working with the otter, found the entire platform tilting on its side. Selka was swimming around cradling a dislodged table leg, having stuffed the accompanying nuts and bolts down the drain.

Almost every photograph of sea otters shows them floating on their backs, often asleep, sometimes holding hands. This creates the deeply misleading impression that they are lazy and sedate. In fact, "they're really fidgety," Strobel tells me. "They're constantly doing things, playing with things, wanting to touch things." This rambunctious quality is something that sea otters share with other mustelids—the mammal group that includes weasels, ferrets, badgers, honey badgers, and wolverines. But sea otters combine what Strobel calls a "general mustelid mojo" with large size—at 3 to 5 feet in length, they're the biggest of the group—and unusually dexterous paws. Consequently, they're infamously hard to house in captivity.* "They're just super-destructive," says Strobel. "They're very curious, and the way they manifest that curiosity is: *How can I break this and figure out what's inside?*"

Inquisitiveness, dexterity, and a penchant for disassembly: These traits serve sea otters well in their native habitat along the western coast of North America. Those frequently cold waters challenge a creature that, though large for a mustelid, is unusually small for a marine mammal. Sea otters have neither the large heat-retaining bodies nor the insulating blubber of seals and whales. They do have the densest fur in the animal kingdom, with more hairs per square centimeter than humans have on our heads, but even that isn't enough to stop heat from rapidly bleeding off their bodies. To stay warm, they need to eat a quarter of their own weight every day; hence their frenetic nature. They're always diving, day and night. Almost everything's on the menu, and almost everything is grabbed by hand. Even when there's not enough light to see by, their paws lead them to food. With the same manual dexterity that Selka displayed in dismantling her table, wild sea otters snag fish, seize sea urchins, and dig out buried clams. Their delicate sense of touch allows them to survive as a small, warm mammal in a big, cold ocean.

The sensitivity of their paws is evident in their brains. As in other species, a region called the somatosensory cortex deals with touch. Different sections of the somatosensory cortex receive inputs from different parts of the body, and the relative size of these sections can hint at an animal's major tactile organs. In humans, the hands, lips, and genitals are most heavily represented. In mice, it's the whiskers; in platypuses, the bill; and in naked mole-rats, the teeth. In sea otters, the part

---

* Orphaned and stranded when she was one week old, Selka was rescued in 2012 and brought to Monterey Bay Aquarium, where she was raised by one of its resident sea otters. After months of learning how to otter, she was released, but after just eight weeks, she was brutally attacked by a shark. The aquarium took her back, fixed her wounds, and released her again. But after a bout of toxic shellfish poisoning and signs that she had become too habituated to humans, the U.S. Fish and Wildlife Service decided that she was "too likely to interact with humans to be safe in the wild." She spent two years in the Long Marine Lab before finally returning to Monterey Bay Aquarium, where she now acts as a surrogate mom to other orphaned pups.

of the somatosensory cortex that receives signals from the paws is disproportionately big compared to those of other mustelids, and even compared to those of other otters.

Those paws don't look like sensitive hands, though. They barely look like hands at all. The skin has the texture of a cauliflower head, and the digits aren't clearly separated. If you held the paw, you could feel the nimble fingers moving underneath, but if you just looked at it, you'd see "knobbly mittens," Strobel tells me. To measure what these mittens are capable of, she put Selka to a test. She trained the otter to recognize the feel of a textured plastic board that was covered in thinly spaced ridges. Selka then had to distinguish that board from others whose ridges were either slightly narrower or slightly further apart. And she did so, reliably and repeatedly, even when the ridges differed in their spacing by a quarter of a millimeter. Her paws really are as sensitive as her brain would suggest.

Sensitivity, however, is not the only metric by which a sense can be judged. As we saw in Chapter 1, humans and dogs can both follow chocolate-scented strings, but the former species labors slowly at the task while the latter does it quickly and assuredly. Likewise, Strobel found that humans are just as sensitive as sea otters at discriminating textures with their hands, but the latter are substantially faster.* In her experiment, human volunteers repeatedly ran their fingertips over the two possible boards, again and again, until they finally made their choice. Selka picked the right board as soon as she laid her paw on it. If the first one she touched was correct, she didn't even bother feeling the alternative. She made her choice in a fifth of a second, 30 times faster than her human rivals. Even her slowest decision times were considerably faster than those of the fastest humans. "They're very confident in whatever they're doing," says Strobel.

Imagine that, right now, a sea otter is about to search for food. Floating on its back on the surface of the sea, it rolls and dives. It will only stay submerged for a minute—roughly the time it will take you

---

* Aristotle once wrote that "in the other senses man is inferior to many animals, but in the sense of touch he far surpasses them all in acuity." He had never heard of sea otters, but he wasn't too far off in his claims.

to read this paragraph. The descent eats up many of the precious seconds, so once the otter reaches the right depth, it has no time for indecisiveness. In a few frantic moments, it presses its knobby mittens over the seafloor, inspecting whatever it can find. The water is dark, but darkness doesn't matter. To some of the most sensitive paws in the world, the ocean is bright with shapes and textures to be felt, grasped, pressed, prodded, squeezed, stroked, and manhandled—or perhaps otterhandled. Hard-shelled prey nestle among the similar hard rocks, but in a split second, the otter feels the difference between the two, and pulls the former from the latter. With its sense of touch, its dexterous paws, and its overabundant mustelid confidence, it snatches that clam, yanks that abalone, grabs that sea urchin, and finally ascends to eat its catches, breaking the water at the end of this sentence.

TOUCH IS ONE OF the mechanical senses, which deal with physical stimuli like vibrations, currents, textures, and pressures. For many animals, touch can operate at a distance. As we shall see later in this chapter, creatures as diverse as fish, spiders, and manatees can all feel the hidden signals that flow, blow, and ripple through air and water. Using tiny hairs and other sensors, they can feel the telltale signals of other animals from afar. Crocodiles can detect the gentlest ripples at the water's surface, crickets can sense the faint breeze produced by a charging spider, and seals can track fish by the invisible currents that they leave as they swim. But most such signals are undetectable to us: I can feel the strong air currents created by my ceiling fan, but little else. For humans (and sea otters), touch is primarily a sense of direct contact.

Our own fingertips are among nature's most sensitive touch organs. They allow us to wield tools with fine precision, to read patterns of raised dots when our vision is impaired, and to control screens with taps, swipes, and touches. Their sensitivity depends on mechanoreceptors—cells that respond to light tactile stimulation. These cells come in several varieties, each of which responds to a different kind of stimulus. Merkel nerve endings respond to continuous pressure: They help you gauge the shape and material properties of this book as you squeeze its pages. Ruffini endings respond to tension and stretch in the skin: They help you adjust your grip, and recognize when objects slip from your grasp. Meissner corpuscles respond to slow vibrations: They produce the feelings of slip and flutter as your fingers move over surfaces, and they allow Braille readers to make sense of raised dots. Pacinian corpuscles respond to faster vibrations: They're useful in assessing finer textures or in sensing objects through tools, like hairs that are gripped by tweezers or soil that crunches beneath a spade. Most of these receptors also exist in a sea otter's paw or a platypus's snout. Collectively, they produce the sensation of touch, just as our sweet, sour, bitter, salt, and umami receptors together define our sense of taste.

At a broad level, we understand how these mechanoreceptors work. Despite their variety, they all consist of a nerve ending enclosed in some kind of touch-sensitive capsule. A tactile stimulus bends or deforms the capsule, causing the nerve inside to fire. But exactly how this happens is still unclear, because touch is one of our least-studied senses. Compared to sight, hearing, or even smell, it inspires less art and fewer scientific devotees. Until very recently, the molecules that allow us to experience touch—the equivalent of opsins for vision, or odorant receptors for smell—remained completely mysterious. We only have a rough sense of the sense that senses roughness.

But touch cannot be ignored. It is a sense of intimacy and immediacy—and it varies just as much as smell or vision. Animals differ widely in how sensitive their touch organs are, what they use those organs to feel, and even the body parts on which those organs are found. And by considering how touch contributes to the Umwelten of different creatures, we will see sandy beaches, underground tunnels, and even internal organs in new ways. Even the true extent of our *own* tactile abilities has only recently come to light. In one experiment, people could distinguish between two silicon wafers that differed only in their topmost layer of molecules, telling them apart thanks to minuscule differences in the way their fingers slid over the two surfaces. In another test, volunteers could tell the difference between two ridged surfaces, even when those ridges differed in height by just

10 nanometers—akin to judging which of two sandpapers is coarser, when the grains are only the size of large molecules.

These incredible feats are possible through movement. If you rest a fingertip upon a surface, you can get only a limited idea of its features. But as soon as you're allowed to move, everything changes. Hardness becomes apparent with a press. Textures resolve at a stroke. As your fingers run over the surface, they repeatedly collide with invisibly small peaks and troughs, setting up vibrations in the mechanoreceptors at their tips. That's how you detect the subtlest of features, even down into the nanoscale.* Movement transforms touch from a coarse sense into an exquisite one. It allows many of nature's tactile specialists to react with incredible speed.

MANY SCIENTISTS SPEND THEIR entire lives studying the same animals. Ken Catania is an exception. In the last 30 years, he has investigated the senses of electric eels, naked mole-rats, crocodiles, tentacled snakes, emerald cockroach wasps, and humans. He is drawn to oddities, and his attraction to weird creatures almost always pays off. "Usually, it's not that, oh, the animal turned out not to be interesting," he tells me. "Usually, it's that the animal is ten times more capable than I could have imagined." No creature taught him that lesson more acutely than the first one he studied: the star-nosed mole.

The star-nosed mole is a hamster-sized animal with silky fur, a rat-like tail, and shovel-like paws. It lives throughout the densely populated eastern parts of North America, but since it dwells in bogs and swamps and spends most of its time underground, few people ever see it. Those who do would recognize it instantly. On the tip of its snout, it has 11 pairs of pink, hairless, finger-like appendages, arranged in a ring around its nostrils. This is the unmistakable star for which the

---

* Mark Rutland, who led the study in which volunteers distinguished between ridges that differed in height by 10 nanometers, said that "if your finger was the size of the Earth, you could feel the difference between houses [and] cars." That's true, but only if you dragged your planet-sized digit down the street—an act that, ironically, would be rather insensitive.

mole is named. It looks like a fleshy flower growing out of the animal's face, or perhaps a sea anemone impaled on its nose.

Scientists have long speculated about what the star might be for, but the answer was obvious to Catania when he first examined it under a microscope in the 1990s. He expected to see a world of different sensors. Instead, he found just one type—a dome-shaped bump called an Eimer's organ, repeated again and again, like the surface of a raspberry. Each bump contains mechanoreceptors that respond to pressure and vibration, and nerve fibers that carry those sensations toward the brain. These were clearly touch sensors, and they constituted the entirety of the star. The star is an organ of touch, and touch alone. Squint at it, and you might mistake it for a set of hands reaching out at the world. More or less, that's what it is.*

Close your eyes and press your hands against the nearest surfaces—the seat or floor beneath you, your own chest or head. With each press, a hand-shaped burst of shape and texture resolves in your mind. Press quickly and often enough, and you start to build a three-dimensional model of your surroundings. This is almost certainly what the star-nosed mole does with its nose. As it scurries through its dark underground world, it constantly presses its star against the walls of its tunnels, a dozen times a second. With every press, its environment comes into focus in a starburst of textures. I imagine that each one adds to a continuous model of the tunnel that builds in the mole's mind, like a pointillist image appearing dot by dot.

The mole's somatosensory cortex—the touch center of its brain—is disproportionately devoted to the star, much as a human's touch center is especially devoted to our hands. And just as our somatosensory cortex has clusters of neurons that represent each of our fingers, the mole's has stripes of neurons that correspond to each ray of the star. "You can

---

* You might think that the rays of the star grow outward from the mole's nose. That's not the case. A star-nosed mole embryo has tiny swellings on the side of its snout, which gradually lengthen into cylinders. These are the future rays of the star. When the mole is born, the cylinders are still attached to its face. Slowly, skin starts to grow beneath them, separating them from the underlying tissue. After roughly a week, the rays break free and spring forward. A star is born.

essentially see the star in the brain," Catania says.* But when he first discovered this mapping, one aspect of it made no sense. The 11th pair of rays, which are smaller than all the rest, is represented by a massive chunk of neurons, which take up a quarter of the brain region that's used for the entire star. Why should the mole devote the largest amount of processing power to the tiniest of its touch sensors?

By filming the mole using high-speed cameras, Catania and his colleague Jon Kaas realized that it *always* ends up investigating a piece of food with the 11th and smallest pair of rays, even if other parts of the star touch the object first. It will often dab an object several times in succession, each one bringing the 11th pair of rays closer. This is remarkably similar to what we do with our eyes. We make tiny adjustments to focus on objects with our fovea, the part of our retina where our vision is most acute. Similarly, the mole's 11th pair of rays is what Catania calls the tactile fovea—the zone where the animal's sense of touch is sharpest. It's no coincidence that this zone lies just in front of the mole's mouth. The instant it decides that an object feels like food, it can part the 11th pair of rays and seize the morsel with its tweezer-like front teeth.

The mole doesn't stroke or rub or palpate with its star. Whatever it's doing occurs through the simplest of actions: press and lift. That is how the animal might be able to recognize its prey through shape, by comparing how neighboring Eimer's organs are dented or deflected. The mole can certainly distinguish textures, since it'll eat bits of dead earthworm but ignore similarly sized chunks of rubber and silicone. And it can do all of this at a speed that puts even the sea otter to shame.

Catania shows me a video that he filmed from below as a star-nosed mole investigated a glass slide containing a piece of worm. When the video is slowed by 50 times, I can see the animal dabbing its star against the glass, detecting the morsel, bringing the tactile fovea across to inspect it more thoroughly, and finally swallowing it. In real time, it's

---

* Around 5 percent of star-nosed moles have mutant noses with either 10 or 12 pairs of rays. Their brains have the corresponding number of stripes.

impossible to work out what is happening. The mole simply appears, and the worm disappears. By analyzing such footage, Catania and his colleague Fiona Remple found that the mole can identify its prey, swallow it, and begin searching for the next mouthful in an average of 230 milliseconds and as little as 120 milliseconds. That's as fast as a human blink. Imagine that your eye starts to close at the exact moment that a foraging mole first touches an insect with its star. Before your lashes cross the midline of your eye, the mole's brain has already recognized what it has touched and sent motor commands to reposition the star. By the time your eye is fully shut, the mole has touched the insect a second time with its supersensitive 11th rays. By the time your eye is half-open, the mole has processed the information from that second touch and decided on a course of action. When your eye is fully open, the insect is gone and the mole is looking for another.

The star-nosed mole seems to be moving as fast as its nervous system will allow, restricted only by the speed at which information can travel between the star and the brain. That trip takes just 10 milliseconds. Within the same time, visual information can't even make it through the retina, let alone reach the brain or complete the return journey. Light may be the fastest thing in the universe, but light *sensors* have their limits, and the star-nosed mole's sense of touch blows past them all. "It's moving so fast that it's almost getting ahead of its brain," says Catania. He shows me another video in which the mole touches a chunk of worm and begins moving away before changing direction and scooping up the briefly missed morsel. "It's on to the next thing before it realizes what it's just touched," he says. Sighted people know what it's like to do a double take after walking past something unexpected. But that's an easy movement—a simple turn of the head. For a star-nosed mole, sensing the world through touch and not vision, and touching with its face instead of its limbs, a double take is a frenetic full-body affair.

Its speed and sensitivity are linked. With its bizarre nose, the mole can detect and capture small prey like insect larvae. But to subsist on such little morsels, it must scoop up a lot of them as quickly as possible.

"They're little vacuum cleaners," says Catania. "They eat things so small that you might think: Why even bother?" They bother because they have no competition. Thanks to the star—a nose that works like a hand and scans like an eye—the underground world appears in glorious detail, and abounds with food that its competitors can't even perceive. A tunnel that might seem like an empty corridor to another mole twinkles with tasty treats under the touch of the star.

LIKE THE STAR-NOSED MOLE, many animals that specialize in touch work in conditions where vision is limited. They're often searching for things that are hidden or hard to find, which forces them to root around with body parts that can probe, press, and explore. Whether we're talking about a sea otter's paw or a human's finger, an elephant's trunk or an octopus's arm, animals discover the world by deliberately moving tactile organs over it. And as the mole shows, those organs don't have to be hands.

The beaks of birds are made of bone and sheathed in the same hard keratin that constitutes your fingernails. They seem inanimate and insensitive—hard, face-mounted tools for grabbing and pecking. But in many species, the tip of the bill contains a smattering of mechanoreceptors, sensitive to vibrations and movements. In chickens, which rely heavily on vision to forage, those mechanoreceptors are relatively rare, and concentrated in a few small clusters on the lower beak alone. But in some ducks, like mallards and shovelers, they're spread all over the bill, upper and lower, inside and out. In some places, these mechanoreceptors are as densely packed as they are in our digits. The mallard's bill may be covered in the stuff of human fingernails, but it's extremely sensitive. Ducks use this sense to find food in murky water. With head submerged and tail aloft, they swirl, strain, and dabble, rapidly opening and closing their bills. They can grab fast-swimming tadpoles in the dark, and filter edible morsels from the inedible mud. "Imagine being given a bowl of muesli and milk to which has been added a handful of fine gravel," wrote Tim Birkhead in his book *Bird*

Sense. "How good would you be at swallowing only the edible bits? Hopeless, I suggest, yet this is precisely what ducks can do."[*]

Many other birds forage by shoving their bills into dark recesses and feeling for food. Such behavior is especially common on shorelines. Even the most deserted beaches are full of buried treasure—worms, shellfish, and crustaceans, all concealed within the sand. To reach this hidden buffet, shorebirds like curlews, oystercatchers, sandpipers, and knots probe among the grains with their beaks. Under a microscope, the tips of their bills are riddled with pits, like corncobs with all their kernels bitten off. Those pits are full of mechanoreceptors that are similar to those in our hands and allow the birds to detect buried prey.

But how does a shorebird know where to stick its bill in the first place? Subterranean prey aren't obvious from the surface, so one might guess that the birds just probe around haphazardly and hope for the best. But in 1995, Theunis Piersma showed that red knots find shellfish up to eight times more frequently than would be expected if they were doing random searches. They must have a technique. To discover it, Piersma trained the birds to inspect sand-filled buckets for buried objects and to indicate if they'd found anything by approaching a designated feeder. This simple experiment revealed that the knots could still detect clams that were buried beyond the reach of their bills. They could even sense stones, so they clearly weren't relying on smells, sounds, tastes, vibrations, heat, or electric fields. Instead, Piersma thinks that they use a special form of touch that works at a distance.

As a knot's bill descends into the sand, it pushes on the thin rivulets of water between the grains, creating a pressure wave that radiates outward. If there's a hard object in the way—say, a clam or a rock—the water must flow around it, which distorts the pattern of pressure. The pits on the knot's bill tip can sense those distortions, detecting sur-

---

[*] Some are especially good at this. Elena Gracheva (the scientist who studies thirteen-lined ground squirrels) and her husband, Slav Bagriantsev, showed that the Pekin duck, an animal that we domesticated from wild mallards and now breed exclusively for meat, is a touch specialist. Compared to other ducks, it has a wider bill, more mechanoreceptors in that bill, and more neurons that carry signals from those mechanoreceptors. More surprisingly, it also has fewer neurons for sensing pain and temperature. Sensory abilities don't come for free, so to become masters of fine touch, mallards had to sacrifice other kinds of tactile sensations.

rounding objects without having to make contact with them. This ability, which Piersma calls "remote touch," is impressive enough, but the knot improves it even further by probing the same areas repeatedly, stabbing its beak up and down several times a second. This stirs up the sand grains, which settle into a denser configuration, heightening the buildup of pressure from the beak and making the distortions more obvious. Every time the knot lowers its head, the food around it becomes more obvious, as if it were using a kind of sonar based on touch instead of hearing.*

The emerald jewel wasp also has a long, probing organ with a touch-sensitive tip, but its goals and methods are far grislier than a red knot's. The wasp—a beautiful inch-long creature with a metallic green body and orange thighs—is a parasite that raises its young on cockroaches. When a female finds a roach, she stings it twice—once in its midsection to temporarily paralyze its legs, and a second time in its brain. The second sting targets two specific clusters of neurons and delivers venom that nullifies the roach's desire to move, turning it into a submissive zombie. In this state, the wasp can lead the roach to her lair by its antennae, like a human walking a dog. Once there, she lays an egg on it, providing her future larva with a docile source of fresh meat. This act of mind control depends on that second sting, which the wasp must deliver to exactly the right location. Just as a red knot has to find a clam hidden somewhere in the sand, an emerald jewel wasp has to find the roach's brain hidden somewhere within a tangle of muscles and internal organs.

Fortunately for the wasp, her stinger is not only a drill, a venom injector, and an egg-laying tube but also a sense organ. Ram Gal and Frederic Libersat showed that its tip is covered in small bumps and pits that are sensitive to both smell and touch. With them, she can detect the distinctive feel of a roach's brain. When Gal and Libersat removed the brain from a cockroach before offering the roach to some wasps, they repeatedly stung it, trying in vain to find the organ that

---

* Inspired by Piersma's discovery, Susan Cunningham showed that distantly related birds also use remote touch. Ibises use the technique when probing through muddy wetlands with their long, sickle-like beaks. New Zealand's kiwis do the same through leaf litter.

was no longer there. If the missing brain was replaced with a pellet of the same consistency, the wasps stung it with the usual precision. If the replacement pellet was squishier than a typical brain, the wasps seemed confused and kept rooting around with their stingers. They know what a brain should feel like.

Both the wasps and their cockroach victims also use their antennae to feel their way around, as most insects do.* Long, sweeping tactile organs are so useful for navigation that many species have independently evolved their own versions.† Humans, ever the tool users, tap the ground in front of them with canes. The round goby, a bottom-dwelling fish, uses supersensitive pectoral fins. The whiskered auklet, a puffin-like seabird, has a large black crest that curves forward from its head, which it uses to feel the walls of the rocky crevices in which it nests.‡

Many other birds have stiff bristles on their heads and faces. These are often wrongly billed as nets that help birds to snag flying insects. It's more likely that they're touch sensors, which the birds use when handling prey, feeding chicks, or maneuvering around dark nests. Such uses might explain why birds have feathers at all. It's clear that birds evolved from dinosaurs, and that many dinosaurs were covered in bristly proto-feathers or "dino-fuzz." These structures were too simple for flight, so they must have evolved for some other reason. The most common explanation is that they provided insulation, but that would only be true if they suddenly appeared in large numbers. Alternatively, and perhaps more plausibly, they could have initially evolved to provide tactile information. As the whiskered auklet shows, an animal

---

* Insects evolved from ancestors that had many body segments, each with its own pair of legs. Over time, several of the frontmost segments fused to create the insect head, and their respective limbs were transformed into either mouthparts or antennae. The antennae are essentially repurposed legs, or sensory limbs.

† Tactile organs don't have to be long or sweeping. The remoras, or suckerfishes, have transformed their dorsal fins into suction cups, which they use to cling to the undersides of larger fish. That sucker is full of mechanoreceptors, which might tell the fish when it has made contact with a host.

‡ When Sampath Seneviratne placed some auklets in a dark maze and taped down their crests and whiskers, they were more likely to bump their heads.

only needs a few bristles to extend its sense of touch in useful ways. Perhaps feathers first appeared as small clumps on the heads or arms of dinosaurs, helping them first to feel and only later to fly.

Mammalian hair might have had a similar start, appearing first as touch sensors that were only later turned into insulating coats. Some hairs still retain that original tactile function. They're called vibrissae, from the Latin word for "vibrate." More commonly, they're known as whiskers. They are typically found on the faces of mammals, and are longer and thicker than other kinds of hair elsewhere in the body. Each one sits in a cup that's full of mechanoreceptors and nerves. When the shaft of the whisker is deflected, its base nudges the mechanoreceptors, which send signals to the brain. (You can get a feel for how this works, no pun intended, by closing your hand around the tip of a pen and deflecting the other end away from you.)

Some mammals continuously sweep their whiskers back and forth, several times a second, as they move. This action, delightfully known as whisking, allows them to explore the zone in front of and around their heads. When I first heard about whisking, I underestimated it. It intuitively felt like what I might do when I stumble down a dark corridor—reaching out with my hands to avoid bumping into a wall or to feel for a light switch. But after talking to sensory biologist Robyn Grant, I realize that a whisking mouse or rat uses its vibrissae in a way that's far closer to what I do with my eyes. The rodent constantly scans and re-scans the area in front of it, building up an awareness of a scene. If it senses something with the long, mobile whiskers on its snout, it investigates further with the shorter, immobile whiskers on its chin and lips, which are more numerous and more sensitive. This behavior is similar to that of a star-nosed mole pressing its nose along a tunnel, detecting objects with its star, and finally bringing the small and most sensitive rays into play. It's also similar to a human sweeping their eyes over a scene, detecting something in their peripheral vision, and focusing on it with their high-resolution foveae.

The similarities to vision don't stop there. If we turn our head, our eyes move first; likewise, a mouse will lead a head turn with its whiskers. Just as we map the world through the pattern of light falling across our retinas, a mouse can map its world by the patterns of touch across its array of whiskers. Each connects to a different part of the somatosensory cortex, so the mouse knows which whiskers have made contact with an object. And since it also knows what orientation those whiskers are in, "it can make maps of what it touches," Grant tells me. The information that builds those maps must flicker in and out as the whisker tips move. But Grant says that a mouse's brain probably interprets these discrete touches in a seamless way. I wonder if whisking for them is like vision for us—an experience that feels uninterrupted even though our eyes are constantly darting and blinking.

Mammals have been using whiskers for almost as long as mammals have existed.* Today, rats and opossums, which share the habits of their small, nocturnal, climbing, scampering ancestors, still whisk. Guinea pigs do it half-heartedly. Cats and dogs don't do it at all, although their whiskers are still mobile. Humans and other apes have lost our whiskers entirely and invested instead in sensitive hands. Whales and dolphins are born with whiskers, but these quickly fall out except around the lips and blowholes. Whisking, after all, is too difficult to do in the water. *Whiskers,* however, can still be useful.

TWO FLORIDA MANATEES LIVE at the Mote Marine Laboratory in Sarasota. As we stare at them, Gordon Bauer tells me that one, Hugh (as in Hugh manatee), is hyperactive. The other, Buffett (after Jimmy, not Warren), is sluggish and a little overweight. I confess to him that I'm struggling to work out which is which. Their 3-meter bodies seem equally rotund and their dispositions equally languid. After a while, though, I notice that one of them is slowly circling around his tank, performing what I guess is the manatee version of a zoomie. That's Hugh.

In the wild, manatees would spend their time trundling along shal-

---

* Grant showed that the opossum—a marsupial—also whisks, and controls its vibrissae using muscles very similar to those used by a mouse. These distantly related species belong to branches of the mammalian family tree that separated shortly after the group first evolved. This suggests that the earliest mammals actively explored their world through whisking.

low seabeds, grazing on underwater plants. In captivity, Hugh and Buffett devour around 80 heads of romaine lettuce every day. Hugh is currently going to town on one of these, slowly rending it apart. Sometimes he holds it between his flippers. Other times he grips it with his face, and specifically with the bit between his upper lip and nostrils. This large area, known as the oral disk, gives manatees the hangdog expression that makes them so endearing. And unlikely though it might seem, it is also an extraordinarily sensitive organ of touch.

The disk is muscular and prehensile, more like an elephant's trunk than a typical lip. By flexing and flaring the oral disk, a manatee can handle and investigate objects with the same dexterity and sensitivity as a hand. This is called oripulation—manipulation done with a mouth. Manatees will oripulate everything in their environment, from anchor lines to human legs. Sometimes this lands them in trouble: Florida manatees, which are endangered, get caught in ropes and crab traps because of their habit of exploring everything face-first. More often, oripulation cements their relationships. "Whenever they meet, they'll oripulate each other's faces, flippers, and torsos," says Bauer.

Reader, Hugh oripulated me. While Buffett took part in an experiment, Hugh was chilling out in a separate part of their enclosure. He lay on his back while a trainer held his flipper and popped beets into his mouth. I leaned over, and he exhaled sweet-scented breath over my face. I put my hand in the water in front of him, and he immediately began exploring it with his oral disk. It felt strange, this meeting of two tactile organs—my hand and Hugh's oral disk, both incredibly different but both devoted to the same sense. I can only imagine what I felt like to him—softer perhaps than the vegetables he eats, but smoother than the skin of his brother Buffett. To me, oripulation felt like being licked by a dog, except with no tongue involved—only prehensile lips, which danced over my palm. My fingertips soon felt like they'd been lightly sandpapered because many of Hugh's whiskers are stubbly.

Those whiskers—vibrissae—are the key to the oral disk's sensitivity. There are around 2,000 of them. Some are long, thin, and bristly.

Others are short and spiky, like broken toothpicks. When the oral disk is relaxed, these whiskers are lost among the fleshy folds. But when it's time to eat or explore, the manatee flares and flattens the disk, extending the whiskers outward. By flexing it in just the right way and moving the whiskers against each other, a manatee can clip grasses and shred lettuces. "They can grab food and bring it into their mouth, but also take things like pebbles out," Bauer says. His colleague Roger Reep once filmed a manatee eating a plant with one side of its mouth while using the other to remove what it didn't want to swallow. By pressing these hairs against an object, a manatee can take the measure of its texture and shape, like a whisking rodent, only much slower. In 2012, Bauer tested Hugh and Buffett to see if they could distinguish between plastic boards with differently spaced ridges, much as Sarah Strobel later did with Selka the sea otter and various human volunteers. The two manatees performed just as well as the other species.* Their faces were the equals of human fingertips.

Manatees are the only known mammals that *only* have vibrissae and no other kinds of hair. Aside from the whiskers on their oral disk, they have another 3,000 scattered all over their large bodies. Thin and widely spaced, they are hard to see at first, but I eventually catch a glimpse of Hugh's, glinting in the daylight. "Every once in a while, when the sun is just right, they look like a field of wheat," says Bauer.†
Manatees use these body-wide whiskers for another purpose—to sense the water flowing around them.

Sensory hairs are versatile structures. They can be actively pressed against surfaces to produce tactile sensations, as whisking rats and oripulating manatees do. But they can also be passively bent and deflected by flowing air or water. By responding to that pressure, an animal can detect the flows created by distant objects, touching things from afar without needing to make direct contact. Manatees can certainly do

---

* Buffett did slightly better, which Bauer attributes to Hugh's shorter attention span.

† A few other mammals have body-wide whiskers, including the naked mole-rat and the hyraxes—small creatures that look like marmots but are in fact the closest relatives of elephants and manatees. These hairs probably help mole-rats and hyraxes to detect the walls of cramped tunnels and rocky crevices, much like the whiskered auklet.

this. Bauer and his colleagues showed that Hugh and Buffett could use their body whiskers to detect the minute vibrations of a sphere that was shaking in the water. The animals were blindfolded, their facial whiskers were covered, and the sphere was positioned a meter away from their flanks. They sensed it nonetheless, even when it was displacing the water by less than a millionth of a meter.

In the wild, they probably use this "hydrodynamic" sense to judge the direction of a current, to work out what other manatees are doing, or to detect the approach of other animals. They successfully keep their distance from snorkelers even though their eyesight is notoriously bad. They often swim upstream from estuaries just as the tide starts to come in. They rest on the seabed in groups and then suddenly rise as one for a breath. Their eyes might be small, and the water around them might be turbid, but they perceive their surroundings through a distributed and distant version of touch. They can tap into the hidden signals that I hinted at earlier—the invisible currents of information that flow around us, and which animals can detect with the right sensory equipment.

AT THE LONG MARINE LAB where Sarah Strobel worked with Selka the sea otter, a harbor seal who goes by Sprouts is floating on his back in a pool. Colleen Reichmuth calls to him, and he hauls his gray, mottled body out of the water. She asks him to speak. He unleashes a startlingly loud noise that sounds like a cross between a roar and a foghorn. "BUH-WAH-WAH-WAH-WAH-WAH-WOOOAAAARRRR," he seems to say. I put my hand on his chest, and I feel the rumble through my entire arm. Underwater, where his song is much louder, it can feel like a punch.

Seals, sea lions, and walruses—the group of animals collectively known as pinnipeds—are often ignored by scientists in favor of more popular marine mammals like whales and dolphins. But Reichmuth has always been fascinated by them, perhaps because they, like her, must split their time between land and sea. "I grew up swimming and I always wanted to be in the water," she says. "I was drawn to these creatures that could just kind of switch back and forth between these two lives." Reichmuth came to the Long Marine Lab in 1990 and has worked there ever since. She has known Sprouts for all of that time: He arrived at the facility a year earlier, shortly after his birth at SeaWorld San Diego. He's approaching his 31st birthday when I meet him, which is well past the life span of male harbor seals in the wild. His old eyes have cataracts, and he can barely see. But that's not a problem: Thanks to their whiskers, blind harbor seals can still thrive, even in the wild.

Sprouts has around a hundred facial whiskers protruding from his snout and his eyebrows. When he looks at me full-on, they form a stiff radar dish around his face. Sprouts can use them to discriminate shape and texture, to sense vibrations in the water, and to avoid obstacles. When he dives back into the water, his whiskers brush along the sides of his tank, allowing him to closely follow the curving wall without ever bumping into it. "But if we were to throw a fish in there, he would have a really hard time finding it," says Reichmuth. "Unless it started swimming."

As a fish swims, it leaves behind a hydrodynamic wake—a trail of swirling water that continues to whirl long after the animal has passed. Seals, with their sensitive whiskers, can detect and interpret these trails.* This ability was only discovered in 2001, by Guido Dehnhardt and his team at Rostock, Germany. They showed that two harbor seals, Henry and Nick, could follow the underwater path of a mini-submarine. They clung to the trail even when their eyes were blindfolded and their ears were plugged by headphones. Only when their whiskers were covered by a stocking did they lose the sub. At the time, most researchers believed that hydrodynamic senses would only work over short distances. The disturbances created by moving underwater objects ought to die away so quickly that beyond a range of a few inches, they would be undetectable. But hydrodynamic wakes can ac-

---

* The seals actively keep the whiskers warm, even when diving in freezing water. This stops the tissues from stiffening and allows the whiskers to move freely. They pay a price for this. Sense organs can't usually be insulated in the same way that internal organs can. They have to be close to the surface, and thus often leak heat. To keep these organs heated in icy water is like powering a radiator that's situated in a doorway. The fact that the animal bothers says something about how valuable these organs are.

tually persist for several minutes. Dehnhardt estimated that a swimming herring should leave a trail that a harbor seal could follow from up to almost 200 yards away.

Sprouts might be getting on in years, but his hydrodynamic sense is still sharp. Reichmuth tests it using a ball that's mounted on the end of a long pole. She walks around the edge of the pool, moving the ball through the water in a sinuous trail. After a few seconds, Sprouts, who was waiting patiently, gets the green light. He searches around, sweeping his whiskers from side to side. As soon as they make contact with the ball's wake, Sprouts instantly turns and follows it. He isn't just heading in the rough general direction. He's following the *exact* path of the ball in minute detail, up and down, in and out, as if pulling himself along an invisible rope. He can't be relying on vision—even if his eyes weren't so old, he's wearing a custom-made blindfold. Instead, he's picking up on a track of invisible whirling vortices temporarily imprinted into the water. When he starts to stray beyond the trail, he moves his head from side to side to find its edge, just as a snake might do with its forked tongue. When the trail crosses a gushing water pipe, he temporarily loses it, but quickly picks it up again on the other side.* When the trail turns back on itself, so does Sprouts. In watching Sprouts, I'm reminded of Finn the dog sniffing his way along odor trails and following the scents of previous passers-by. To us, touch is rooted in the present, in the instants when a sensor makes contact with a surface. But to Sprouts, touch extends into the recent past, just as smell does to Finn. His whiskers can feel what was, rather than simply what is.

This ability seemed impossible back when Dehnhardt first discovered it. As a seal swims, its whiskers should produce their own swirling vortices of water. These ought to vibrate the whiskers and drown out the subtler signals produced by the wakes of distant fish. But harbor seals have an answer to this problem, which becomes clear when

---

* For obvious reason, the U.S. military funds studies like these, in the hope of creating instruments that can also track stealthy objects that are moving underwater. "Can you build devices that mimic the biological capabilities of an animal like this one?" says Reichmuth, pointing at Sprouts. "The answer so far is no."

Sprouts sticks his head out of the water. Looking closely at his whiskers, I can see that they're slightly flattened and angled so that the bladed edge always cuts into the water. They aren't smooth, either. At first glance, they look like they're covered in beads of water. But as I run my finger over them, I realize that they're dry, and that the "beads" are part of the whiskers' actual structure. They have an undulating surface that repeatedly widens and narrows along their entire length. The Rostock team showed that these shapes dramatically reduce the vortices left by the whiskers themselves. Through this quirk of anatomy, seals can tone down the signals from their own bodies and enhance those left by their prey. These flattened, undulating whiskers aren't found in walruses, which use their numerous vibrissae to feel out buried shellfish. They aren't found in sea lions, which are still strongly guided by vision. They're unique to seals, which are consequently better at following hydrodynamic wakes than other pinnipeds.*

Having shown off his skills, Sprouts sinks to the bottom of his tank and lies there, waiting. Harbor seals do this in the wild, too. They'll lurk in the darkness of a kelp forest using their radar dish of erect whiskers to detect the wakes of passing fish. From those impressions alone, the seal can tell in which direction a fish was swimming. It can discriminate between the wakes left by objects of different sizes and shapes, which might help it to pursue only the largest and most nutritious individuals. It might not need a wake at all. In one experiment, Henry and other seals at Rostock could detect gentle currents rising from the seabed, as might be produced by the gills of buried flatfish. Those fish might be camouflaged and lying perfectly still, but a seal can still feel their breaths with its face. A seal's tactile world is attuned to flow and motion, and their prey cannot help but move. It would seem like an unfair contest, if those prey didn't have incredible hydrodynamic powers of their own.

---

* Bearded seals are an exception that proves the rule. Their many whiskers are also simple and cylindrical, because they, like walruses, are bottom-feeders that root around for prey. They don't need a particularly strong hydrodynamic sense.

WHEN SEALS AND OTHER underwater predators charge at a group of fish, the school moves as one. The fish don't flee in random directions. They don't collide with each other. They seem to flow around their attackers like the very water in which they're immersed. This miraculous feat of coordination depends partly on vision. But it also depends on a system of sensors called the lateral line.

The lateral line is found in all fish (and some amphibians). It usually includes a smattering of visible pores on a fish's head and flanks, along with fluid-filled canals running just below its skin. After describing the pores in the seventeenth century, scientists spent 200 years thinking that they mostly secreted mucus. But on closer inspection, they noticed small groups of pear-shaped cells, capped in a gelatinous dome. These structures, now called neuromasts, were obviously sensors. In the 1930s, the biologist Sven Dijkgraaf showed that blind fish can use their lateral lines to detect the currents produced by objects moving nearby.[*] More impressively, he showed that they could also detect stationary objects by analyzing the currents that they themselves produce.

A swimming fish displaces the water in front of it, creating a flow field that envelops its body. Obstacles distort that field, and the lateral line can detect those distortions, providing the fish with a hydrodynamic awareness of its surroundings. If it swims toward an aquarium wall, the wall "prevents the water particles giving way as freely as in unobstructed water," Dijkgraaf wrote, and "the fish will experience an 'unexpected' rise of water resistance." This is similar to the technique that red knots use to locate buried clams, and it's likely how manatees perceive whatever's in the turbid water around them. But fish had been using their lateral lines to feel at a distance for hundreds of millions of years before either manatees or knots existed, and they are far more sensitive to water movements.[†]

---

[*] In 1908, the ichthyologist Bruno Hofer came close to working out what the lateral line did. He noticed that a blind pike could still avoid collisions and react to water currents as long as its lateral line was intact. Hofer correctly deduced that the organ allowed the pike to "feel at a distance" by sensing the flow of water. Unfortunately, he published his claim in an obscure and short-lived journal that he himself had founded and that hardly anyone read.

[†] In 1963, Dijkgraaf summarized his work in a seminal paper, which argued that the lateral line is a "specialized organ of touch," analogous to the vibrissae of mammals. In a nice bit of

With the lateral line, fish can feel the rich sources of information that are literally flowing around them. This awareness extends in almost all directions, for up to a body length or two away, which Dijkgraaf described as "touch at a distance." Humans can feel strong water currents flowing over our skin, but "I don't think that even gets close to the rich perceptions that fish must have through their lateral line," says Sheryl Coombs, who has been studying this system for decades. When we walk down the street, patterns of brightness and color move over our retinas, and we perceive our surroundings flowing past us. Perhaps a fish gets a similar experience from the patterns of water moving over its lateral line. They can certainly use those patterns to orient in flowing water, find prey, escape from predators, and keep tabs on each other. Schooling fish use their lateral lines to match the speed and direction of their nearest neighbors. When a predator lunges, the rush of incoming water triggers the lateral lines of the nearest individuals, which dart away. Their startled movements trigger the lateral lines of their neighbors, which trigger their neighbors, and so on. Waves of panic spread outward, and the school seamlessly parts around the predator. Each fish only attends to the small volume of water around it, but the sense of touch connects them all and allows them to act as a coordinated whole. Blind fish can still school.

Though all fish share the same basic neuromast structure, many of them have expanded and tweaked the lateral line in unusual ways. Surface-feeding fish have flattened heads loaded with neuromasts, which detect the vibrations of insects falling on the water's surface. Halfbeaks have massive underbites, and the neuromasts that line their protruding lower jaws can tell them if prey are swimming in line with their mouths. Blind cavefish have lost their sight and use exceptionally large, numerous, and sensitive neuromasts to find their way around.*

---

conceptual turnabout, when the hydrodynamic abilities of the manatee's body vibrissae were first discovered, they were billed as a mammalian equivalent of the lateral line.

* Some blind cavefish have evolved a unique style of swimming where they alternate between rapidly kicking forward and gently gliding along. The kicks provide propulsion but swamp the lateral line. The glide is slower but generates a stable flow field that makes surrounding objects easier to discern.

And some fish, unexpectedly, have almost lost their lateral lines altogether.

In 2012, Daphne Soares, a lover of both caves and unusual animals, traveled to Ecuador to see a blind catfish called *Astroblepus phoeleter*, which lives in a single cave and is so obscure that it has no common name. Examining it under a microscope, she expected to find giant and exceptionally sensitive neuromasts, like the ones found in many cave-dwelling fish that have dispensed with vision. Soares was shocked to find barely any neuromasts at all. Instead, the animal's skin was covered in what looked like little joysticks, the likes of which she had never seen before. "That's the reason I'm in science—that feeling of: I wonder what this is," she says.

Soares showed that the joysticks are mechanosensors. More unexpectedly, she learned that they're *teeth*. They're not tooth-like structures—they're actual teeth, made of enamel and dentine, with nerves coming out of their bases. While most catfish have expanded their taste buds to cover their bodies, this cave species has done the same with its teeth, turning them into a body-wide coat of flow sensors. That seems like a strange innovation for an animal whose ancestors would already have had a fully functioning lateral line. But Soares notes that these catfish live in a cave that experiences torrential floods on an almost daily basis. Those raging currents, she thinks, might have overwhelmed the lateral line, forcing the fish to evolve stiffer sensors. They now use their skin-teeth to find calm zones, where they can wait out the torrents by sticking to rocks with their sucker-like mouths. Soares is now studying other cavefish to see if they, too, have strange touch sensors.* "I like weird animals," she tells me. "The more extreme or ancient or unique, the better."

---

* One of these is a Chinese fish called *Sinocyclocheilus*. Between its long, upturned snout and a mysterious forward-pointing hump on its back, it looks like a cross between a fish and an iron. Its lateral line is normal, but Soares suspects that the horn might somehow sensitize the neuromasts by creating a bow wave ahead of the fish. It'll take more work to confirm that idea, but Soares is keen to start.

IN THE SUMMER OF 1999, before cavefish came into her life, Soares was sitting in the back of a pickup truck, next to a large alligator that had been collected by the U.S. Fish and Wildlife Service. During the long ride, she got a good look at her companion's taped mouth. That's how she first noticed the bumps.

Alligators have rows of dark, raised domes along the edges of their jaws, as if they're wearing beards made of blackheads. Scientists first described these bumps in the nineteenth century, but no one knew what they were for. "I thought they have got to be some sort of sensory things," Soares says. Back in her lab, she found that the bumps contained nerve endings. But she couldn't find any hairs, pores, or other obvious sensory structures that might stimulate those nerves. Working with sedated alligators that were lying in water, Soares tried exposing the bumps to light, electric fields, or bits of smelly, tasty fish. The nerves didn't react. Then, one day, she reached into the water to retrieve a tool that she had dropped. As her hand broke the surface, it caused ripples. And when these ripples hit the alligator's face, the nerves in its bumps finally started to fire. "I called my friends over to confirm that I wasn't hallucinating," Soares tells me.

The bumps, she discovered, are pressure receptors that can detect vibrations at the water's surface. They might work like little buttons, akin to the Eimer's organs on moles. They're so sensitive that if Soares let a single drop of water fall into an (unsedated) alligator's tank, the animal would turn and lunge toward the disturbance, even when its eyes and ears were covered. But if Soares covered its snout in a plastic sheet, the drops went unnoticed. The animals use the bumps to scan the thin horizontal layer where air and water meet. They sit in ambush in that layer, waiting for something to land in the water or to arrive at its edge for a drink. This strategy demands stillness, so they can't engage in the comparatively hectic explorations of moles, mice, or even manatees. Unmoving, they use their touch sensors to monitor everyone else's movements.*

---

* Crocodilians—alligators, crocodiles, and their relatives—weren't always aquatic. They and their extinct relatives have been around for some 230 million years, and many of those ancient species were land-living creatures that prowled like cats or galloped like horses. It's hard to

These bumps might detect more than the ripples of prey. When male alligators want to attract mates, they produce deep-throated bellows. These vibrate the water above their backs, causing it to dance and sputter like oil on a sizzling pan. Other alligators might be able to sense these vibrations through their delicate faces. The bumps are also found around the teeth and inside the mouth, so crocodilians might use them to assess their food or adjust their bites. When they forage underwater by sweeping their jaws around, the bumps could tell them when they've hit upon something edible. When a mother croc hears the cries of infants about to hatch, she might use the bumps to deliver just enough force to crack the eggs. When she carries her hatchlings around in her jaws, her fine sense of touch might help her to distinguish between prey she should bite and babies she should not.

This goes against every stereotype one might have about crocodiles as brutish, unfeeling animals. With jaws that can crush bone and thick skin that's heavily armored with bony plates, they seem like the antithesis of delicacy. And yet, they are covered head to tail in sensors that, as Ken Catania and his student Duncan Leitch showed, are 10 times more sensitive to pressure fluctuations than human fingertips.

What other organs of touch might people have missed because they exist in creatures that seem insensitive? Many snakes have thousands of touch-sensitive bumps on the scales of their heads. These bumps are especially common and prominent in sea snakes, which might use them as hydrodynamic sensors much as crocodilians seem to do. *Spinosaurus,* an enormous sail-backed dinosaur, had pores at the tip of its snout that resemble the holes in a crocodile's skull and that might have also allowed nerves to pass into pressure-detecting bumps. *Spinosaurus* had a crocodile-like face and has often been depicted as a semi-aquatic fish-eater; perhaps it also used touch sensors to feel for rippling prey. *Daspletosaurus,* a close relative of *Tyrannosaurus,* also had telltale holes in its

---

know what senses these prehistoric animals possessed, but their skulls provide a clue. If they had the same ripple-detecting bumps as modern crocodilians, they would also have had telltale holes in their jaws through which nerves would have passed. Some of them did—but not all of them. Crocodilians only evolved the pressure-sensitive bumps when they started transitioning to life in the water.

jaws and might well have been covered in sensory bumps. These dinosaurs didn't live in water, but perhaps they rubbed their sensitive faces during courtship, or used them to carry their young in their mouths. Such speculations might sound far-fetched, but perhaps they shouldn't when we think about the bumps of crocodiles, the lateral lines of fish, or the whiskers of seals. Science has a long track record of underestimating or overlooking touch and flow sensors—including ones that were sitting in full view.

FEW BIRDS ARE MORE recognizable or ostentatious than the peacock. But ignore, if you can, the gaudy iridescent tail. Focus instead on the stiff, spatula-like feathers that form a crest on their heads. These are utterly conspicuous, but often ignored. To find out if they have a purpose, Suzanne Amador Kane acquired several of them from aviaries and breeders, plus one unfortunate zoo peacock that flew into a polar bear enclosure. Her student Daniel Van Beveren then mounted the crests on a mechanical shaker, and watched as they wobbled to and fro. When shaken at exactly 26 Hz—that is, 26 times a second—they moved with exceptional vigor. That's their resonant frequency. It happens to be the *exact* frequency at which a courting male peacock shakes his tail feathers. That, Kane tells me, "couldn't possibly be a coincidence." Van Beveren played different recordings to his mounted peacock crests. When he put on a clip of an actual peacock rattling his tail, the crest feathers resonated. When he put on other recordings, including "Staying Alive" by the Bee Gees, they did not.

These results suggest that a peahen that stands in front of a courting male might be able to detect the air disturbances produced by his tail. As well as seeing his efforts, she might *feel* them. (This also works in reverse, since females will sometimes display back to males.) Kane now wants to prove this idea by filming the crests of living courting peacocks to see if they actually shake at the same frequencies.* If they do,

---

* That's easier said than done since a female's crest is green and is usually in front of green foliage. But Kane knows some breeders who have white peacocks, and discussions are afoot.

it would mean that a peacock's display, despite its flamboyance, has always had a secret component that was inconspicuous to human observers. We just don't have the right equipment to fully appreciate it. And if we're missing something in one of the most flamboyant exhibitions in the animal kingdom, then what else are we missing?

A clue can be found at the base of each peacock crest feather, where there's a smaller companion feather called a filoplume. It's just a simple shaft with a tufted tip, and could act as a mechanosensor. When moving air shakes the crest feather, the crest feather could nudge the filoplume, and the filoplume could trigger a nerve. Filoplumes are found in most birds, and are almost always associated with another feather. Birds can use them to monitor the position of their feathers, perhaps to sense when their smooth plumage has become ruffled and needs to be preened. But filoplumes are especially important during flight.

Bird flight looks so effortless that it's easy to forget just how demanding it is. To stay aloft, birds continuously adjust the shape and angle of their wings. If they get everything right, air flows smoothly over the contours of each wing, producing lift. But if they hold their wings at too steep an angle, the smooth flows form turbulent vortices and the lift disappears. This is called stalling, and if the bird can't avoid or correct it, it will drop out of the sky. This rarely happens, in part because filoplumes provide birds with the information they need to rapidly adjust their wings and stay in the air. Which is, frankly, incredible. I remember once standing on a boat and watching a gull fly alongside me. It was windy, and we—the boat and the bird—were moving fast. As I held my hand out and felt the air blow over and between my fingers, I marveled that the gull's wing could shape those same currents and keep it aloft. But I didn't realize all the bird was doing—that it was also using its filoplumes to read the air around it and make tiny adjustments to its flight. The French ophthalmologist André Rochon-Duvigneaud once wrote that a bird is a "wing guided by an eye," but he was wrong—the wings also guide themselves.

The same could be said about bats. Their membranous wings are very different from the feathered ones of birds, but are no less sensitive. They are covered with a smattering of touch-sensitive hairs, protruding from small domes and connected to mechanoreceptors.* Susanne Sterbing showed that most of these hairs react only to air that flows from the back of the wing to the front, which typically occurs when the wing is about to stall. Bats, like birds, can sense those moments and take corrective measures. Thanks to their hairs, they can bank steeply, hover, backflip to catch insects in their tails, and even land upside down. When Sterbing treated bat wings with hair removal creams and flew the animals through obstacle courses, the effects were obvious. They never crashed, but they kept a wide distance from the objects around them, and their turns were wider and clumsier. By contrast, with their hairs intact, they could fly within inches of obstacles and pull off hairpin turns. For them, airflow sensors make the difference between flying and flying acrobatically.

For other animals, however, such sensors mean the difference between life and death. Perhaps that is why they have evolved into some of the most sensitive organs in the world.

IN 1960, A SHIPMENT of bananas arrived at a marketplace in Munich, Germany. It had come from somewhere in Central or South America, and had brought with it a few hitchhikers—three large spiders, each as big as a hand. The spiders were sent to the University of Munich, where a scientist named Mechthild Melchers began studying and breeding them. The species, now known as the tiger wandering spider for the black and orange stripes on its legs, has since become the most thoroughly studied spider in the world.

The tiger wandering spider doesn't spin a web to catch food; instead, it sits in wait for its prey. Its legs are covered in hundreds of thousands of hairs, which are packed so densely that there can be 400 in a square millimeter. Almost all of them are connected to nerves and are sensitive to touch. Prod just a few on a single leg, and the spider

---

* Too short and thin to be seen with the naked eye, these hairs are not for insulation. In 1912, scientists suggested that they could be airflow sensors that allowed bats to fly in darkness. But once people realized that bats use a kind of sonar to navigate, interest in their tactile sense dropped, until Susanne Sterbing reignited it in 2011.

will either withdraw its limb or turn to investigate. If it is running and its hairs brush against an object—say, a wire strung across its path by a curious scientist—the spider will arch its body and scurry over the obstacle. During courtship, a male might stimulate a female's hairs in just the right way to prevent her from eating him.

Most of these hairs only respond to direct contact, but some are so long and sensitive that they will also be deflected by the wind. These are called trichobothria, from the Greek words for "hair" (*trichos*) and "cup" (*bothrium*). Like a bird's filoplumes or a fish's neuromasts, they're flow sensors—albeit exceptionally sensitive ones. Even air that's moving at just an inch per *minute*—a breeze so gentle it could hardly be called a breeze—will deflect them. Watch them under a microscope, and you'll see them fluttering away under the influence of imperceptible currents, while everything around them is still. With a hundred trichobothria on every leg, the tiger wandering spider can tune in to the airflow around its body, in every possible direction. It uses this sensitivity for lethal ends.

In its rainforest home, the spider spends the day hiding within the leaf litter and only emerges half an hour after sunset. It walks onto a leaf and waits. As the darkness intensifies, gusts of wind become rare, and the steady ambient airflow is dominated by low frequencies that the spider ignores. Its trichobothria are tuned instead to the higher frequencies produced by airborne insects, like a fly zooming toward the spider. The fly might be minuscule, but it still pushes air ahead of it. At first, the spider can't distinguish that moving air from the background flow. But once the fly is about 1.5 inches away, its air signal becomes noticeable, like a silhouette emerging from a fog. The trichobothria on the leg closest to the fly start to move before those on the other seven, and sensing this difference, the spider turns to face its incoming prey. As soon as the fly moves over one of its legs, it deflects the trichobothria from straight overhead, and the spider jumps. It grabs the fly from the air with its front legs, drags it to the ground, and delivers a venomous bite. "It's even able to correct its path while jumping," says Friedrich Barth, who has been studying the spider since 1963 and has watched its jumps many times over. "I've always thought about how difficult it would be to build a robot to do this."

Insects aren't helpless, though. Many have airflow sensors of their own. Wood crickets have a pair of spines called cerci that protrude from their rear ends. These are covered in hundreds of hairs that are just as sensitive as a spider's trichobothria, if not more so. These so-called filiform hairs can detect the current produced by a wasp's wing-beats. And, as Jerome Casas has shown, they can detect the infinitesimal wind created by a charging spider.

The wolf spider is the cricket's major predator and runs down its prey. On the uneven, leaf-strewn floor of a forest, it must launch its attacks while standing on the same leaf as its target. It is fast, but Casas found that the cricket's hairs can sense it almost as soon as it starts to run. Indeed, the faster the spider moves, the more detectable it becomes. Its only hope is to sneak up on the cricket, moving so slowly that it barely disturbs the air in front of it, and getting close enough for a final lunge. Even then, its odds of success are just 1 in 50. "The cricket almost always wins," Casas tells me. "As soon as it jumps away from that leaf and lands somewhere else, the game is over. It's on another world."*

The filiform hairs of crickets and the trichobothria of spiders are almost inconceivably sensitive. They can be deflected by a fraction of the energy in a single photon—the smallest possible quantity of visible light. These hairs are a hundred times more sensitive than any visual receptor that exists, or could *possibly exist*. Indeed, the amount of energy needed to shift a cricket's hairs is very close to thermal noise—the kinetic energy of jiggling molecules. Put another way, it would be almost impossible to make these hairs more sensitive without breaking the laws of physics.

So why doesn't everything in the world set them off? Why aren't spiders constantly leaping at imagined insects, or crickets constantly fleeing from phantom spiders? Partly, the hairs only respond to biologically meaningful frequencies—the kind produced by predators or

---

* This ability resembles Spider-Man's spider-sense, which warns him of danger. In some movies, the spider-sense is represented by small hairs that stand up on Peter Parker's arm. But as Roger Di Silvestro wrote on the National Wildlife Federation's blog, "Spiders can detect danger coming their way with an early-warning system called eyes."

prey, and not by the environment. The mechanoreceptors at the base of the hairs are also less sensitive than the hairs themselves and need stronger stimulation before they fire. Finally, no single hair will send the spiders into action. Animals rarely respond to the excited buzz of a single mechanoreceptor. Instead, they listen to the entire chorus.

Why, then, is each hair so sensitive? The obvious explanation is that long arms races between predators and prey have led to the evolution of sensors that detect the faintest possible signals. "But that's a bit of an easy answer, and I'm not totally convinced," says Casas. As a biologist, he's used to talking about optimization, where animals make the best of what they've got given the many constraints they face. But the cricket hairs are a rare example of *maximization,* he says. "They almost couldn't be better than they are, and that's surprising. No one really knows why."*

Most arthropods—the diverse group that includes insects, spiders, and crustaceans—have hairs that detect the flow of either water or air. The implications of this widespread sense are profound, in ways we have barely begun to grapple with. For example, in 1978, Jürgen Tautz showed that caterpillars can use hairs on their midsection to sense the air movements produced by flying parasitic wasps. They react by freezing, throwing up, or falling to the ground. Thirty years later, Tautz showed that flying honeybees can trigger the same effect. Simply by moving the air around the plants that they visit, bees can reduce the amount of damage that very hungry caterpillars might inflict. Few groups of insects matter more to plants than bees and caterpillars. And yet no one appreciated that these groups—the pollinators and the despoilers—are connected by the slightest gusts of wind and the minuscule deflections of hairs. The air around us is full of signals that we don't detect. And so is the ground below us.

---

* Does this airflow sense count as touch at a distance, as it is often described? Is it some version of hearing, which also relies on hairs that respond to movements of air? Opinion is divided. Casas thinks it has elements of both. Barth feels that it's a distinct sense in its own right. I personally find it hard to categorize without knowing more about what the animals are actually experiencing. How does the airflow of a distant fly feel to a spider compared to a wire directly brushing its leg? Do these feel as distinct as, say, hot or cold to us, or are they two ends of the same spectrum of tactile sensations?

7.

The Rippling Ground

Surface Vibrations

IN 1991, KAREN WARKENTIN WAS LIVING THE DREAM. THEY LOVED frogs and snakes, and as a new PhD student, they had somehow ended up in a place with plenty of both—Costa Rica's Corcovado National Park. Sitting by a pond, they'd observe the abundant red-eyed tree frogs with their lime-green bodies, orange toes, electric-blue thighs, yellow-striped flanks, and bulging, tomato-red eyes. In just one evening, each female would lay around a hundred eggs, which she'd encase in jelly and stick to leaves overhanging the water. But around half of the clutches were devoured by cat-eyed snakes. The others would hatch after six or seven days, releasing their tadpoles into the water—or, occasionally, onto Warkentin. "It was pretty common, in the field, to have tadpoles falling in your hair, tadpoles falling in your notebook," they tell me. "I also had the experience of bumping into a clutch and seeing a few embryos hatch out very quickly."

That was weird. The tadpoles weren't passively spilling out of eggs that Warkentin had broken. It looked like they were actively making a run for it. If they could do that when Warkentin bumped them, could they also flee from an attacking snake? Could they sense the motion of chewing jaws and decide to take their chances in the water? Warkentin presented this idea at a scientific meeting and was met with skepticism. Frog embryos were meant to be passive entities that hatch on a fixed schedule and are oblivious to their environment. "Some people thought it was a crazy idea," Warkentin says. "I thought it was a testable one."

They collected batches of eggs and housed them in outdoor cages along with cat-eyed snakes. The snakes are nocturnal, so Warkentin had to check on them throughout the night. They'd sleep on a couch in an adjacent building, suffer through the clouds of mosquitoes, and wake up every 15 minutes to groggily inspect the eggs. It was rough, but they were right: Embryonic tadpoles can hatch early when attacked. Warkentin even saw them bursting out of eggs that were held in a snake's mouth.

Warkentin has been studying this behavior ever since. Fortunately, their research now involves fewer itchy all-nighters and more infrared video cameras. They show me one recent video in which a cat-eyed snake lunges at a tree frog clutch and grabs several eggs in its jaws. As it tries to pull its mouthful free from the jelly, the surrounding embryos wriggle furiously, releasing an enzyme from their faces that quickly disintegrates their eggs. One of them plops into the water. A second later, another joins it. Soon, tadpoles are tumbling down too quickly to count, and the snake, still chewing its first mouthful, is left with a smear of empty jelly. "I never get tired of watching this," Warkentin tells me.

Their experiments showed that frog embryos are neither as helpless nor as unaware as people thought. The embryos' sensory bubble extends beyond the actual bubble in which they're trapped. Light can pass through the translucent eggs, and chemicals can diffuse into them. But vibrations are what really matter. They pass into the eggs and into the embryos, which can distinguish between bad vibes and benign ones without any previous experience of either. A bite from a snake will trigger hatching. Rain, wind, and footsteps will not. Even when a mild earthquake rattled Warkentin's pond, the embryos didn't react. By recording different vibrations and playing them back at the eggs, Warkentin showed that they're attuned to pitch and rhythm. Falling raindrops produce a steady pitter-patter of short, high-frequency vibrations. Attacking snakes produce lower frequencies and more complicated patterns, with prolonged bouts of chewing punctuated by periods of stillness. If Warkentin edited gaps of stillness into rainfall recordings to make them feel more snake-like, the tadpoles found them scarier and were more likely to hatch. They can clearly sense the world before entering it, and they can use that information to defend themselves. They have agency. They have an Umwelt.

"As they develop, they get more and more senses, and more and more information," says Warkentin. At two days old, the embryos can detect the oxygen levels around them, which tells them if their eggs have accidentally fallen into water. But they don't respond to snakes until they are just over four days old because, as Warkentin's student Julie Jung discovered, that's when the vibration sensors in their inner ears come online. They can escape from danger before then, but they have no way of sensing it.* Snakes are not yet part of their Umwelt. But in a matter of hours, everything changes: A new sense kicks in, and a realm of vibrations to which they were once oblivious transforms their lives.

Once the tadpoles have transformed into frogs and are ready to make tadpoles of their own, males compete for access to mates. By watching them with infrared cameras, Warkentin and their colleague Michael Caldwell saw that males would square off along a branch, raise their bodies, and vigorously shake their backsides. These displays are meant to be visually captivating, but males will also perform when their lines of sight are obscured. They might not be able to see each other, but they can still *feel* the vibrations created by their rival's quivering bum and use those vibrations to assess size and motivation. In these contests, the victors are usually those that shake for more time and create longer-lasting vibrations.†

---

* When the tadpoles' bodies are shaken, small crystals in their inner ears push against touch-sensitive hair cells, which send signals to their brains. This same inner ear system also controls a reflex that steadies the tadpoles' gaze by moving their eyes in the opposite direction to their head. So Jung built a jury-rigged tadpole rotator. By placing the tadpoles in tubes, gently turning them, and watching if their eyes swiveled, she could work out exactly when their inner ears become sensitive to vibrations.

† Caldwell even provoked males with a model frog mounted on an electric shaker. When this Robofrog vibrated, other males responded with their own aggressive signals. When it made visual signals without accompanying vibrations, the other males didn't care.

Many other animals probably communicate in this way. Male fiddler crabs attract mates by thumping their gigantic claws on the sand. Termite soldiers drum their heads against the walls of their mounds to create vibrational alarms that attract more soldiers. Water striders—insects that skate along the surface of ponds and lakes—can coerce partners into sex by making ripples that summon vibration-sensitive predators. All of these creatures create and respond to vibrations that travel along the surfaces around them, whether branch or beach. Scientists call these substrate-borne vibrations. Everyone else might just call them vibrations, or perhaps tremors or surface waves.*

To some people, these surface vibrations (and the airflow patterns that excite wandering spiders and crickets) count as "sound." By that logic, everything I described in the second half of the previous chapter and everything I'm about to describe in this one falls within the rubric of "hearing." I have no horse in this race and don't care to pick one. If you're a lumper, feel free to read these as a single continuous chapter, and if you're a splitter, think of them as three discrete ones. Either way, it's worth noting that while these stimuli have a lot of overlaps, they do also have important differences in their physical properties that, in turn, determine which animals pay attention to them and what those species do with the information.

For example, airborne sounds are waves that oscillate in the direction of travel—imagine stretching and compressing a Slinky. Surface waves, by contrast, oscillate perpendicularly to the direction of travel—imagine shaking the Slinky up and down. Those oscillations are obvious as ripples on the surface of water. They also occur on solid ground to a less visible extent. Throw a rock on the ground, and a subtle wave will ripple along the surface. If an animal is sensitive enough, it could feel the rise and fall of the ground beneath its feet. Many animals *are* sensitive enough, but most humans are not. Aside from the bass of a speaker or the shake of a cellphone, most of us miss out on the lush

---

* The vocabulary gets a little difficult, even for scientists. Many of them use *vibrations* in a colloquial way to specifically refer to substrate-borne vibrations, even though the term technically also encompasses sounds. I'm going to do the same here, with apologies to engineers who are now surely recoiling in disgust.

vibrational landscape that other species are privy to. It doesn't help that surface vibrations can be hard to separate from airborne sounds. Animals often produce both at the same time, shaking earth and air simultaneously. And animals often detect both kinds of waves with the same receptors and organs, like hair cells and inner ears. We certainly talk about them using a shared vocabulary: Creatures are said to be "listening" for vibrations, even when those are inaudible.

Perhaps the most important distinction between surface vibrations and sounds is that the former are largely ignored, including by scientists who study the senses. For the longest time, researchers saw all kinds of drumming, thumping, shaking, and quivering body parts, and interpreted them as visual or auditory signals, while ignoring the surface waves that those movements produce. Every red-eyed tree frog cues into that sensory world from four and a half days of age, but generations of scientists ignored it. "We have encountered it, but we were not looking for it," wrote ecologist Peggy Hill. It's a lesson that sensory biologists, and everyone else, should heed: By giving in to our preconceptions, we miss what might be right in front of us. And sometimes what we miss is breathtaking.

I'M IN A LAB in Columbia, Missouri, staring at a tick-trefoil plant. A dot of red light is shimmering on one of its leaves, as if someone planned to assassinate it. The dot is coming from a device called a laser vibrometer. It converts the vibrations moving over the surface of the leaf, which we cannot hear, into audible sounds, which we can. When I touch the table, I shake the entire plant and hear a loud roar. When I speak, the sound waves from my mouth set up surface waves in the leaf, which are converted back into sound waves by the speaker. I hear my own voice, as channeled through the plant. No one's interested in the sound of my voice, though. Rex Cocroft and his student Sabrina Michael are more interested in the song of the minute creature on the leaf. It's a treehopper—a kind of sap-sucking insect. It has large orange eyes, legs tucked so closely under its head that they resemble a beard, and black-and-white textures that look like a seashell. This species is known as *Tylopelta gibbera,* and though it has no official common name, Cocroft makes one up on the spot—the tick-trefoil treehopper.

We met Cocroft in the introduction, when he took his mentor Mike Ryan to meet some treehoppers in the Panamanian rainforest. That encounter took place more than 20 years ago, but Cocroft is still fascinated by these insects and the messages they exchange. By rapidly contracting muscles in their abdomen, they can create vibrations that move along the plants on which they stand, and up the legs of other treehoppers. These vibrations are normally silent, but a vibrometer can convert them into audible sound. Cocroft, Michael, and I all lean in toward the tiny tick-trefoil treehopper with almost comical expectation. And then we hear a rumbling noise, which sounds entirely unlike what an insect would produce. It's a purr but a startlingly deep one, more lion than house cat.

"Here we go," says Cocroft, beaming.

"Good job, buddy," says Michael.

Plants are strong, flexible, and springy, which makes them fantastic carriers of surface waves.* Insects exploit that property, filling plants with their vibrational songs. Between treehoppers, leafhoppers, cicadas, crickets, katydids, and more, Cocroft estimates that around 200,000 species of insects communicate through surface vibrations. Their songs aren't normally audible, and so most people are completely unaware that they exist. Those who become aware often get hooked.

Cocroft remembers his first time. He was a young student interested in animal communication and had decided to focus on treehoppers because they were obscure and understudied. In a field in Ithaca, he found a goldenrod plant that was covered in the species *Publilia concava*. He clipped a contact microphone onto the stem of the plant and listened through headphones. "Very shortly, I heard this *woo-woo-woo-woo*," he tells me, mimicking a noise that sounds like a plaintive bullfrog. "It was a crazy sound that nobody had ever heard before, and it

---

* "Surface waves" isn't strictly accurate here. When a wave travels along a long, thin structure, like a plant stem or a strand of spider silk, it's not that the surface ripples. Instead, the structure itself bends and flexes, which is properly known as a bending wave. I have relegated this to a note so that we aren't drowning in terms.

was right in my backyard. And that was it. I think that everyone who learns about this vibrational world can't help but be charmed by it, but there's a certain fraction of people who become so amazed that they have to go out and record from more species. There's so much out there. It's really endless."

Cocroft now has a library of treehopper recordings. When he plays them to me, I'm dumbfounded. The songs are haunting, mesmerizing, and surprising. None of them sound remotely like the familiar, high-pitched chirping of crickets or cicadas, but instead sound more like birds, apes, or even machinery and musical instruments. They're often deep and melodic, and they likely sound that way to the insects themselves. The song of *Stictocephala lutea* resembles a scratchy didgeridoo. *Cyrtolobus gramatanus* melds a hooting monkey with mechanical clicks. *Atymna* sounds like the warning that a truck makes when it's backing up, combined with a drum. *Potnia* lures me into a false sense of security with a mundane *brum-brum-brum* train, which then ends with a shocking half moo, half scream. When Cocroft first heard that, he tells me, "I sat back in my chair and thought: No way! *Is that an insect?*"

These vibrational songs are so strange because they're not subject to the same physical constraints as airborne sounds. In the air, an animal's pitch is normally tied to its size, which is why mice don't bellow and elephants don't squeak. That constraint doesn't exist for surface waves, so small animals can make low-frequency vibrations that seem like they're coming from much larger bodies. A treehopper can produce a mating call that's as low as that of an alligator, even though the latter is millions of times heavier.

Airborne sounds have another limitation: They radiate outward in three dimensions, and so lose energy very quickly. Insects compensate for this by concentrating all their efforts in a narrow range of frequencies, producing simple chirps. But surface waves only have to travel along flat paths, so they retain their energy over longer distances. Insects that signal along this channel can afford to get more creative. They can produce melodic upsweeps and downsweeps, stacks of tones, and percussive backdrops. That's why they sound more like birds.

There are more than 3,000 species of treehoppers, and they use surface waves in a variety of ways.* Some babies produce synchronized vibrations to summon their mothers when they sense a predator. Some mothers produce vibrations that silence the youngsters, lest their panicked tremors attract even more predators. Tick-trefoil treehoppers, like the one I saw in Cocroft's lab, use surface waves to congregate in groups. One will purr, and if another is within legshot, it responds with a sharp tick. The duo repeatedly move toward each other while purring and ticking, like children shouting "Marco" and "Polo" until finally they meet. They court each other in a similar way. A male makes a vibratory whine, followed by a train of high-pitched pulses. If a female hears him and is receptive, she makes a hum as soon as he finishes. He uses that hum to gauge her direction, walks a little closer, and makes another whine. She hums again, and slowly, the two duetters find each other. But if a second male is on the same plant, he'll unleash his own whine in the final moments of the first male's call; this shuts down the female's response. The first male retaliates by timing his next call to interrupt the second male, and the two go back and forth, repeatedly jamming each other. "If there's more than one male, it takes them a long time to find a female," Cocroft says.†

Treehoppers can gather on a single plant in the hundreds, and many of them might be vibrating away at the same time. A single stem might be as raucous as a busy street, full of cries for help, calls for silence, invitations to hang out, and literal booty calls. Even if you've never heard of treehoppers until now, if you spend any time outdoors, you will almost certainly have sat next to one, oblivious to the vibrational serenade it was performing. And these are just some of the many animals taking part in the full vibrational chorus. Masked birch caterpillars scrape their anuses on leaves to invite other caterpillars to social gatherings. Acacia ants vigorously defend their home trees from browsing

---

* Cocroft often tries to work out what different vibrations are for by recording them, playing them back to the treehoppers, and seeing how the insects react to the artificial noises. His sister once told a friend about this, and the friend said, "He lies to bugs?"

† Many duetting insects will jam each other's signals, and scientists can exploit this behavior to control agricultural pests. By playing the right vibrations along wires that run through vineyards, they can shut down the sex lives of leafhoppers that spread diseases.

mammals if they sense the vibrations created by chewing mouths. Even species whose calls we can hear are often sending vibrational signals that we can't. Cocroft plays me more recordings, made through plant stems, in which chirping cicadas sound like cows, and katydids sound like revving chainsaws. "I'm just amazed at the unbelievable richness of nature that already seemed so rich," he says.

It is surprisingly easy to tap into that extra richness, even without a laser vibrometer. In 1949, three decades before such instruments were invented, a pioneering Swedish entomologist named Frej Ossiannils-son heard the vibrations of leafhoppers by putting them on grass blades, sticking the blades in test tubes, and holding the tubes to his ear. As a trained violinist, he transcribed what he heard in musical notation. To hear them today, Cocroft simply uses a cheap speaker and a digital re-corder connected to a clip-on microphone that a guitarist might use. With this kit, he spends his spare time prospecting for vibrations, mik-ing random stems, leaves, and branches in nearby parks, or even in his backyard. Most times, he'll hear something new. I ask him to show me.

We drive to a park just a few minutes away from his lab. In a sunny spot, next to a wall of long grass, Cocroft and his students kneel down and begin clipping their microphones onto plants. For a while, we hear nothing. It's late September, and the season for vibrational song is drawing to a close. Strong gusts of wind are drowning everything else out. I can hear the footsteps of a walking caterpillar, and a beetle land-ing heavily on a leaf, but nothing like the haunting melodies I had hoped to experience firsthand. After a disappointing half hour, Co-croft apologizes. But just as we decide to call it a day, one of his stu-dents, Brandy Williams, calls to us. "There's something really cool here," she says.

We walk over, and from her speaker, we hear what sounds like . . . sniggering? "Eh, eh, eh, eh, eh," it seems to say. It is more hyena-like than insect-like. "Eh, eh, eh, eh, eh." Williams has clipped her micro-phone onto the bottom of a random blade of grass, and we cannot see any insects upon it. And yet, there's definitely an insect there. "Eh, eh, eh, eh, eh." So few people have listened to the vibrational world of treehoppers and other insects that on any attempt, there's always a chance of experiencing something that no other human ever has. I ask Cocroft if he's heard the mysterious sniggering before. "I've heard things *like* that," he tells me, "but whether I've heard that one . . . I really don't know. There are so many species out there."

Satisfied, we head back to his car. I'm suddenly aware of the choruses that might be vibrating through all the plants we walk past. I think about the vibrations that we ourselves are making with every step—the seismic surface waves that ripple out from each footfall. Although we hear the crunch of twigs underfoot and the soft squelches as shoes meet mud, we don't detect the tremors our footsteps send out. But other creatures do.

AS NIGHT FALLS ON the Mojave Desert, so does silence. Aside from the occasional howl of a coyote or the distant roar of a passing plane, the air is soundless. The dunes, however, thrum with vibrations. As insects emerge to forage, their petite feet create tremors that course along the sand. These waves are extremely faint and short-lived. But they're strong enough for the sand scorpion to sense.

Sand scorpions are some of the Mojave's most common residents and will eat anything they can successfully grab and sting, including other sand scorpions. In the 1970s, Philip Brownell and Roger Farley realized that the scorpions would readily attack anything that walks or lands within 20 inches of them. "Gentle disturbances of the sand with a twig also triggered a vigorous attack," Brownell later wrote in *Scientific American,* "but a moth held squirming in the air a few centimeters from the scorpion did not attract its attention." It seemed to track its prey using surface waves.

Brownell and Farley tested this idea by placing scorpions in a cunningly designed arena. It looked smooth and continuous on the surface, but a buried air gap blocked vibrations from traveling between the two halves. If a scorpion stood on one half, it was completely oblivious when the researchers prodded the other half with a stick, even at a point just an inch away. But if even one of the scorpion's legs straddled the gap, it became aware of the entire arena and would turn to face any disturbance.

Its sensors lie in its feet. On the joint that could be loosely described as an "ankle," there's a cluster of eight slits, as if the exoskeleton had been scored by a sharp knife. These are the slit sensilla—vibration-detecting organs common to all arachnids. Each slit is spanned by a membrane and connected to a nerve cell. When a surface wave reaches the scorpion, the rising sand pushes against its feet. This compresses the slits by an infinitesimal amount, but enough to squeeze the membrane and cause the nerves to fire. By sensing the tiniest changes in its own exoskeleton, the scorpion can feel the steps of passing prey.

The first time this happens, it shifts into its hunting stance. It raises its body, opens its pincers, and arranges its eight feet into a near-perfect circle. In this position, it can work out where surfaces waves are coming from by noting when those waves hit each of its feet. It turns and runs before pausing and waiting for another wave. When one arrives, it turns and runs again, getting closer to its target with each successive tremor. If its pincers collide with something, the scorpion seizes and stings. If it arrives at the source of the waves and can't find anything, it knows that its prey is underground, and digs it out.

Fittingly, these discoveries were earthshaking. They were made over a decade before Karen Warkentin found their frog-filled pond and Rex Cocroft started listening to treehoppers. At the time, the study of surface vibrations was even more niche than it is now. Scientists knew that animals can feel such vibrations, but few believed they could track down a source, any more than a human can locate an earthquake's epicenter without equipment.* It seemed especially preposterous that an

---

* Can animals sense earthquakes before they happen? It seems likely that many species could detect the incoming seismic waves, but whether they can parse that information and take appropriate evasive action is unclear. For millennia, there have been many anecdotal reports of creatures acting strangely before a quake, but such behaviors aren't consistent, and it's hard to know if human observers are simply remembering unusual activity in hindsight. In a few cases where elephants and other animals had been coincidentally fitted with tracking collars before an earthquake struck, they didn't seem to move any differently in the period before the shaking began.

animal could do so on sand, whose loose grains ought to damp and absorb vibrations rather than transmit them. But Brownell and Farley's meticulous experiments showed that these assumptions were wrong. Sand, soil, and solid earth are surprisingly good at transmitting surface waves, which are strong enough for animals to detect and informative enough for them to use. They were also interesting enough for scientists to study. Others began to look for seismic senses in other animals. They didn't have to look very far.

THE LARVAE OF ANTLION INSECTS, which are known as doodlebugs in North America, also hunt using surface waves that travel along sand. But rather than running down their victims, they bring their prey to them. They dig conical pits in dry sand and lurk at the bottom with their plump bodies buried and their gigantic jaws agape. The pits are precisely constructed traps. Their sides are shallow enough that they don't spontaneously collapse, but steep enough that any ant that walks into them will start to slip. The footfalls of an ant, even a struggling one, are hardly heavy, but the antlion is covered in bristles that can detect vibrations of less than a nanometer. It can sense when an ant is walking outside the pit, and can definitely tell when one is inside it. It reacts by tossing sand at the thrashing creature, creating an avalanche that further destabilizes the already slippery ground beneath it. Eventually the ant falls into the antlion's jaws, and is pulled under and injected with venom. Its vibrations then cease.

Other predators hunt by exploiting the seismic senses of their prey. Every April, the town of Sopchoppy, Florida, hosts a festival to celebrate the old tradition of worm grunting. Since the 1960s, several local families have ventured into the woods, pounded stakes into the ground, and created strong vibrations by scraping the stakes with iron. Soon, hundreds of large earthworms rise up, where they are easily collected by the bucketful and sold as bait. Some worm grunters believe that their vibrations mimic the sound of rainfall. Ken Catania—the same man who studied the star-nosed mole—proved otherwise. While attending the Sopchoppy Worm Gruntin' Festival in 2008, he showed

that worms barely react to the patter of raindrops, but they hightail it
to the surface if they detect the vibrations of a digging mole, or even a
recording of those vibrations. This is usually a sensible strategy since
moles don't pursue their prey aboveground. But several surface predators have learned that they can summon worms by deliberately shaking
the ground. Herring gulls and wood turtles do this, as, apparently, do
Floridians. For decades, worm grunters have been unknowingly mimicking mole-quakes.*

Animals have likely been able to sense seismic vibrations from the
moment they ventured onto the land from the oceans. The first backboned creatures to make that move—early amphibians and reptiles—
probably laid their large heads on the ground, allowing surface waves
to travel through the bones of their jaws and into their inner ears. In
the ancestors of mammals, three of those jawbones became repurposed
for transmitting airborne sounds. They shrank and moved, turning
into the small bones of the middle ear—the hammer, anvil, and stir-
rup. Now, instead of transmitting surface vibrations from the ground
via the jaw, they transmit sounds from the air via the outer ear and
eardrum.

But the ancient bone-conduction pathway still works: Vibrations
can pass directly to the inner ear via the bones of the skull, bypassing
the outer ear and eardrum altogether. Cyclists and runners can use
bone-conduction headphones to listen to music while keeping their
ears free. People with hearing difficulties can use bone-conduction
hearing aids, while deaf dancers can use special vibrating dancefloors.
And everyone who can hear does so partly through bone conduction,
which is why people often think they sound strange on recordings.
Those recordings reproduce the airborne components of our voices,
but not the vibrations traveling through our skulls.

Other mammals have tweaked their own anatomy to better sense
vibrations through bone conduction and restore their ancestral seismic
sense. Among the sands of southwestern Africa lives the Namib Desert

---

* In 1881, Charles Darwin wrote that "if the ground is beaten or otherwise made to tremble,
worms believe that they are pursued by a mole and leave their burrows." Over a century later,
Catania confirmed his statement.

golden mole. It is mostly insensitive to airborne sounds, because its outer ear is tiny and hidden in fur. But it is highly sensitive to vibrations, thanks to its malleus—the hammer bone of its middle ear. This bone is relatively enormous: Even though the golden mole weighs just an ounce and would fit in your palm, its malleus is bigger than yours.*

The golden mole forages at night, either by trundling over the Namib's dunes or by "swimming" through the loose sand with its paddle-like feet. It searches for sparse mounds of dune grass, where delicious termites might nest. Peter Narins has suggested that wind blowing over these mounds produces gentle low-frequency vibrations through the dunes, which the golden mole can detect by periodically dipping its head and shoulders into the sand. Every time it does, vibrations pass into its inner ear via its malleus, and humming beacons of dune grass resound around it.† The golden mole's seismic sense is so acute that, though blind, it can walk between distant mounds in virtually straight lines.

Golden moles, sand scorpions, antlions, and earthworms all have poor eyesight and all live either very close to the ground or within it. It seems plausible, and perhaps even obvious with hindsight, that they should be attuned to vibrations in the ground. But a seismic sense is harder to intuit in creatures that stand higher off the ground. Cats, for example, have a lot of vibration-sensitive mechanoreceptors in the muscles of their bellies. When a cat crouches down during a stalk, is it doing more than lying low? Is it also sensing the vibrations of potential prey? Could a lion pinpoint distant antelope herds? "The lying about that nature documentaries attribute to innate laziness of lions may actually be a period of astute assessment," wrote Peggy Hill in her book about vibrational communication. Hill herself admits that such ideas could be "greeted with applause or derision," but her point is that the

---

* Golden moles, despite their name and appearance, are not moles. They independently evolved the same physique and lifestyle, but they're more closely related to a motley menagerie of mammals that includes manatees, aardvarks, and elephants.

† The malleus normally picks up sound vibrations from the eardrum, and moves to transmit them to the incus (anvil). The golden mole's version is so big that it works in a slightly different way. When seismic waves reach the mole's head, the malleus mostly stays in the same place, and the rest of the skull, including the incus, vibrates around it.

questions are worth asking. Seismic senses have been long neglected, and biologists always seem to be one stray observation away from uncovering an unseen side to even the most familiar creatures.

IN THE EARLY 1990S, Caitlin O'Connell spent weeks at a time sitting in a dank, cramped, half-buried cement bunker, gazing through a narrow slit at a waterhole. She had come to Etosha National Park in Namibia to study elephants and to find ways of keeping them away from croplands. In the meditative confines of her bunker, she got to know the local herds, and certain behaviors began to leap out. Sometimes, she noticed, an elephant seemed to sense something in the distance, freeze midstride, and lean forward with a foot propped up on its toenails. To O'Connell, that pose seemed strangely familiar. As a master's student, she had studied the vibrational communication of planthoppers, which are related to treehoppers, and which also lean forward and press down on their feet when trying to detect each other's signals. Could the elephants really be doing the same? It surely wasn't a coincidence that whenever one of them adopted this pose, other elephants soon appeared in the distance. The animals seemed to be listening with their feet, but no one seemed to have noticed.

In 2002, O'Connell returned to her waterhole to test her idea. She had previously recorded the alarm call of local elephants that were being threatened by lions. The original call was audible, but O'Connell transformed it into a mostly seismic signal by cutting off the higher frequencies and playing it through shakers buried in the ground. When she did this, entire herds would freeze. They'd fall silent, become wary, and bunch up into defensive formations. Watching them through night-vision goggles, O'Connell was thrilled. "All these years of planning, hoping, and dreaming of this moment. We were finally showing that my original hunch so long ago was true," she wrote in her book *The Elephant's Secret Sense.* "Elephants were detecting and responding to our seismic cues."

A few years later, she repeated the experiment, but with an extra anti-predator rumble recorded in Kenya. This time, the Etosha elephants responded to the vibrations of the familiar local alarm, but not to the unfamiliar Kenyan one. They not only paid attention to vibrations but could tell if they were coming from elephants they knew. More recently, O'Connell has shown that elephants can respond to other kinds of seismic signals. In one video, a sexually active bull named Beckham searches fruitlessly for a fertile female after hearing her rumbles through a hidden speaker.*

What of the other elephant-like creatures, like mammoths and mastodons, that used to roam the planet? What about the giant ground sloths, the short-faced bears that would have towered over modern grizzlies, the armadillos the size of cars, or the hornless rhinos that were 10 times heavier than modern ones? These megafauna are now all extinct, and humans and our prehistoric relatives are to blame. As we spread around the globe, the biggest animals blinked out. That trend continues today. The three remaining species of elephants—two in Africa and one in Asia—are all endangered. The next-biggest land animals—white and black rhinos, giraffes, and hippos—are in trouble, too. Great herds are also diminished. Between 30 and 60 million bison once roamed North America in groups that were thousands strong, but European colonists slaughtered them in a bid to also exterminate the Indigenous peoples who depended on them. Now just 500,000 bison remain, and most are confined to private lands. Imagine how much quieter the ground is now without all those hooves and paws. Six continents that once would have thundered with the footsteps of titans now reverberate with sparse gurgles.

Can humans, the cause of that seismic silencing, even feel the loss? Western societies have largely cut themselves off from the ground beneath their feet with shoes, seats, and floors. If they spent more time

---

* As we saw in Chapter 1, doing experiments with animals as big, powerful, and intelligent as elephants is not easy, and their seismic sense remains largely mysterious. O'Connell has shown that elephants produce surface waves when they call and walk, but do they do so deliberately, or are such waves incidental? The vibrations can travel over several miles, and elephants could potentially use them to coordinate their social groups over long distances—but do they? Can they use that information to tell which elephants are nearby, or whether they are distressed or aggressive? Seismic signals are likely part of their Umwelt, but it's not yet clear if they're an important part.

sitting upon instead of standing above the ground, what might they sense? Luther Standing Bear, an Oglala Lakota chief and author, offered a clue. "The Lakota . . . loved the earth and all things of the earth, the attachment growing with age," he wrote in 1933. "The old people came literally to love the soil, and they sat or reclined on the ground with a feeling of being close to a mothering power. . . . This is why the old Indian still sits upon the earth instead of propping himself up and away from its life giving forces. For him, to sit or lie upon the ground is to be able to think more deeply and to feel more keenly; he can see more clearly into the mysteries of life and come closer in kinship to other lives about him. The earth was full of sounds which the old-time Indian could hear, sometimes putting his ear to it so as to hear more clearly."

That direct connection to the natural vibratory world may be in decline, but a different vibroscape has arisen. Modern cellphones buzz against our skin and fingertips, alerting us of breaking news, upcoming events, and social attention. Our devices use vibrations to connect us to the world beyond our bodies, extending our Umwelt beyond the reach of our anatomy. As usual, though, another group of animals got there first.

"IT IS PRETTY GROSS in here, just to warn you," Beth Mortimer warns me. And yet, I am unprepared.

I had asked to see her colony of *Nephila* spiders, which I assumed would be housed individually in a row of cages. Instead, we walk through a heavy door and a curtain of wide plastic slats into a large room that used to be an aviary but now houses a few dozen free-range spiders. Mortimer and I stand in the middle of this arachnarium to avoid blundering into the messy, meter-wide webs. They are hard to see, but I can easily sense where they are by looking for the large spiders at their centers. Each is the size of an ear. In the wild, *Nephila* webs can be big and strong enough to catch bats. In this room, they are fed on flies, which are also allowed to roam freely. That's the gross bit: The flies are bred from a compost bin in the corner, full of rotting bananas and milk powder. As Mortimer tells me about this, and about her work on spider silk, I try to ignore the large blowflies landing on my hair, notepad, and pen. "I bring undergrads in here and they're disappointingly squeamish," she says.

To humans, whose eyes can scan the entire scene and are sharp enough to just about make out the silk of the webs, the room is a labyrinth of death traps waiting to ensnare the flies. To the spiders, which have very poor eyesight, the room doesn't really exist: There is only the web, and whatever vibrates it. To the flies, the thin webs are imperceptible until they are ensnared in one. I almost feel sorry for them. "I don't," says Mortimer. "I hate flies." She adores spiders, though, and *Nephila* most of all. She studies other vibration-sensing animals, including water striders, planthoppers, and elephants. But *Nephila,* the first creatures she worked with when she started her scientific career, "will always be my first love," she says. "I really respect the elephants. But I *love* the spiders. The fact that they're so misunderstood by so many people just really makes me want to sing their praises so much more."*

Spiders have been around for almost 400 million years, and they've likely been producing silk for all that time. Their silk is a marvel of engineering. Though light and elastic, it can be stronger than steel and tougher than Kevlar. Spiders use it to wrap their eggs, construct shelters, hang in the air, and soar through the skies (more on that later). Most famously, many species fashion it into a flat, circular shape—the orb web.

The orb web is a trap, which intercepts and immobilizes flying insects. It's also a surveillance system, which extends the range of the spider's senses well beyond the reach of its body. That body is covered in thousands of slit sensilla—vibration-sensing cracks similar to those that sand scorpions use to detect the seismic activity of their prey. On spiders, these slits are also concentrated around the joints, where

---

* It's striking to me that many scientists who study vibrational senses are also musicians. Frej Ossiannilsson, who pioneered the field, was a violinist. Rex Cocroft was originally going to major in piano before he was seduced by biology. Beth Mortimer is a singer who also plays the French horn and piano.

they're grouped into clusters called lyriform organs. Using these exquisitely sensitive organs, all spiders can sense the vibrations coursing through whatever they're standing upon. For the tiger wandering spider of the previous chapter, that surface is the ground. For orb-weavers like *Nephila,* it's the web. These spiders construct the surfaces that they then sense vibrations through. For that reason, the orb web isn't just another substrate, like soil, sand, or plant stems. It is built by the spider and it is part of the spider. It is as much a part of the creature's sensory system as the slits on its body.

Like the *Nephila* in Mortimer's arachnarium, most orb-weavers sit in the middle of their webs and rest their legs on the radial spokes that funnel vibrations toward them. From this position, they can distinguish the vibrations generated by rustling wind or falling leaves from those created by struggling prey. They can probably work out where those struggles are coming from by comparing the strength of the vibrations hitting each of their legs. They can assess the size of their prisoners, and will approach the larger ones more carefully or not at all. If the prey stops moving, they can find it by deliberately plucking the silk and "listening" to the returning vibrational echoes. When it comes to capturing prey, vibrations supersede other stimuli. If a tasty fly buzzes above an orb-weaver, the spider will simply wave it away with its legs. The fly only becomes recognizable as food if it shakes the web.

This dependency on vibrations is so absolute that many animals can exploit orb-weavers by camouflaging their footsteps. The small dewdrop spider *Argyrodes* is a thief, stealing from larger spiders like *Nephila* by hacking their webs. From a nearby hiding place, it runs several lines of silk over to the hub and spokes of a *Nephila* web, effectively plugging its sensory system into that of the bigger spider. It can tell when *Nephila* has caught something and is wrapping it in silk for storage. It then runs over and eats the insect itself, often after cutting it free from the main web so that the host spider can no longer detect it. *Argyrodes* acts carefully to avoid creating its own telltale vibrations. It runs only when *Nephila* is moving and treads more slowly when *Nephila* is still. It also holds on to any strands it cuts to avoid any sudden releases in tension. Through such subterfuge, this thief is almost never caught. As many as 40 of them might be plugged into a single *Nephila* web.

Other creatures have more lethal intentions than pillaging food. Some assassin bugs walk so stealthily that they can creep right up to a spider and kill it on its own web. *Portia*, a jumping spider that eats other spiders, will violently twang a web to mimic the impact of a twig and use this vibrational smoke screen to charge at its prey. Both *Portia* and the assassin bugs can pluck webs to mimic the vibrations of ensnared prey and lure spiders to them. These predators are all visually conspicuous, but as long as their vibrations feel like those of an insect, a twig, or a breeze, an orb-weaver can't tell the difference. It lives in what Friedrich Barth calls "a small woven world full of vibrations."

An orb-weaver not only builds its own vibrational landscape but also can adjust it as if tuning a musical instrument. The range of that instrument is immense. By using gas guns to fire projectiles at individual silk fibers and analyzing the threads with high-speed cameras and lasers, Mortimer concluded that some silks can transmit vibrations over a wider range of speeds than any known material. A spider can theoretically change the speed and strength of those vibrations by altering the stiffness of its silk, the tension in the strands, and the overall shape of the web. It can do this every time it builds a new web, by pulling silk out of its body at different speeds, by creating fibers of different thicknesses, or by adding tension to the new strands. It can adjust webs that have already been spun by adding, removing, or tugging on specific threads. It can rely on silk's natural tendency to contract in humidity, and then stretch out these tightened threads to just the right degree. It's not clear when orb-weavers might decide to do any of this, but they certainly have the option of tuning their own senses and defining their own Umwelt according to their needs.

Zoologist Takeshi Watanabe showed that the Japanese orb-weaver *Oclonoba sybotides* changes the structure of its web when it is hungry. It adds spiral decorations that increase the tension along the spokes, improving the web's ability to transmit the weaker vibrations transmitted by smaller prey. When it is famished, every morsel counts. To capture such morsels, the spider expands the range of its senses by changing the nature of its web.

But here's the truly important part: Watanabe found that a well-fed spider will *also* go after small flies if it is placed onto a tense web built by a hungry spider. The spider has effectively outsourced the decision about which prey to attack *to its web*. The choice depends not just on its neurons, hormones, or anything else inside its body, but also on something outside it—something it can create and adjust. Even before vibrations are detected by its lyriform organs, the web determines which vibrations will arrive at the leg. The spider will eat whatever it's aware of, and it sets the bounds of its awareness—the extent of its Umwelt—by spinning different kinds of webs.* The web, then, is not just an extension of a spider's senses but an extension of its *cognition*. In a very real way, the spider thinks with its web. Tuning the silk is like tuning its own mind.

A spider can also tune its body. Biophysicist Natasha Mhatre showed that the infamous black widow can adjust the lyriform organs on its joints to different vibrational frequencies by changing its posture. The widow spins a messy horizontal web, and normally hangs upside down from it with legs outstretched. But when it's hungry, it can also draw its legs into a "crouch"—a sensory power pose that retunes its joints to higher frequencies. Like the tense web of Watanabe's orb-weaver, this stance might shift the spider's Umwelt toward the movements of smaller prey. It might also help it to ignore the low frequencies of wind. It's like a postural squint, which allows the spider to focus its attention. The analogy isn't exact, though, since squinting helps us to focus on particular parts of space. Here, the black widow's posture focuses on different parts of *information space*. It's as if a human could emphasize the red parts of our vision by squatting, or single out high-pitched sounds by going into downward dog.

The black widow's crouch reminds me of the hunting stance of the

---

* Orb-weavers will also pull on spokes that lead to areas where prey are repeatedly caught, focusing their attention on parts of the web most likely to yield food.

sand scorpion, the dipped head of the golden mole, and the forward-leaning, tipped-toe posture that clued Caitlin O'Connell in to the seismic sense of elephants. It seems only right that animals that parse the vibrations moving beneath them might have special ways of interacting with whatever they're standing on. For us, sitting down will suffice.

Since getting a puppy, I've been spending a lot more time on the floor than I used to. From that position, I can feel surface vibrations that I hadn't ever noticed before. I can feel the footsteps of my neighbors as they come in and out. I can feel the rumbles of garbage trucks as they drive past outside. This is a world I can lower myself upon, but it's one that Typo always resides in. Being a corgi, he is usually five feet closer to the rippling ground. I wonder what he feels. I also wonder what he hears. Typo will often perk up from a rest, his Yoda-esque ears picking up something that mine did not. He reminds me of what I'm missing: not just the surface waves traveling through the floor below us but also the pressure waves—sounds—moving through the air around us.

8.

All Ears

Sound

ROGER PAYNE USED TO BE SCARED OF THE DARK. WHILE IN high school, he tried to overcome that phobia by going on long night-time walks through a nature reserve near his home. During these solitary strolls, he often heard (and occasionally saw) an owl that lived in a nearby building. And as his fear of the night subsided, his interest in owls grew. In 1956, when he got a chance to study the birds as an undergraduate student, he leapt at it.

Owls have large eyes, but they can catch prey in darkness so total that even they can't see. Payne suspected that they used their ears. To test this idea, he taped black plastic sheets over the windows of a large garage and carpeted the floor with a thick layer of dry leaves. On a perch in the corner, he placed a hand-raised barn owl, who was named Wol after the character from *Winnie-the-Pooh*. Then, sitting in the dark, Payne released a mouse. "I couldn't see anything, but once the mouse started moving, I could hear the sounds of rustling," he tells me. So could Wol. For the first three nights of the experiment, the bird did nothing. But on the fourth night, Payne heard the sound of a strike. He flicked on the lights and saw Wol with the mouse in his talons.

Over the next four years, Payne did more experiments with Wol and other barn owls, all of which confirmed how adept they are at finding their prey through sound. The mice seemed aware of the danger, and would skulk at a glacial pace when Payne introduced them into a leaf-strewn room. As soon as they started rustling, they were done for. Watching through an infrared scope, Payne saw that owls would react to the first rustle by leaning far forward. On the second, they would swoop headfirst toward the rodent and, at the last moment, rotate their body by almost 180 degrees to place their talons where their faces had been. They were so accurate that they could not only land on a mouse but strike it along the long axis of its body. If Payne dragged a mouse-sized wad of paper through the leaves, the owls struck that, too. If he tied a single leaf to the tail of a mouse and allowed it to scamper over a foam floor, the owls attacked the leaf. These tests confirmed that the birds couldn't be using smell, vision, or any other sense. They were unquestionably using their ears to guide their strikes. And if Payne plugged one of those ears with cotton, the once-unerring birds would miss their rodents by more than a foot. "It was a thrill," he tells me. "The evidence was so clear."

If a mouse rustles, a dog barks, or a tree falls in a forest, it produces waves of pressure that radiate outward. As these waves travel, the air molecules in their path repeatedly bunch up and spread out. These movements, which occur in the same direction as the wave's line of travel, are what we call sound. The number of times the molecules compress and disperse in a second determines the sound's frequency—its pitch, which is measured in hertz (Hz). The extent to which they move determines the sound's amplitude—its loudness, which is measured in decibels (dB). Hearing is the sense that detects those movements.

Your ear consists of three parts—the outer, middle, and inner ears. Your outer ear greets incoming sound waves, collecting them with a fleshy flap and sending them down the ear canal. At the end of the canal, they vibrate a thin, taut membrane called the eardrum. Those vibrations are amplified by the three small bones of the middle ear, which we met in the last chapter, and transmitted to the inner ear—specifically, into a long fluid-filled tube called the cochlea. There, the vibrations are finally detected by a strip of movement-sensitive hair cells, which send signals to the brain. A sound is heard.*

The barn owl's ear shares the same basic structure: The outer ear collects, the middle ear amplifies and transmits, and the inner ear detects. But while your outer ears are a pair of fleshy flaps, the owl's are effectively its entire face.† The feathers of the conspicuous facial disc that makes owls look owlish are thick, stiff, and densely packed. They act like a radar dish that collects incoming sound waves and funnels them toward the ear holes. These enormous openings are found behind the owl's eyes, hidden among its feathers. In some species, they're so wide that if you part the overlying feathers and look into the ears, you can see the back of the owl's eyeball. These features, combined with an eardrum and a cochlea that are much bigger than you'd expect for a bird of its size, contribute to the exceptional sensitivity of a barn owl's hearing.

The owl excels not only at detecting sounds but also at working out exactly where they're coming from.‡ As we saw in the chapter on vision, if you make a thumbs-up sign with your arm outstretched, your nail represents roughly one degree of space. Masakazu Konishi and Eric Knudsen showed that at best, barn owls can localize a sound's

---

* These hair cells are similar to those in the lateral lines of fish, because both the ear and the lateral line likely evolved from the same ancestral sensory system.

† Some other differences: The owl's cochlea is curved like a banana, while yours is coiled like a snail shell, and its middle ear has just one bone instead of three. Also, unlike mammals, barn owls and other birds have ageless ears. Their hair cells regenerate, and the sensitivity of their hearing barely decreases with age. Confusingly, the prominent tufts of the long-eared owl, the short-eared owl, and their relatives are just ornaments that aren't actually part of the ear and aren't involved in hearing.

‡ Even a barn owl can't hear everything, though. Like humans and every other animal, it can only detect sounds within a certain range of frequencies, or pitches. That range is determined by the hair cells in its cochlea, which are arrayed on a long strip called the basilar membrane. The base of that membrane vibrates at lower frequencies, while the tip vibrates at higher ones. Based on which parts of the strip are vibrating, and thus which hair cells are being stimulated, the owl's brain can work out which frequencies are hitting its ear. The length, thickness, shape, and stiffness of the membrane determine the upper and lower limits of its hearing range. On average, humans can hear sounds between 20 Hz and 20 kHz, while owls have a slightly narrower hearing range, between 200 Hz and 12 kHz. Within that range, they're especially sensitive to anything between 4 and 8 kHz, which not coincidentally covers the frequencies that mice make when they scamper through leaf litter.

source to within 2 degrees. That's better than most land-living animals. For comparison, cats, whose ears are roughly as sensitive as a barn owl's, can only localize sounds to within 3 to 5 degrees.

Humans are almost as good as owls in the horizontal direction, but considerably worse in the vertical, where our accuracy falls to between 3 and 6 degrees. That's because our ears are level with each other, so sounds hit both at roughly the same time whether they're coming from above or below.* An owl's ears, however, are uniquely asymmetric, with the left being higher than the right. If you think of an owl's face as a clock, its left ear opens at two o'clock and its right ear at eight o'clock. If a sound comes from above *or* from the left, it arrives a little sooner and a little more loudly at the higher left ear than the lower right one. If the sound comes from below or from the right, the opposite is true. The owl's brain uses these differences in timing and loudness to work out the position of a sound's source in both the vertical and horizontal. If I go on a hike and hear a rustling noise nearby, I can tell roughly where it's coming from, and I turn my head so that my eyes can spot the source. But an owl perched overhead can tell *exactly* where the noise is coming from with its ears alone. A great gray owl can pluck a lemming from within its snow-covered tunnel or accurately bust through the roof of a gopher burrow, solely by listening to the chewing or scurrying sounds coming from beneath the ground. These feats are remarkable, and they hint at why hearing can be such a useful sense.

AMONG THE TRADITIONAL FIVE SENSES, hearing is most closely related to touch. That might be counterintuitive, since the latter is concerned with surfaces, which are solid and tangible, and the former deals with sounds, which seem airborne and ethereal. But both hearing and touch are mechanical senses, which detect movements in the out-

---

* We localize sounds without consciously thinking about it, which conceals how hard that task actually is. An eye comes with an inbuilt sense of space, because light from different parts of the world falls on different parts of the retina. But ears are set up to capture qualities like frequency and loudness that have no intrinsic spatial component. For an animal to take that information and turn it into a map of the world, its brain has to work really hard.

side world using receptors that send electrical signals when they're bent, pressed, or deflected. In touch, those movements occur when fingertips (or whiskers, bill tips, and Eimer's organs) are pressed or stroked against a surface. In hearing, the movements occur when sound waves reach the ear and deflect small hair cells within it.

But unlike touch, hearing can operate over long distances. Unlike vision, hearing functions in darkness and through solid, opaque barriers. Unlike the vibrational sense from the previous chapter, hearing doesn't need a surface and can work through all-encompassing media like air or water. And unlike smell, which is limited by the slow diffusion of molecules, hearing works at the considerably faster speed of sound. Some senses have a few of these qualities, but hearing has them all, which is why some animals rely so heavily upon it. William Stebbins once encapsulated this beautifully: "Very different from other forms of stimulation, [sound] can impart information on current events at an unseen distance," he wrote.

Compare an owl to a rattlesnake. Both are nocturnal. Both hunt rodents. The rattlesnake doesn't need to eat very often and is an ambush hunter. It can use its sense of smell to find the right spot for a lengthy stakeout, and wait for victims to run within the short range of its infrared sense. The owl has no such luxury. To sustain its high metabolism, it must find prey more regularly, which means scanning a wide swath of forest and accurately localizing the rustles of fast-moving but unseen rodents. Hearing—long in range, fast in speed, and precise in resolution—is naturally its primary sense.

But hunting by sound has one major disadvantage—interference. A visually guided predator like an eagle doesn't emit light when it moves, but an owl can't help but make noise with its own wingbeats. Those noises, which are close to the owl's ears, could potentially drown out the faint and distant sounds of its prey. Fortunately, the owl has soft feathers on its body and serrated edges on its wings that make its flight almost imperceptibly quiet. The noise it does make is mostly below the range to which its ears are most sensitive and below the lower limit of what small rodents can hear. The owl can hear a mouse just fine, but a mouse can barely hear an owl coming.

Kangaroo rats can. These little hopping rodents have relatively huge middle ears, which are larger than their brains. These chambers specifically amplify the low frequencies produced by an owl's wings and allow kangaroo rats to hear incoming danger that most other rodents can't perceive. So they're especially difficult for barn owls to catch. They can even hear the sounds that rattlesnakes make when they strike, with enough time to jump away, turn in midair, and kick the lunging snakes in the face. (Rulon Clark, the snake expert whom we met in the chapter on heat, describes them as a "particularly obnoxious prey item.")

All of these creatures are connected by sound. Their lives and deaths are determined by the frequencies they can hear, how sensitive they are to those frequencies, and their skill at localizing the source of sound. Every species has its own strengths and weaknesses. An owl is maximally sensitive to the frequencies produced by scurrying mice and can locate those sounds with almost unmatched accuracy, but it's oblivious to the highest and deepest notes that human ears can detect. Mice can't hear the low wingbeats of an owl, but they can make high-pitched alarm calls that the owl can't hear. As with other senses, an animal's hearing is tuned to its needs. And some animals don't need to hear at all.

OUR ROUNDED EARS MIGHT look very different from the pointy triangles of fennec foxes, the giant flaps of elephants, or the simple holes of dolphins, but these differences are superficial. Most mammals have very good hearing, and most mammalian ears are very similar. They always exist, for a start. There are always two of them. They're always found on the head. None of these absolutes is absolutely true for insects. They have also evolved ears, but those ears come in a dazzling variety that offers three important lessons about why animals hear at all.

The first lesson: Hearing is useful, but not universally so in the way that touch or nociception is. After all, the first insects were deaf. They had to evolve ears, and over their 480-million-year history, they did so on at least 19 independent occasions, and on almost every imaginable body part. Ears exist on the knees of crickets and katydids, the abdomens of locusts and cicadas, and the mouths of hawkmoths. Mosquitoes hear with their antennae. Monarch caterpillars hear with a pair of hairs on their midsection. The bladder grasshopper has six pairs of ears running down its abdomen, while mantises have a single cyclopean ear in the middle of their chests.* Insect ears are so diverse because most of them evolved from movement-sensitive structures called chordotonal organs, which are found throughout an insect's body. These organs consist of sensory cells that lie just beneath the hard outer cuticle and respond to vibrations and stretching motions. They tell insects about the position of their own body parts—beating wings, moving limbs, swelling guts. But since chordotonal organs can also react to very loud airborne sounds, they're almost predisposed to becoming ears. They just need to become more sensitive, and that's easily done by thinning the cuticle lying over them to create an eardrum.† Since this can happen almost anywhere on the body, insects can conjure ears from the unlikeliest of places. It's as if their entire surfaces are primed for hearing.

But many insects haven't exploited this evolutionary gimme. As far as anyone knows, mayflies and dragonflies don't have ears. The majority of beetles don't, either. Indeed, most insects seem to be deaf, and since they handily outnumber all other animal species, it follows that *most animals might be deaf.* This might seem odd, especially since sound seems so omnipresent to those of us who can hear. And yet millions of deaf people do just fine without it, and many animals don't bother with it at all. If you look at our fellow mammals and other vertebrates,

---

* In 1968, a zoologist named David Pye published a delightful five-verse poem about insect ears in *Nature,* one of the world's top-ranking science journals. By 2004, scientists had learned so much more about these ears that Pye was compelled to publish a sequel with 12 extra verses. "In later years some further ears / Were found in other forms. / The more we know just goes to show / There are no real norms," he wrote.

† Not all insect ears have eardrums. The antennae of mosquitoes and the hairs of monarch caterpillars act more like the airflow-sensing hairs of spiders and crickets, which we met in Chapter 6.

you might be forgiven for thinking that hearing is invaluable. If you look at insects, you realize that it is decidedly optional.

As with vision, to think about how animals hear, you have to understand how animals use their ears. Hearing is specifically useful in that it offers fast, precise, long-range, and 24-hour information that allows animals to sense both rapidly moving prey and rapidly approaching threats. Accordingly, many insects seem to have evolved ears to listen out for predators. Many butterflies, including the striking blue morpho, have ears on their wings. These species are silent, so they're certainly not listening to each other. Instead, Jayne Yack has shown that their wing-ears are tuned to the same frequencies produced by predatory birds. From several feet away, they can hear wingbeats, territorial calls, and probably other relevant sounds like feathers swishing through grass or feet hopping on branches. They're likely using their ears in the same way that a kangaroo rat uses its ears.*

The qualities that make hearing good for detecting predators also predispose it to communication. By producing sounds, and listening out for them, animals can exchange signals over longer distances than surface vibrations would allow, in dark and cluttered spaces that obscure visual cues, and with greater speed than pheromones can achieve. This may explain why, millions of years ago, crickets and katydids started to sing.

The males are the noisy ones. They have a ridge on one of their wings, and a comb-like row of teeth on the other. When they rub these together, they produce a *thrrrrp* sound, which females hear with eardrums on their front legs. Fossilized insects that have the same ridges and combs on their wings suggest that these songs have filled the air for at least 165 million years, and likely much longer. But around 40 million years ago, another group of insects started eavesdropping on the singers: parasitic tachinid flies. Most tachinids track their victims

---

* The ear's aptitude for detecting predators might explain why some insect groups haven't bothered evolving them. Perhaps mayflies, which have no ears, take to the air in such large numbers that they find safety from predators without needing an early-warning system. Maybe dragonflies, which are also earless, rely on their excellent eyesight to spot incoming danger, and their aeronautical acumen to evade even close-range attacks.

through sight or smell, but *Ormia ochracea*—a yellow, half-inch-long species that's found throughout the Americas—uses sound. Like female crickets, it listens out for the song of a male. Homing in on those dulcet *thrrrrp*s, it lands either on or near the singer, and deposits maggots. These burrow into the cricket and slowly devour him from within.

Ormia's ears are not obvious. But Daniel Robert is so familiar with insect ears that when he first looked at the fly under a microscope in the early 1990s, he instantly recognized a pair of eardrums—two thin oval membranes just below its neck. ("Maybe I'm too much of a nerd," Robert tells me.) These ears are very different from those of most flies, which are usually feathery and found on the antennae. They're much closer to those of a female cricket, and they're similarly tuned to the frequency of a male's song. *Ormia* has tapped into the female cricket's auditory Umwelt and uses it for the same goal: pinpoint an unseen male from afar. If you've ever been plagued by a cricket singing somewhere inside your house, you'll know how hard it is to find the source of the infernal chirping. *Ormia* has no such problem. It can turn toward a singing cricket with an accuracy of 1 degree, which is better than humans, barn owls, and almost every other animal that's been tested.*

Despite this superlative acuity, *Ormia*'s ears control a very simple behavior: *Find cricket*. That's true for many insect ears, and Jayne Yack thinks this might also explain why they've evolved in such a wide variety of body parts. Ears, she says, tend to appear near the neurons that control the actions for which those ears evolved. Female crickets turn and walk toward singing males, so their ears are on their legs. Mantises and moths execute evasive dives and rolls when they hear predators, so

---

* The barn owl showed us that animals can work out where a sound is coming from by comparing the time at which it arrives at each ear. But as animals get smaller, their ears get closer together, and sounds reach both of them almost simultaneously. *Ormia*'s ears are less than half a millimeter apart—the width of the dot on this *i*. At such tiny distances, a cricket's song should hit the two eardrums no more than 1.5 microseconds apart—a time window so narrow that it might as well not exist. (For comparison, human ears need separations of at least 500 microseconds to accurately localize a sound.) But Robert and his mentor Ron Hoy showed that *Ormia*'s eardrums, unlike ours, are connected. Within the fly's tiny head, they are linked by a flexible lever that looks like a coat hanger. When sound vibrates one eardrum, the lever transmits those vibrations to the opposite one—but with a slight delay of around 50 microseconds. This greatly extends the time difference between the two ears, and makes the difference between *Ormia* hearing a cricket and *Ormia* hearing a cricket over there.

their ears are on or near their wings. (Blow on a dog whistle next to an eared moth, and it will start doing loops and spirals.)

This is the second lesson that insect ears can teach: Hearing can be incredibly simple. One might think that a listening cricket creates a mental representation of what it hears, and compares that against some internal template of an ideal male song. None of that is necessary. Through several painstaking studies, Barbara Webb showed that the female cricket's ears, and the neurons connected to them, are wired so that she *automatically* recognizes a male's song and turns toward it. Her actions are built into her sensory system.* As the sense that underpins most of our music and language, hearing can be hard to separate from sophistication of thought, emotionality, and creativity. But it can be akin to the reaction of a human who kicks out when their knee is stimulated by a hammer.

Even simple behaviors can have big consequences. *Ormia*'s acoustic prowess is so acute that on Hawaii, it once infested a third of male crickets and was seriously suppressing their numbers. In response, the crickets acquired a mutation that warped the comb-like structure on their wings and muted their songs. To avoid the grave, they became as silent as one. This happened within 20 generations, making the "flat-wing" crickets one of the fastest cases of evolution that has ever been documented in the wild. The newly silent males are undetectable to *Ormia,* but also to females. The silent males are reduced to loitering around the few males who can still sing, in the hopes of sneakily mating with approaching females. They also still go through the motions of singing, rubbing their wings together as if they could still *thrrrrp* away.

Here, then, is the third lesson from insect ears: Animal hearing can drive the evolution of animal calls, and vice versa. Just as eyes define nature's palette, ears define its voices.

----

* Webb even built a simple robot that behaves exactly like a female cricket and can track a singing male even though it has no internal conception of his song.

IN THE SUMMER OF 1978, after a long flight, a train journey, and a boat ride, a young graduate student named Mike Ryan finally arrived at Panama's Barro Colorado Island to study frogs. He had been hooked on the amphibians ever since he had witnessed an older biologist identifying one species after another from their calls alone. If another human could hear so much in what his own ears perceived as a formless cacophony, Ryan wondered, what might the frogs themselves hear? He knew that males called to attract mates, but what parts of the song were the females listening to? What sounds beautiful to a frog?

Initially, Ryan's plan was to study the Panamanian red-eyed tree frog, the same species that his future student Karen Warkentin would focus on two decades later.* But these animals stuck to the canopy and weren't very talkative. When Ryan tried to record their calls, he would instead pick up a much louder species that was shouting at his feet— the túngara frog. "I kept kicking them away to get them to shut up," he tells me. "And then I said: Duh, what if I just study *them*? There are tons of them and they're right in front of me."

Picture, in your mind, an average frog. The túngara frog looks like that. It's about the size of a quarter, with bumpy skin and drab, mossy colors. But what it lacks in visual flamboyance, it makes up for in acoustic flair. After sunset, the males inflate their huge vocal sacs and force air through voice boxes larger than their brains. The result is a short whine that falls in pitch, like a tiny, receding siren. After that, the male might add one or more short, staccato embellishments that are known as chucks. To some human ears, the combined call sounds like "tún-ga-ra"—hence the name. To Ryan, it resembles a sound effect from an old video game.† To a female frog, it sounds like an invitation. She'll sit in front of various males, compare their whines and chucks, choose the most attractive-sounding specimen, and allow him to fertilize her eggs. Courting males might call 5,000 times in a single evening

---

* Rex Cocroft, the treehopper aficionado whom we met in the last chapter, was also one of Ryan's students.

† Ryan does a very good túngara frog impression, but to my disappointment, he has never tried playing his own rendition through a speaker to see if he can fool an actual female. "I should do that," he tells me.

before they're chosen. Ryan knows this because he spent 186 consecutive nights at Barro Colorado, recording the serenades and escapades of a thousand individually marked túngara frogs from dusk to dawn. It was a marathon of voyeurism, from which he learned one crucial fact: Chucks are *very* sexy.

Females almost always go for males who embellish their whines with chucks over males who merely whine. The chucks are so desirable that if a male is reluctant to make them, a female will sometimes body-slam him until he does. Ryan recorded the males' songs and spliced their whines and chucks into different combinations. In a soundproof room, he played pairs of these remixes to females through different speakers and noted which they hopped toward. He learned that a whine is attractive on its own, but a chuck makes it five times more appealing. More chucks are sexier than fewer chucks. Deeper chucks are sexier than higher-pitched ones. These preferences are straightforward. The reasons for them are not.

Ryan found that the frog's inner ear is especially sensitive to frequencies of 2,130 Hz, which is just under the dominant frequency of an average chuck.* Even at a noisy pond, where several species might be calling simultaneously, a female can easily find her own males, because she can hear their calls more acutely than those of other frogs. Larger males sound especially loud and clear since their lower-pitched chucks are closer to the ideal frequency of her inner ear. Perhaps, Ryan reasoned, that's why the túngara ear is tuned in that specific way. Larger males can also fertilize more eggs, so in past generations, females who preferred lower frequencies would have been drawn to males who provided them with more offspring. Their predilections became more common and the species ended up with ears that were tuned to the male's voice. This narrative is perfectly plausible. It's also completely wrong.

Ryan discovered the actual story by studying the túngara frog's close relatives. These other species all whine, but only a few chuck.

---

* Technically, the frog has two hearing organs in its inner ear. One, the amphibian papilla, is most sensitive to the pitch of the whine—700 Hz. The other, the basilar papilla, is tuned to the frequency of the chuck.

And yet, all of them have inner ears that are tuned to the same chuck-adjacent frequency as the túngara frog's. These other frogs are predisposed to find chucks attractive, without ever actually hearing them. Ryan demonstrated this by traveling to Ecuador and studying the Colorado dwarf frog—one of the túngara's chuck-less cousins. He recorded the male's whine, added túngara chucks after them, and played the hybrid calls to the females. "I thought it would scare the hell out of them," he tells me. Instead, the females hopped toward the unfamiliar chimeric sounds. The chucks, which the females had never heard before, proved irresistible because they tapped into a preexisting quirk of their senses.

This discovery flipped Ryan's narrative on its head. The túngara frog's hearing didn't change to match its call. It was the other way around. The frog's ancestor already had ears that were tuned to 2,130 Hz, and the chucks evolved to exploit that bias. The reasons for that ancestral tuning are still unclear: Perhaps that's the pitch produced by a rustling predator, or some other important aspect of the frog's environment. Regardless, the female's aesthetic preference came first, and the male's calls changed to fit her conception of beauty. Ryan calls this phenomenon "sensory exploitation," and he and others have shown that it is common throughout the animal kingdom.* Nature's ears really do define its voices.

Male túngara frogs, for their part, get an easy way to earn their partner's attention. A chuck takes very little effort, and enhances their attractiveness fivefold. "Think of all the stuff we do to make ourselves more attractive—and this is for free," Ryan says. They ought to chuck as frequently and repeatedly as possible, but they're strangely unwilling to do so. While some individuals have been heard slapping up to seven chucks onto their whine, most add just one or two. Many refuse

---

* Sensory exploitation works across the senses. In swordtail fish, the bottom half of the male's tail fin is unusually long. The longer this sword, the more attractive the male is to females. But Alexandra Basolo found that this same preference exists in the closely related and swordless platyfish. If she glued artificial swords onto the tails of platyfish males, they became more attractive. The sword, then, is like the túngara frog's chuck—a trait that evolved to exploit a preexisting preference.

to chuck at all. Their reticence was puzzling, until Ryan realized that females aren't the only ones listening to their calls.

A year before Ryan arrived at Barro Colorado, his colleague Merlin Tuttle caught a bat with a half-eaten túngara frog in its mouth. This species, the fringe-lipped bat, turned out to be a voracious frog-eater. Tuttle and Ryan showed that it tracks its prey by eavesdropping on its courtship calls, much as *Ormia* does with cricket songs. And the bat, just like female túngara frogs, is particularly drawn to males that add chucks to their whines. The females hear a mate, the bats hear a meal, but both are listening for the same qualities. This leaves the male frogs with an unenviable choice. Their chucks court both females and death. No wonder they sometimes stick to whines.*

I find it astonishing to consider how these creatures have been bound together through their senses. For whatever reason, an ancestral frog had ears that were partial to frequencies of 2,130 Hz. Túngara frogs took advantage of that sensory quirk by adding chucks to their whines. Fringe-lipped bats took advantage of those chucks with an auditory add-on that expanded their hearing into unusually low frequencies for a bat. The frog's Umwelt shaped the frog's calls, which then shaped the bat's Umwelt. The senses dictate what animals find beautiful, and in doing so, they influence the form that beauty takes in the natural world.

FEW ANIMAL SOUNDS ARE as beautiful to human ears as the songs of birds. And few bird songs have been studied as intensely as those of

---

* Ryan remembers that after he first presented his bat findings at a seminar, a very senior researcher told him that he was wrong. Bat ears are tuned to the exceptionally high frequencies of their own calls, Professor Bigshot said, and should be deaf to the lower notes of a túngara chuck. Undeterred, Ryan showed that they are not. Their inner ears are wired up to more neurons than those of almost any other mammal, and uniquely among bats, a subset of these are sensitive to the low frequencies found in frog calls. It's as if they've added a special frog-detecting module onto what is otherwise basic bat hardware. One of Ryan's students, Rachel Page, later showed that under some circumstances, the bats find it easier to locate the frogs if they're chucking as well as whining. They aren't the only eavesdroppers, either. Another of Ryan's students, Ximena Bernal, showed that bloodsucking midges are drawn to frog calls, and especially to those with chucks.

zebra finches. Visually, these Australian birds are striking, with gray heads, white chests, orange cheeks, red beaks, and black stripes beneath their eyes that resemble running mascara. Vocally, the males are equally flamboyant, singing complicated and raucous songs. To my ears, they sound like melodic printers. But I also wonder if a zebra finch's song sounds to another zebra finch like it does to me. In terms of pitch, the answer is yes. The frequency range of bird hearing is roughly similar to that of humans, so birds generally hear the same range of pitches that we hear. But their songs can also be incredibly fast. The notes that emerge from a zebra finch's beak fly by so quickly that I can barely distinguish them. Even in the notes I *think* I can hear, there seems to be something more, some intricacy I cannot fully discern, lurking at the edges of my awareness. Surely, the birds can hear something in these songs that I cannot.

Bird enthusiasts have long suspected that bird hearing works on a faster timescale than ours. Some birds prove their temporal prowess by singing dazzlingly synchronized duets, slotting their notes in and around each other's with such precision that the two songs can sound like one. Others, including zebra finches, learn their songs from listening to each other, and so must be able to hear the acoustic minutiae that they then reproduce. The same goes for mimics like mockingbirds. To our ears, the song of the whip-poor-will comprises three notes, but it actually has five, which becomes clear if we slow it down. A mockingbird doesn't need the help: When mimicking the whip-poor-will, it gets all five notes.

In the 1960s, before his work on barn owls, Masakazu Konishi found direct evidence that the processing speed of bird hearing is exceptionally fast. He played strings of rapid clicks to sparrows, while recording the electrical activity of neurons in the hearing centers of their brains. The neurons fired once per click, even when the clicks were just 1.3 to 2 milliseconds apart. At such speeds—between 500 and 770 clicks *per second*—a cat's auditory neurons can only keep to the same tempo around 10 percent of the time. The sparrows' neurons kept pace perfectly. Even pigeons, whose songs don't contain rapid sounds, had ears that seemed to resolve them.

Later studies were less clear. From the 1970s onward, Robert Dooling repeatedly failed to find any differences between the ways birds and humans perceived the temporal nature of sounds. For example, he showed that humans can tell if a silent gap of just 2 milliseconds is inserted into an otherwise continuous noise. Birds, surprisingly, don't do any better. Test after test, "nothing popped out as being different," Dooling tells me. "We measured birds in a gazillion different ways over the years, but their hearing always looked like a human's." It took him a long time to realize the problem: He had been testing birds with simple sounds like pure tones, which are nowhere close to the rich complexity of actual songs. You can visualize a pure tone as a smooth curve that undulates up and down, representing increases and decreases in pressure over time. A bird's song, when visualized in the same way, looks more like the skyline of a city or the ridgeline of a mountain range. It's full of jagged bumps, which represent extremely fast shifts that occur within the span of a single note. Those details are known as the temporal fine structure. They're missing from the pure tones that are typically used to study hearing. And as it happens, they're what songbirds are actually listening for.

Dooling confirmed this through an elegant experiment, in which he asked various songbirds to discriminate between sounds that differed only in their temporal fine structure. This isn't intuitive, so let's use a visual analogy. Imagine taking a movie and reversing the order of every three frames. The color palette would stay the same, the scenes would be composed in the same way, and the plot would still be comprehensible. But *something* would feel off, and you'd likely notice the difference. This is roughly what Dooling did with his birds. He presented them with pairs of buzzy sounds. One consisted of repeated chunks in which the pitch rose over a few milliseconds before falling again. In the other, the pitch of the chunks *fell* over the same range of frequencies, and over the same time period. To a slow ear, both sounds would average out to the same pitch, and seem identical. To a fast ear, they'd be completely different. Dooling found that humans could only distinguish between these sounds if the chunks were longer than 3 to 4 milliseconds. Canaries and budgerigars hit their limit at between

1 and 2 milliseconds. And zebra finches weren't even slightly duped by the shortest 1-millisecond chunks. This experiment clearly showed that birds can hear complexities that are imperceptibly fast to humans. And it so thoroughly contradicted Dooling's previous work that "it kind of freaked me out," he says. Indeed, further tests showed that "our electronics couldn't handle the fine detail that the birds are capable of discriminating." That was the first of many surprises.

A zebra finch's song consists of several distinct syllables that it always sings in the same sequence—A-B-C-D-E. When Beth Vernaleo and a team of Dooling's students reversed one of these syllables—A-B-Ɔ-D-E—zebra finches almost always noticed the change. Human listeners couldn't, even after a lot of practice. But when the team doubled the gap between two of the syllables, humans could easily tell—it sounded like a glitch in the recording—and the *finches* were completely oblivious. They couldn't hear the differences between two songs that were obviously different to human ears.

Two students, Shelby Lawson and Adam Fishbein, went even further. They completely shuffled the order of the syllables—C-E-D-A-B. The finches *still* couldn't discriminate between them. The two sequences are patently different, but not different *in a way that matters to the finches*. Even though these birds learn their individual sequence of syllables in their youth, and sing that same unchanging sequence for the rest of their life, "they don't give a crap about the sequences," Dooling says. "They care about what's inside the individual notes." It's as if two conversing humans were paying close attention to the nuances of each other's vowels, while blithely disregarding the order of each other's words.

The answer to my question is clear: A zebra finch's song must sound entirely different to a zebra finch than to us. Their disregard for sequence is especially unexpected, and flies in the face of our intuitions about bird songs. The sequences in those songs are both beautiful and useful to human ears. Birders use them to identify particular species. Neuroscientists study them because of their similarities to human languages. And yet, they might be utterly irrelevant to the birds that produce them. Not all species behave this way: Budgies seem sensitive to the sequence of notes as well as their fine structure. But many others, including Bengalese finches and canaries, mostly care about the latter. To them, the beauty and significance of the song lie in its minutiae. They ignore the big acoustic picture in favor of the details. They can't—or don't care to—hear the forest for the trees.

Humans have the opposite tendency. To our ears, each delivery of a zebra finch's song sounds the same as the last, and we could be forgiven for thinking that they all carry the same information. But Dooling's colleague Nora Prior showed that the fine structure of seemingly identical renditions can sound very different to a finch. If she swapped syllable B from one recording with syllable B from another, the birds could hear that something had changed. Their songs must be full of subtle nuance that we simply cannot detect. While we might hear repeated iterations of the same unwavering tune, they could conceivably hear information about sex, health, identity, intention, and more. Zebra finches sing to establish lifelong bonds with their partners, to find each other when they're apart, to stay together while traveling, and to coordinate their parenting responsibilities. Perhaps they accomplish all of this through information encoded in their songs' fine structure.

Part of the thrill of listening to animals comes from wondering what they are saying to each other. Writers have conjured up characters like Dr. Dolittle who can understand the meaning of the tweets, bleats, and hisses of other species. Naively, we might imagine this to be a problem of vocabulary, as if there might exist some word-chirp dictionary that would suddenly allow us to speak bird. There isn't, and Dooling's work reminds us why: The communication barrier between species is also a sensory one. Birds encode meaning in aspects of their songs that our ears can't pick out and our brains don't pay attention to. "Now, when I hear birdsong, I think it's amazing that it sounds so complex but I'm *still* missing most of it," Dooling tells me. "There's a lot in there that another bird is appreciating that I can't."

IN THE EARLY 2000S, while Robert Dooling was running the first of his fine structure experiments, Jeffrey Lucas stumbled upon another

unexpected side to bird hearing. He and his colleagues placed electrodes on the scalps of six North American bird species to record how
their auditory neurons responded to different sounds. This simple
technique is called the auditory evoked potential (AEP) test. Doctors
use it to check hearing levels in human patients. Biologists use it to
work out what animals can hear. Lucas used it to see if species with
more complex songs hear differently than those with simpler tunes.
More through accident than planning, he happened to test birds in two
waves—one in the winter, and a second in the spring. And when he
compared those snapshots in time, he saw that they were very different. Birds, Lucas realized, hear differently across the seasons.

Their hearing changes because of an important trade-off that's inherent to all ears. Let's say I played you two musical notes—one with a
frequency of 1,000 Hz, and another with a frequency of 1,050 Hz.
These roughly correspond to two adjacent keys at the high end of a
piano, which should be easy to tell apart. But if I played 10-millisecond
snippets of the two notes, they'd be indistinguishable. Why? Because
within that short timeframe, *both* notes would oscillate 10 times each,
and sound the same. If I increased the snippet length to 100 milliseconds, the notes would oscillate 100 times and 105 times, respectively,
and sound different. For this reason, animal ears become more adept at
discriminating between similar frequencies if their neurons integrate
sound information over longer periods of time. But in doing so, they
also become *less* sensitive to fast changes that occur within those periods. We saw a similar trade-off in the chapter on vision: Eyes can have
exceptional resolution or exceptional sensitivity, but not both. Likewise, ears can have exceptional *temporal resolution* or exceptional *pitch
sensitivity*, but not both. "The auditory system that does fast stuff is
completely different from the auditory system that does frequency
stuff," Lucas tells me. And he found that birds don't have to settle for
one or the other. They can flip between the two, as the situation demands.

Consider the Carolina chickadee—a small, inquisitive songbird
that graces much of eastern America. Its signature *chick-a-dee-dee* call
rapidly changes in pitch and volume, much like the songs of zebra

finches. That call can be heard all year round, but it's especially important during the fall, when the sociable chickadees form large flocks. At
that time, the birds need to parse all the information encoded within
the fine structure of their calls, so their hearing needs to be as fast as
possible—and it is. Lucas found that in the fall, their temporal resolution goes up, but their pitch sensitivity goes down. When spring rolls
around, everything changes. The flocks begin to break up, as females
and males pair up to establish their own breeding territories. To attract
mates, the chickadee males start singing their courtship songs, which
are much simpler than their year-round calls. There are four notes—
fee-bee-fee-bay—and each is close to a pure tone. The male's attractiveness depends on how consistently he can sing these notes, and
specifically on whether he can maintain the exact drop in pitch between the *fee* and the *bee*. Now the chickadees need to hear the frequencies of their songs as sharply and precisely as possible—and they do.
While speed takes all in the fall, pitch is king in the spring.

The hearing of the white-breasted nuthatch changes in the opposite direction. Its courtship song—a nasal, fast-paced *wha-wha-wha*—
has a fine structure that includes fast changes in volume. So, unlike the
chickadee, its hearing becomes *faster* during the breeding season, and
less sensitive to pitch. Both birds completely retune their sense of hearing from one season to the next to process the information that matters
most in that season. Their voices and their needs change with the cal-
endar. So do their ears.

These changes are driven by sex hormones like estrogen, which can
directly influence the hair cells in songbird ears. This might explain
why in some species, the hearing of males and females changes in different ways. Lucas and his colleague Megan Gall showed that female
house sparrows have seasonal hearing that shifts in the same way as the
chickadees': It gets better at handling pitch in the spring at the expense
of speed. Male hearing, however, stays fast all year round. So, while
Robert Dooling showed that humans experience bird songs in a different way than birds, Lucas showed that birds can also experience *their
own songs* in different ways, depending on their sex and the season. In
the fall, all house sparrows hear in the same way. In the spring, males

and females get different experiences of the same tunes. Their Umwelten converge and diverge throughout the year.

These cycles influence more than their sense of aesthetics. As we saw with both owls and *Ormia,* animals can calculate where sounds are coming from by noting if those sounds reach one ear slightly later than the other. If ears become worse at detecting small time differences, their owners become worse at mapping sounds. So when a female sparrow's sense of acoustic timing becomes slightly slower in the spring, her acoustic *space* also becomes slightly fuzzier.

These seasonal cycles shocked Lucas when he first discovered them in 2002. Other researchers didn't believe his early results, either. At the time, people thought that hearing was mostly static. It might get duller with age in some species—humans, sadly, among them—but it wasn't thought to change over shorter timescales. But as we've repeatedly seen, an animal's senses are finely tuned to its environment and have evolved to extract whatever information is relevant. When the environment fluctuates from one season to the next, the information that's relevant also changes.* For a North American bird, spring often means sex. The air fills with courtship calls that are absent in other times of year and must now be carefully judged. Fall brings openness: Bare branches make little birds more visible to predators. The ability to localize the sound of approaching danger, which is inextricably linked to fast hearing, becomes paramount. An animal's Umwelt cannot be static, because an animal's world isn't static.

Bird songs don't lie beyond the reach of human senses, like the circularly polarized patterns of mantis shrimps or the vibrational songs of treehoppers. We can very much hear them. The *fee-bee-fee-bay* of chickadees and the *wha-wha-wha* of nuthatches are obvious enough that we can transcribe them. And yet, we still don't appreciate these signals in the same way as their intended audiences can. To us, a chickadee song sounds the same whether we listen to it in October or March.

---

* Males of the plainfin midshipman fish attract females by making long and very deep hums, and during the breeding season, the females' ears become several times more sensitive to the main frequencies. Green tree frogs become more sensitive to their own calls after just two weeks of listening to a chorus.

To a chickadee, it does not. If so much mystery can exist within sounds that we can hear, how much more are we missing in sounds that we can't?

IN THE 1960S, AFTER his seminal work on barn owls, Roger Payne switched his attention to whales. In 1971, he published two historic papers. One, based on recordings that Payne analyzed with his wife, Katy Payne, revealed for the first time that humpback whales sing haunting songs. It prompted decades of research, turned whale song into a cultural phenomenon, spawned a bestselling album, and helped to spark the Save the Whales movement. The second showed that fin whales—the second-largest animals after blue whales—make extremely low-pitched calls that can be heard across entire oceans. It nearly destroyed Payne's career.

That controversial paper was born of the Cold War. To listen for Soviet submarines, the U.S. Navy installed chains of underwater listening posts in the Pacific and Atlantic. This network, known as the Sound Surveillance System, or SOSUS, picked up a deluge of oceanic noises. Some were clearly biological. Others were more mysterious. One especially enigmatic sound was monotonous, repetitive, and low, with a frequency of 20 Hz—an octave below the lowest key on a standard piano.* This hum was so loud that people doubted it could be coming from an animal. Did it have a military origin? Was it produced by underwater tectonic activity? Did it come from waves crashing on some distant shoreline? The actual source only became clear when Navy scientists started following the sounds to their sources, and often found a fin whale at the end.

Human hearing typically bottoms out at around 20 Hz. Below those frequencies, sounds are known as infrasound, and they're mostly inaudible to us unless they're very loud. Infrasounds can travel over

---

* Hearing ranges don't have sharp boundaries. Instead, it just becomes harder and harder to hear sounds at a specific volume. Humans, for example, can hear some infrasonic frequencies if they're loud enough.

incredibly long distances, especially in water.* Knowing that fin whales
also produce infrasound, Payne calculated, to his shock, that their calls
could conceivably travel for 13,000 miles. No ocean is that wide. Together with oceanographer Douglas Webb, Payne published his calculations, speculating that the largest whales "may be in tenuous acoustic
contact throughout a relatively enormous volume of ocean." The response was brutal. Leading whale researchers told him that his paper
was pure fantasy. Colleagues hinted that critics had been questioning
his mental health behind his back. "When you get to distances like
that, people just refuse to believe that it's true," Payne tells me.

Payne's work made a more positive impression on Chris Clark. A
young acoustician and former choirboy, Clark was recruited by Roger
and Katy Payne to be a sound technician on a 1972 trip to Argentina to
study right whales. It was a thrilling and formative time. Camped on a
beach beneath the Southern Cross, with penguins bumbling past and
albatrosses wheeling overhead, Clark began listening to whales. He
placed hydrophones in the water to eavesdrop on their songs and found
ways of assigning specific recordings to individual whales. He went on
to compile libraries of whale calls, recorded all over the world, from
Argentina to the Arctic. And all the while, Payne's idea of giant whales
talking over oceans stuck with him.

In the 1990s, with the Cold War over and the threat of Soviet subs
diminished, the Navy offered Clark and others a chance to observe
real-time recordings from their SOSUS hydrophones. Amid the
spectrograms—visual representations of the sounds that SOSUS
picked up—Clark saw the unmistakable signal of a singing blue whale.
On his first day, Clark saw that more blue whale vocalizations had been
recorded from a single SOSUS sensor than had been described before
in the entire scientific literature. The ocean was awash with their calls,
and those calls were coming in from enormous distances. Clark calculated that one individual was 1,500 miles from the sensor that recorded
it. He could listen to whales singing in Ireland with a microphone situ-

---

* Humans exploited this property during World War II, when aircraft were armed with
explosive charges that went off if the planes sank. Listening posts could detect the locations of
the wrecks, and rescue teams could be deployed.

ated off Bermuda. "I just thought: *Roger was right,*" he says. "It is physically possible to detect a blue whale singing across an ocean basin." For Navy analysts, these sounds were regular parts of their workday, irrelevancies to be marked on the spectrograms and promptly ignored. For Clark, they were mind-blowing epiphanies.

Although blue and fin whale songs can traverse oceans, no one knows if the whales actually communicate at such ranges. It's possible that they're signaling to nearby individuals with very loud calls, which just happen to extend further afield. But Clark points out that they repeat the same notes, over and over again, and at very precise intervals. A singing whale will stop calling when it surfaces for air, and come back on the beat when it submerges. "That's not arbitrary," he says. It reminds him of the redundant and repetitive signals that Martian rovers use to beam data back to Earth. If you wanted to design a signal that *could* be used to communicate across oceans, you'd come up with something similar to a blue whale's song.

Those songs might have other uses, too. Their notes can last for several seconds, with wavelengths as long as a football field. Clark once asked a Navy friend what he could do with such a call. "I could illuminate the ocean," the friend replied. That is, he could map distant underwater landscapes, from submerged mountains to the seafloor itself, by processing the echoes returning from the far-reaching infrasounds. Geophysicists can certainly use fin whale songs to map the density of the ocean crust. But can the whales do so?

Clark sees evidence in their movements. Through SOSUS, he has seen blue whales emerging in polar waters between Iceland and Greenland and making a beeline—a whaleline?—for tropical Bermuda, singing all the way. He has seen whales slaloming between underwater mountain ranges, zigging and zagging between landmarks hundreds of miles apart. "When you watch these animals move, it's as if they have an acoustic map of the oceans," he says. He also suspects that the animals can build up such maps over their long lives, accruing sound-based memories that lurk in their mind's ear. After all, Clark recalls veteran sonar specialists telling him that different parts of the sea had their own distinctive sounds. "They said: If you put a pair of headphones on me, I can tell you if I'm near Labrador or off the Bay of Biscay," says Clark. "I thought that if a human being could do this in 30 years, what could an animal do with 10 million years?"

The scale of a whale's hearing is hard to grapple with. There's the spatial vastness, of course, but also an expanse of time. Underwater, sound waves take just under a minute to cover 50 miles. If a whale hears the song of another whale from a distance of 1,500 miles, it's really listening back in time by about half an hour, like an astronomer gazing upon the ancient light of a distant star. If a whale is trying to sense a mountain 500 miles away, it has to somehow connect its own call with an echo that arrives 10 minutes later. That might seem preposterous, but consider that a blue whale's heart beats around 30 times a minute at the surface, and can slow to just 2 beats a minute on a dive. They surely operate on very different timescales than we do. If a zebra finch hears beauty in the milliseconds within a single note, perhaps a blue whale does the same over seconds and minutes.* To imagine their lives, "you have to stretch your thinking to completely different levels of dimension," Clark tells me. He compares the experience to looking at the night sky through a toy telescope and then witnessing its full majesty through NASA's spaceborne Hubble telescope. When he thinks about whales, the world feels bigger, stretching out in space and time.

Whales weren't always big. They evolved from small, hoofed, deer-like animals that took to the water around 50 million years ago. Those ancestral creatures probably had vanilla mammalian hearing. But as they adapted for an aquatic life, one group of them—the filter-feeding mysticetes, which include blues, fins, and humpbacks—shifted their hearing to low infrasonic frequencies. At the same time, their bodies ballooned into some of the largest Earth has ever seen. These changes are probably connected. The mysticetes achieved their huge size by evolving a unique style of feeding, which allows them to subsist upon tiny crustaceans called krill. Accelerating into a krill swarm, a blue

---

* There's a running joke in Pixar's *Finding Nemo* where the protagonist Dory speaks whale by saying the usual things loudly and slowly. Talking to Clark, I wonder if that's surprisingly accurate.

whale expands its mouth to engulf a volume of water as large as its own body, swallowing half a million calories in one gulp. But this strategy comes at a cost. Krill aren't evenly distributed across the oceans, so to sustain their large bodies, blue whales must migrate over long distances. The same giant proportions that force them to undergo these long journeys also equip them with the means to do so—the ability to make and hear sounds that are lower, louder, and more far-reaching than those of other animals.

Back in 1971, Roger Payne speculated that foraging whales could use these sounds to stay in touch over long distances. If they simply called when fed and stayed silent when hungry, they could collectively comb an ocean basin for food and home in on bountiful areas that lucky individuals have found. A whale pod, Payne suggested, might be a massively dispersed network of acoustically connected individuals, which seem to be swimming alone but are actually together. And as his partner Katy later showed, the largest animals on land might use infrasound in the same way.

IN MAY 1984, KATY PAYNE found herself in the company of several Asian elephants at Washington Park Zoo in Portland, Oregon, 16 years after she and Roger Payne learned that humpback whales sing. She was searching for another species to study, and elephants, which were also intelligent and sociable, seemed like good candidates. As she observed them, she occasionally felt a deep shuddering sensation in her body. "It had been like the feeling of thunder but there'd been no thunder," she later wrote in her memoir, *Silent Thunder*. "There had been no loud sound at all, just throbbing and then nothing." The feeling stirred a memory from her teens, of singing in a chapel choir while the pipe organ shook her body as it played its deepest notes. Maybe, Payne reasoned, the elephants had affected her in the same way because they were also producing imperceptibly deep notes. Maybe they were conversing in infrasound, just as some whales were said to do.

Payne returned to the zoo in October with two colleagues and some recording equipment. They left the recorders running while keeping round-the-clock notes on the animals' behavior. Payne didn't listen to the tapes until Thanksgiving eve, and she began with a recording from one especially memorable event. She had felt that familiar silent throbbing at a time when two elephants—Rosy, the matriarch, and Tunga, a male—were facing each other on opposite sides of a concrete wall. At the time, they seemed silent. But when Payne sped up the recordings from that encounter, raising their pitch by three octaves, she heard what sounded like mooing cows. Across their concrete divide, and unbeknownst to the nearby humans, Rosy and Tunga had been having an animated chat. That night, she had a dream in which she was visited by a group of elephants. The matriarch said, "We did not reveal this to you so you would tell other people." Payne interpreted this not as a call for secrecy but as an invitation: *We revealed it to you not to make you famous among people, but to give you access to us.*

Payne's discovery, which she published in 1984, made perfect sense to Joyce Poole and Cynthia Moss, who had been studying African elephants in Kenya's Amboseli National Park. They'd noticed that elephant families would often move in the same directions for weeks at a time, even though they were separated by several miles. In the early evenings, different groups would also converge on the same waterholes at the same time, but from different directions. Infrasound carries over long distances, even in air, and if elephants use it to communicate, that would explain how they can synchronize their movements across a savannah. Poole and Moss invited Payne to join them. She accepted, and in 1986, the team showed that African elephants use infrasound just like their Asian counterparts—and in every conceivable context. There are contact rumbles that help individuals find each other. There are greeting rumbles that they make when reuniting after a separation. Males make rumbles when in heat, and females make rumbles in response to them. There's a "let's go" rumble, and an "I just had sex" rumble. At close range, most of these rumbles contain frequencies audible to human ears, but some became apparent only when the team sped up their recordings, or visualized them.

These infrasonic rumbles are airborne sounds, so they're partly distinct from the surface-borne signals that Caitlin O'Connell more recently identified, and that we encountered in the last chapter. Both are mostly imperceptible to us, and both can be detected by other elephants over long ranges. The low-frequency parts of the rumbles range between 14 and 35 Hz—about the same as a large whale's. Those calls don't carry as far in the air as underwater, and atmospheric conditions dictate how far they can travel: The colder, clearer, and calmer the air, the greater the range. In the heat of midday, an elephant's auditory world shrinks. A few hours after sunset, it expands tenfold, theoretically allowing elephants to hear each other over several miles.* "But we really don't know how far these animals are listening to each other, or what they're listening for," Payne says. "That's a very important question, and no one can answer it."

The same applies to whales. Much of what Roger Payne, Chris Clark, and others have theorized is still speculative, based on little snapshots of whale behavior and educated guesses about what they should be capable of. When it comes to the largest animals that live or have ever lived, actual data are hard to come by, and experiments are nigh impossible. Birds, by contrast, can be easily housed in cages, and bird songs have been analyzed for centuries. And yet, it was 2002 before Robert Dooling discovered that some species pay attention to temporal fine structure at the expense of qualities we can hear. If it's that hard to understand the Umwelt of a bird, no wonder scientists barely understand what giant whales are really listening for in each other's calls. Are those songs courtship displays? Territorial calls? Dinner bells? Assertions of identity? No one knows. Even if you could find a blue whale and play a recorded song to it, how would you expect the animal to behave?

No one even knows for sure what a mysticete whale's hearing range is. The AEP method, where researchers play sounds to an animal and record its neural responses through electrodes on its scalp, is impossible to use on a free-swimming blue whale. Researchers have managed to

---

* Other land animals experience the same expansions and contractions, which is why songbirds sing at dawn and wolves howl at night. Nightfall also increases the range over which a predator might pick up on a call, which might be why elephants call most often in the late afternoon, when their sounds travel reasonably far but lions are still snoozing.

use AEP on smaller whales and dolphins that either strand or live in captivity, but mysticetes rarely do the former and never do the latter. In lieu of direct measurements, scientists like Darlene Ketten have estimated what these giants hear by analyzing their ears with medical scanners. Her work strongly suggests that they hear the same infrasonic frequencies that are found in their calls. What they do with that sense is another matter.

There are still holes in Payne's and Clark's ideas. Only male blue whales seem to sing, so if they're really navigating or communicating with their calls, then what are females doing? There's also the matter of proportions. A 20 Hz note has a wavelength of 75 meters, which means that the distance between two peaks of pressure is two to three times as long as the longest blue or fin whale. These superlatively big animals have the same problem as the tiny *Ormia* fly: Their calls should sound the same to both ears, so it shouldn't be possible to track their source. "It may be impossible, but watch that fly!" Clark says. "I don't believe in spirits or astrology, but don't underestimate evolution. I've been more than chastised in scientific meetings for proposing all these preposterous things that I can never prove. But I'd much rather be open-minded. And I constantly try to put myself in the space of the animal."

WHILE ELEPHANTS AND WHALES produce calls that are below the range of human hearing, other species go above it. In the winter of 1877, Joseph Sidebotham was staying in a hotel at Menton, France, when he heard what sounded like a canary singing on his balcony. He soon discovered that the singer was actually a mouse. He fed it with biscuits, and it reciprocated by singing for hours by the fireplace, cranking out a tune as beautiful as that of any bird. His son suggested that all mice might sing similar melodies at pitches too high for humans to hear. Sidebotham disagreed. "I am inclined to think the gift of singing in mice is but of very rare occurrence," he wrote to the journal *Nature*.

He was wrong. Roughly a century later, scientists realized that mice, rats, and many other rodents do indeed make a wide repertoire of "ultrasonic" calls, with frequencies too high to be audible to humans. They make these sounds when playing or mating, when stressed or cold, when aggressive or submissive. Pups that are separated from their nests make ultrasonic "isolation calls" that summon their mothers. Rats that are tickled by humans make ultrasonic chirps that have been compared to laughter. Richardson's ground squirrels produce ultrasonic alarm calls when they detect a predator (or a tan fedora repeatedly thrown by a scientist to mimic a predator). Male mice that sniff female hormones produce ultrasonic songs that are remarkably similar to those of birds, complete with distinctive syllables and phrases. Females attracted to these serenades join their chosen partners in an ultrasonic duet. Rodents are among the most common and intensively studied mammals in the world and have been fixtures of laboratories since the seventeenth century. All that time, they've been spiritedly talking to each other without any human realizing, exchanging messages that slipped beneath the senses of the oblivious researchers and technicians milling around them.

Like *infrasound,* the term *ultrasound* is an anthropocentric affectation. It refers to sound waves with frequencies higher than 20 kHz, which marks the upper limit of the average human ear. It seems special—*ultra,* even—because we can't hear it. But the vast majority of mammals actually hear very well into that range, and it's likely that the ancestors of our group did, too. Even our closest relatives, chimpanzees, can hear close to 30 kHz. A dog can hear 45 kHz; a cat, 85 kHz; a mouse, 100 kHz; and a bottlenose dolphin, 150 kHz. For all of these creatures, ultrasound is just sound. Many scientists have suggested that ultrasound offers animals a private communication channel that other ears can't eavesdrop upon—the same claim that was made about ultraviolet light. We can't hear these sounds, so we bill them as "hidden" and "secretive," even though they're patently audible to many other species.

Rickye and Henry Heffner have a different explanation for why so many mammals can hear ultrasound: It helps them work out where that sound is coming from. Like barn owls, mammals do this by comparing when a sound arrives at their two ears. But as the space between those ears goes down, such comparisons only become possible for higher frequencies with shorter wavelengths. As a general rule, the smaller a mammal's head, the higher its hearing range. The boundaries of our auditory worlds are set by the physics of sound hitting our skulls.*

High-frequency sounds may be easier to locate, but they have an important limitation. They lose energy quickly, and can be easily scattered and reflected by obstacles like leaves, grasses, and branches. This means that ultrasonic calls can only spread over short ranges. A singing blue whale might be heard across an ocean, but a singing mouse is only audible to its immediate neighbors. This limited range might explain why relatively few mammals—rodents, toothed whales, small bats, domestic cats, and a few others—use ultrasound to communicate even though they can hear those frequencies. The sounds just die off too quickly. (This is also why devices that claim to repel pests with ultrasound don't really work: Their range is far too limited to be of much practical use.)

A limited range might be beneficial, however, if animals want to limit their audience. The isolation call of a helpless mouse pup can alert a nearby parent without also alerting more distant predators. In this way, ultrasound really can provide a secret communication channel, not because it lies in an inaccessible frequency range but because it doesn't travel very far. Annoyingly, that limited range makes ultrasound even harder to study: We can't hear it, and even if we could, we might not be close enough to do so. Given how long it took to learn that rodents use ultrasound extensively in their social lives, it's entirely possible that such communication is far more abundant among animals than we currently appreciate.

Many examples of ultrasonic communication were only discovered when scientists noticed that animals seemed to be screaming silently, going through all the motions of making a call but without unleashing

---

* Subterranean animals are a striking exception. Their hearing range is much lower than expected for the size of their heads, perhaps because they don't need to localize sounds, and instead use surface-borne vibrations.

any actual noise. That's what Marissa Ramsier noticed while watching Philippine tarsiers—fist-sized, big-eyed primates that look like gremlins. They would open their mouths, but no sound would emerge. Ramsier only heard what they were saying by placing them in front of an ultrasound detector. Their calls, she learned, have frequencies of 70 kHz—well above the ultrasonic boundary and higher than any mammal aside from bats or cetaceans. What are they saying? What are they listening for, besides each other?

Hummingbirds are even more mysterious. As with Ramsier's experience with tarsiers, many observers have noticed hummingbirds opening their beaks and fluttering their chests without seeming to sing. The blue-throated hummingbird of North America sings an elaborate song that we can partly hear, but that also extends up to 30 kHz—well into the ultrasonic range. That was surprising since, as Carolyn Pytte showed in 2004, it can't hear above 7 kHz. It can still perceive the lower registers of its song, but much of what it sings is inaudible to its own ears. Several other hummingbirds, like the black jacobin and the violet-tailed sylph, make calls beyond the hearing of most birds, and the part of these songs that people can perceive sounds like crickets. The Ecuadorian hillstar goes even further, singing entire phrases in an ultrasonic register. Birds tend to have similar hearing ranges that top out before 10 kHz. So either these hummingbirds have very unusual ears or they can't actually hear what they're saying.* And if the latter is true, then why are their songs so high-pitched? Calls demand listeners. If the hummingbirds' tunes lie beyond their own Umwelten, who's the audience?

Maybe it's insects? Even though most insects can't hear at all, many of those with ears can hear ultrasonic frequencies. More than half of the 160,000 species of moths and butterflies are so equipped. The greater wax moth can even hear frequencies near 300 kHz—the high-

---

* It might seem absurd to think that an animal couldn't hear its own calls, but there's at least one clear example where that's the case—the pumpkin toadlet of Brazil. This orange frog is insensitive to the frequencies in its calls, but calls all the same, perhaps because the sight of its inflating vocal sacs is attractive to mates.

est limit of any animal by some margin. Hummingbirds eat insects as well as nectar, so perhaps they produce ultrasonic calls that they can't hear to flush out the insects that can.

But why did so many insects evolve ultrasonic hearing, especially since most of them can't hear at all? It certainly wasn't to hear hummingbirds, which are relatively recent evolutionary arrivals. It probably wasn't to hear each other, since many of them are silent.* The most likely answer is that their ears were tuned to extremely high pitches to listen out for their nemeses, which appeared around 65 million years ago—bats. Bats evolved the ability both to call and to hear at ultrasonic frequencies, and they combined these traits into one of the most extraordinary animal senses of all.†

---

* Some moths do make ultrasonic courtship calls. Males will follow a female's pheromone trail, land next to her, and vibrate their wings to produce a volley of ultrasound. These calls are very quiet, almost like whispers. Like other ultrasonic communicators, these moths are probably making use of ultrasound's limited range so that they'll be heard by a prospective mate sitting nearby, but not by a hungry bat flying overhead. But unlike most songs, ultrasonic or otherwise, these calls aren't meant to be attractive. They're meant to sound dangerous. They mimic the calls of bats, prompting the females to freeze and allowing the males to mate more easily.

† For years, hundreds of textbooks and scientific papers have claimed that echolocating bats drove the evolution of ears in moths and other insects. But halfway through writing this book, I (and the wider scientific community) learned that this narrative is false. Moth ears almost always evolved before bat ultrasound, by at least 28 and as many as 42 million years. They only shifted toward higher frequencies once bats arrived on the scene. As sensory biologist Jesse Barber tells me, "Most of the introductions I've written in my papers are wrong."

9.

A Silent World Shouts Back

Echoes

AS I LOOK THROUGH THE WINDOW OF A HEAVY DOOR, A gloved hand on the other side holds up a ball of brown fur with long ears and a dark Chihuahua-like face. This is Zipper. She's a big brown bat—one of seven spending the summer at Boise State University under the care of Jesse Barber. Big browns are certainly brown, but, at roughly the weight of a mouse, they're only big relative to other small bats. They thrive in attics throughout the United States, but since they're nocturnal and quiet, people rarely see them, and certainly not at this distance. They emerge at dusk to chase after moths and other night-flying insects, and Zipper was so named because she's especially good at maneuvering. Some of her roommates have been given food-related epithets like Ramen, Pickles, and Tater. Others were named for their personalities: Casper (after the ghost) is friendly; Benny (after a character in *Rent*) is vocal. All these bats will be released in October in time for hibernation, but until then, they're in for a cushy summer, dining on juicy mealworms, snuggling up in warm cages, and going on regular "flight walks." "We take them out of their cages to let them exercise," Barber tells me. "It's like having 16 dogs."

As I watch Zipper through the window, she opens her mouth, exposing her surprisingly long teeth. This isn't an aggressive display. She's trying to make sense of her surroundings. She's unleashing a stream of short, ultrasonic pulses from her mouth. By listening for the returning echoes, she can detect and locate objects around her—a form of biological sonar. Only a few animals have this skill, and only two groups have perfected it: toothed whales (like dolphins, orcas, and sperm whales) and bats. Currently, Zipper's sonar is telling her that there's a solid barrier in front of her, even though she can see large creatures standing beyond it. (Despite the common idiom, bats aren't blind.) It must be a little confusing, but in fairness, Zipper's ability didn't evolve to detect windows. It evolved to find small insects at night when vision is limited. During the day, sharp-eyed predators like birds have their way with bugs. At night, those prey belong to bats. Since we rarely see bats, it's easy to mistake them for ecological B-listers that dine on the nocturnal scraps that birds leave behind. It's actually the other way round: In some rainforests, bats devour twice as many insects as birds. And when Zipper's handlers take her into an adjacent flight room and release moths into the air, I start to understand why.

The flight room, which is completely dark, is watched by three infrared cameras. The handlers inside can only hear the sounds of flapping. Everyone outside—Barber, his student Juliette Rubin, and I—can see what's happening on a monitor. And what we see is Zipper, for whom darkness is no impediment, slicing through the air and catching moth after moth. Outside, Rubin and Barber whoop and cheer like excited sports fans.

RUBIN: Did she get it? No, she just touched it.
BARBER: There it is. . . . OOOOOHHHHH.
RUBIN: Second interaction. Third. She's gonna get it. This bat is so good.
BARBER: That moth's pretty good too. . . .
RUBIN: OH, got it. Knew it!
HANDLERS, ON THE WALKIE-TALKIE: Did she get it?
RUBIN: Yeah, she did, the badass.
BARBER, TO ME: She'll need a minute to consume it.
RUBIN: This one's eaten two lunas, some wax moths, plus mealworms. She's an empty pit.

[The team gives Zipper a break, takes Poppy—another bat—into the room, and releases another moth.]

RUBIN: Okay, we're rolling. Ooooh, nice. Whoa! Oh my gosh. She is . . . OH, did you *see* that acceleration she just did?
EVERYONE, INCLUDING ME: WHOAAAAAA!

The images on the monitor are monochrome and grainy, but, on his laptop, Barber shows me several videos that he captured with much better cameras. In slow motion and high definition, a red bat does a double backflip, catching a moth in its tail and then flipping it into its mouth. A leaf-nosed bat tackles another moth in an explosion of scales. A pallid bat descends upon a scorpion like a dragon. These are bats in their element—and they are glorious. "For many people, when I talk about my research, their first reaction is: Oh, how could you work with those things?" Rubin says. "I forget that most humans think bats are gross, because they're so incredible at what they do—and they look good doing it." They are so misunderstood, so often used as symbols of evil, and so separate from us in altitude and time of day that "some of their most basic biology is unknown," Barber adds. "Bats might as well be in the deep ocean. We know more about their sonar than any other aspect of their lives."

For a long time we didn't know about their sonar, either. In the 1790s, the Italian priest and biologist Lazzaro Spallanzani realized that bats could still navigate in spaces too dark for a captive owl. In a series of cruel experiments, he showed that bats could orient when blinded, but would blunder into objects when deafened or gagged. He never fully grasped the meaning of these curious findings and could only write that "the ear of the bat serves more efficiently for seeing, or at least for measuring distances, than do its eyes." His contemporaries scoffed at that idea; one philosopher ridiculed it by asking, "Since bats see with their ears, do they hear with their eyes?"

The meaning of these observations remained unclear for more than a century, until a young undergraduate named Donald Griffin came up with a clever idea.* Griffin had spent many hours studying migrating bats and marveled at how they flew through dark caves without face-planting into stalactites. He heard about an untested hypothesis that bats listen out for echoes from high-frequency sounds. And he knew that a local physicist had invented a device that could detect ultrasonic sounds and convert them into audible frequencies. In 1938, Griffin showed up at the man's office with a cage of little brown bats, which he placed in front of the detector. "We were surprised and delighted to hear a medley of raucous noises from the loudspeaker," Griffin wrote in his classic book *Listening in the Dark*.

A year later, Griffin and his fellow student Robert Galambos confirmed that bats make these same ultrasonic cries as they fly, that their ears can detect such frequencies, and that both skills are necessary for them to avoid obstacles. With mouths and ears unimpeded, they could effortlessly negotiate around a labyrinth of fine wires hung from the ceiling. If their ears were blocked or their mouths were gagged, they were loath to take wing and quick to collide with walls, furniture, and even Griffin and Galambos themselves. The animals were clearly finding their way around by listening to the echoes of their own calls. Others thought this preposterous. As Griffin later recounted, "One distinguished physiologist was so shocked by our presentation at a scientific meeting that he seized Bob [Galambos] by the shoulders and shook him while expostulating: You can't really mean that!" But the duo did mean it, and in 1944, Griffin gave a name to the bat's astonishing skill. He called it echolocation.†

---

* For over a century, scholars claimed instead that bats feel their way through the night by sensing air currents playing along their wings. In 1912, Hiram Maxim (hot off inventing a fully automatic machine gun) modified this idea by suggesting that bats feel the reflections of low-frequency sounds produced by their wingbeats. It wasn't until 1920 that the physiologist Hamilton Hartridge correctly speculated that they were listening for echoes from high-frequency sounds. This was the idea that Griffin heard.

† The Dutch scientist Sven Dijkgraaf had being doing similar studies. But with Germany occupying the Netherlands and war disrupting scientific communication across the Atlantic, Dijkgraaf had no idea what Griffin and Galambos were up to, and didn't have access to an ultrasonic detector.

Even Griffin underestimated echolocation at first. He saw it merely as a warning system that alerted bats of possible collisions. But his views changed in the summer of 1951. Sitting by a pond near Ithaca, he began to record wild echolocating bats for the first time. Pointing his microphone at the skies, he was shocked by how many ultrasonic cries he heard, and how different these were from those he had witnessed in enclosed spaces. When bats were cruising through open skies, their pulses were longer and duller. When they swooped after insects, the steady *put-put-put*s would quicken and fuse into a staccato buzz. By using a slingshot to launch pebbles in front of the bats, Griffin confirmed that they go through the same sequence of quickening pulses every time they pursue an airborne object. Echolocation, he was staggered to realize, wasn't just a collision detector. It's also how bats hunt. "Our scientific imaginations had simply failed to consider, even speculatively, [this] possibility," he later wrote.

To study wild bats, Griffin had to stuff a station wagon with microphones, tripods, parabolic reflectors, radios, a generator with a car muffler welded onto it, gasoline tanks, and around 200 feet of extension cord. Technology has progressed since then, and so has the study of echolocation. Back in 1938, the ultrasound detector that Griffin used was one of a kind (and he was appalled when he and Galambos temporarily broke it). When I visit Cindy Moss's state-of-the-art lab in Baltimore 80 years later, I count 21 ultrasonic microphones dotting the walls of just one of two flight rooms. Infrared cameras film the bats as they fly. Laptops represent the bats' inaudible sounds as visible spectrograms, and these displays are precise enough that experienced researchers can use them to identify individual bats. One might have a stutter. Another might have an unusually low voice—a bat baritone.

These gadgets mean that bat echolocation, which was once undetectable to human ears and implausible to human minds, is one of the most accessible of all senses. Of course, "what bats perceive is not yet known," Moss tells me. "That's a really important problem." I mention that this is the same philosophical dilemma that Thomas Nagel discussed in "What Is It Like to Be a Bat?"—that the conscious experiences of other animals are inherently hard to imagine.

"Right," Moss says. And with a wry smile, she adds, "Except he thought you would never know."

THERE ARE MORE THAN 1,400 species of bats. All of them fly. Most of them echolocate.* Echolocation differs from the senses we have met so far, because it involves putting energy into the environment. Eyes scan, noses sniff, whiskers whisk, and fingers press, but these sense organs are always picking up stimuli that already exist in the wider world. By contrast, an echolocating bat creates the stimulus that it later detects. Without the call, there is no echo. As bat researcher James Simmons explained to me, echolocation is a way of tricking your surroundings into revealing themselves. A bat says, "Marco," and its surroundings can't help but say, "Polo." The bat speaks, and a silent world shouts back.

The basic process seems straightforward. The bat's call is scattered and reflected by whatever's around it, and the animal detects and interprets the portion that rebounds. But to successfully do this, a bat must cope with many challenges. I count at least 10.

First, distance is an issue. A bat's call must be strong enough to make the outward journey to a target and the return journey back to its ears. But sounds quickly lose energy as they travel through air, especially

---

* The origins of echolocation are still unclear, because the origins of bats themselves are unclear. Bat skeletons tend to be small and delicate, which means they leave behind few fossils that might hint at their ancestry. And modern bats, despite their variety, are more physically similar than they are different, which makes it hard to work out how different groups are related. For these reasons, there's still vigorous debate about when bats first started to echolocate, whether they could already fly at that point, whether they initially used the ability to avoid obstacles or find prey, and how many times that ability evolved. Traditionally, the bat family tree has two main branches—one containing the smaller echolocating species, and another containing the larger fruit bats that (with one exception) do not echolocate. We now know this is wrong. The most recent tree, which includes genetic data, shifts several of the smaller bats, including horseshoes and false vampires, over to the fruit bat branch. That's huge news in the world of bat academia. If correct, it means either that echolocation evolved once in the common ancestor of all bats and was subsequently lost in the fruit bats or that it evolved on two separate occasions.

when they're high in frequency, so echolocation only works over short ranges. An average bat can only detect small moths from around 6 to 9 yards away, and larger ones from around 11 to 13 yards. Anything farther away is probably imperceptible, unless it's very large, like a building or a tree. Even within the detectable zone, objects on the periphery are fuzzy. That's because bats concentrate the energy of their calls into a cone, which extends from their heads like the beam of a flashlight; this helps the sounds to carry farther before petering out.*

Volume helps, too. Annemarie Surlykke showed that the sonar call of the big brown bat can leave its mouth at 138 decibels—roughly as loud as a siren or jet engine. Even the so-called whispering bats, which are meant to be quiet, will emit 110-decibel shrieks, comparable to chainsaws and leaf blowers. These are among the loudest sounds of any land animal, and it's a huge mercy that they're too high-pitched for us to hear. If our ears could detect ultrasound, I would have recoiled in pain while listening to Zipper, and Donald Griffin probably would have fled from the unbearable hubbub of his Ithaca pond.

But bats can hear their own calls, which creates an obvious second challenge: They must avoid deafening themselves with every scream. They do so by contracting the muscles of their middle ears in time with their calls. This desensitizes their hearing while they shout and restores it in time for the echo. More subtly, bats can adjust the sensitivity of their ears as they approach a target so that they perceive the returning echoes at the same steady loudness, no matter how loud the echoes actually are. This is called acoustic gain control, and it likely stabilizes the bat's perception of its target.

The third problem is one of speed. Every echo provides a snapshot. Bats fly so quickly that they must update those snapshots regularly to detect fast-approaching obstacles or fast-escaping prey. John Ratcliffe showed that they do so with vocal muscles that can contract up to 200

---

* The big brown bat actually produces a forked sonar beam with two horns—one pointing ahead and another pointing downward. The bat might use the forward horn to scan for insects and obstacles and the downward one to keep track of its altitude. This is reminiscent of the eyes of birds of prey, which have two foveae, one for scanning the horizon and another for tracking prey.

times a second—the fastest speeds of any mammalian muscle.* Those muscles don't always contract so quickly. But in the final moments of a hunt, when bats are bearing down upon their targets and need to sense every dodge and dive, they produce as many pulses as their super-fast muscles will allow. This is the so-called terminal buzz. It is what Griffin first heard at his Ithaca pond. It is the sound of a bat sensing its prey as sharply as possible, and of an insect likely losing its life.

Fast pulses address the third challenge while creating a fourth. For echolocation to work, a bat must match every outgoing call to its re-spective echo. If it's calling very quickly, it risks creating a jumbled stream of overlapping calls and echoes that can't be separated and thus can't be interpreted. Most bats avoid this problem by making their calls very short—a few milliseconds long for the big brown. They also space their calls, so that each goes out only after the echo from the preceding one has returned. The air between a big brown bat and its target is only ever filled by a call or an echo, and never both. The bat's control is so fine that even during its rapid terminal buzz, there's no overlap.

After receiving the echoes, the bat must now make sense of them. This fifth challenge is the hardest yet. Consider a simple scenario where a big brown bat is echolocating on a moth. It hears its own call on the way out. After a delay, it hears the echo. The length of that delay tells the bat about its distance to the insect. And as James Simmons and Cindy Moss have shown, the bat's nervous system is so sensitive that it can detect differences in echo delay of just one or two millionths of a second, which translates to a physical distance of less than a millimeter. Through sonar, it gauges the distance to a target with far more preci-sion than any human can with our sharp eyes.†

But echolocation reveals more than just distance. A moth has a complex shape, so its head, body, and wings will all return echoes after

---

* The bat's flashlight, however, pulses on and off several times a second, offering a series of stroboscopic snapshots. It seems likely that the bat's brain knits these snapshots into something smooth and continuous, much as our brains do when we watch a movie where static frames appear in quick succession.

† This is another reason why bats keep their calls short: Since they compute distance from time, a shorter call provides a more precise estimate of range.

slightly different delays. Complicating matters further, a hunting big brown bat produces a call that sweeps across a broad band of frequencies, falling over an octave or two. All of these frequencies bounce off the moth's body parts in subtly different ways, and provide the bat with disparate pieces of information. Lower frequencies tell it about large features; higher frequencies fill in finer detail. The bat's auditory system somehow analyzes all this information—the time gaps between the call and the various echoes, *at each of their constituent frequencies*—to build a sharper and richer acoustic portrait of the moth. It knows the insect's position, but maybe also its size, shape, texture, and orientation.

All of this would be hard enough if the bat and the moth were staying still. Usually, both are in motion. Hence, the sixth challenge: A bat must constantly adjust its sonar. To even find a moth in the first place, it must scour wide expanses of open air. During this search phase, it makes calls that carry as far as possible—loud, long, infrequent pulses whose energy is concentrated within a narrow frequency band. Once the bat hears a promising echo and approaches the possible target, its strategy changes. It broadens the frequencies of its call to capture more detail about the target and to more accurately estimate its distance. It calls more frequently to get faster updates about the target's position. And it shortens each call to avoid overlapping with the echoes. Finally, once the bat goes in for the kill, it produces the terminal buzz to claim as much information as possible as quickly as possible. Some bats will also broaden the beam of their sonar at this point, widening their sensory zone to better catch moths that try to bank to the side.

The entire hunting sequence, from initial search to terminal buzz, might occur over a matter of seconds. Again and again, bats adjust the length, number, intensity, and frequencies of their calls to strategically control their perception. Handily, this means that a bat's voice reveals its intent. If its call is long and loud, it's focusing on something far away. If the call is soft and short, it's homing in on something close. If it produces faster pulses, it is paying more attention to a target. By measuring these calls in real time, researchers can almost read a bat's mind.

This approach has helped to explain how bats cope with their seventh challenge—cluttered environments. Bats can race through rugged caves, tangled branches, and even mazes of hanging chains. These messy spaces pose special problems for sonar that don't apply to vision. Imagine that a bat is flying toward two branches that are the same distance away. If it could see them, it could easily tell them apart, because light reflecting off each branch would fall on different parts of its retina. A sense of space is baked into the anatomy of its eye. That's not true for ears. The bat must *compute* space from the timing of its echoes, and since echoes returning from the two equidistant branches would arrive after the same delay, they might sound like the same object.

Cindy Moss showed how bats solve this problem by training big browns to zoom through a hole in a net. She saw that the animals would aim the center of their sonar beams onto the edges of the hole, scanning it before hurtling through. "Just as we can scan different objects in a room with our eyes, the bat can do the same by directing its sonar beam," Moss tells me. She also found that whenever the bats are doing something demanding, like flying around obstacles or chasing erratically moving targets, they shorten their calls and broaden their frequency range to wrest as much detail from the echoes as possible. They also tend to group their calls into distinctive clusters that Moss calls sonar strobe groups (*buh-buh-buh-buh . . . buh-buh-buh-buh . . . buh-buh-buh-buh*). Bats may process each group as a unit, summing up the detail from all the constituent echoes to build a sharper representation of their surroundings.*

Echolocation suffers from another problem—the eighth in our series—that vision does not. Eyes have no problem picking out objects against a background, unless that object is camouflaged. But for sonar, small objects on large backgrounds are automatically camouflaged. If a moth is flying in front of a leaf or sitting upon it, the strong echoes from the leaf would drown out the fainter ones from the moth. Of

---

* If that scene is especially complex, big brown bats can get even more detail by shifting the frequencies of the individual calls within the strobe groups, so that each is lower than the last. Several species do this kind of "frequency-hopping": The chestnut sac-winged bat produces triplets of ascending frequencies, and is also known as the do-re-mi bat.

several solutions to this problem that bats have developed, the common big-eared bat's is the most impressive. Using sonar, and sonar alone, it can grab dragonflies and other insects right off a leaf, even when they're still and silent—a feat that scientists had long considered impossible. Inga Geipel found that the bat pulls off its amazing trick by approaching its prey from a sharp angle, so that echoes from the insect bounce toward it while those from the leaf bounce away. The bat accentuates this effect by hovering upward and downward in front of the insect, with its head fixed upon it. Initially, it probably hears something fuzzy and indistinct—the merest hint of possible prey. But as it slides up and down, gathering information from different angles, the shape of its meal sharpens, and to the insect's misfortune, an impossible feat becomes all too possible.

The ninth challenge arises when bats fly in groups, as they often do. Now they must somehow distinguish the echoes of their own calls from those of other individuals. Big browns do this by aiming their calls away from other bats, shifting the frequencies of their calls to avoid overlapping with other bats' sounds, or taking turns to fly silently.* But such strategies are less useful for Mexican free-tailed bats, which gather in the millions. When 20 million bats are flooding out of a cave together, how on earth does each one pick out its own echoes? Researchers have called this the "cocktail party nightmare," and it's not clear how bats wake from it. They might only process echoes that arrive within a certain timeframe, or from a specific direction. They might also ignore echolocation altogether, relying instead on other senses or their memories. Mexican free-tailed bats probably know the path in and out of their caves, and can just follow the right trajectory without needing to consult any echoes. This explains the many historical incidents in which people barricaded the entrances to caves for

---

* When bats want to communicate with each other, they tend to make types of calls very different from the ones they use as sonar. The difference between communication and echolocation isn't clear-cut, though. Some bats can recognize the sonar calls of familiar individuals and will eavesdrop on each other's feeding buzzes. The greater bulldog bat can also modify its sonar call into a message: It'll add a deep warning honk at the end of the pulse if it's about to hit another bat.

safety reasons, only to later find that bats had fatally crashed into the doors.

These tragic mishaps illustrate the 10th challenge of echolocation: It takes a lot of effort to solve the other nine. Echolocation is mentally demanding, especially since bats do everything they do at speed. Often they simply don't have the time to use their sonar to its fullest capacity, which is why they often make ridiculous mistakes that seem beneath them.* They can distinguish two grades of sandpaper whose grains differ by half a millimeter, but will also plow headlong into a newly installed cave door. They can discern flying insects by shape, but will go after a pebble launched into the air. Bats are fully capable of avoiding such errors. They're just not paying attention. They're relying on memory and instinct. Humans behave in the same way: Most car accidents occur close to home, in part because drivers are less watchful when going down familiar routes. In both cases, perception is influenced not just by information from sense organs but also by what brains decide to do with that information. Those brains, and their workings, are still mysterious. For all we have learned about echolocation, Nagel was still right: We might never fully know what it is like to be a bat. But if we dared to take an educated guess, it might be something like this.

It is dark, and you, a big brown bat, are hungry. Easily sensing trees and other large obstacles, you zip around them, searching for insects by lobbing strong, infrequent, and narrow-pitched calls into the intervening air. Most of those calls disappear into the distance, but some return, revealing the presence of something flying at one o'clock. A moth? You turn your head and then your body to keep the target within the cone of your sonar. You know precisely how far away the

---

* In *Listening in the Dark,* Donald Griffin devoted an entire section to "bumbling bats." In it, he noted that the miraculous feats for which these animals are rhapsodized, like flying through a curtain of thin wires, are only performed by "the most alert and wide-awake" individuals. Under some conditions, Griffin wrote, bats "are quite clumsy and they sometimes blunder headlong into obstacles which they dodge without the slightest difficulty at other times. Perhaps I have become a trifle sensitive about this point for whenever a bat is seen to bump into anything I am very likely to hear about it, often in slightly accusing tones."

target is by now, but your perception of it is still blurry. That changes as you draw closer. As your calls shorten, speed up, and broaden in pitch, your sense of the target sharpens—it *is* a moth, a large one, flying away. As you bear down upon the insect, the incredible muscles in your throat unleash the fastest possible barrage of sonar pulses, snapping the moth into sharp focus. Head, body, and wings all become richly detailed even as you scoop the lot into your mouth with your tail. And you accomplish all of that in the time between you reading *this* word . . . and you reading *this* one.

It is no wonder that bats are so successful. They're found on every continent except Antarctica, and they account for one in every five mammal species. There are bats that pluck insects from the air and bats that pluck fruit from trees. There are bats that catch frogs, bats that drink blood, and bats that sip nectar with tongues more than twice as long as their bodies. There are bat-eating bats. There are bats that go fishing by echolocating on ripples. There are bats that pollinate plants by echolocating on dish-shaped leaves that are adapted to reflect sonar pulses. And there are bats that have solved the challenges of echolocation in a way fundamentally different from what we've already seen, and have developed the most specialized form of sonar in the world.

MOST BATS ECHOLOCATE IN a way broadly similar to that of the archetypal big brown. They send out short sonar pulses that last between 1 and 20 milliseconds and are separated by relatively longer silences. Those pulses also sweep down across a broad band of frequencies, which is why these bats are known as FM, or frequency-modulated, bats. But around 160 species—the horseshoes, hipposiderids, and Parnell's mustached bat—do something very different. Their calls are much longer, lasting for many tens of milliseconds in some species, and separated by much shorter gaps. And instead of covering a range of frequencies, these species hold one particular note. For that reason, they are called CF, or constant-frequency, bats. And they are listening out for a very specific kind of echo.

When a sonar pulse hits an insect's flapping wing, the echo strength varies as the wing moves up and down. But at one particular moment, when the wing is exactly perpendicular to the incoming sound, an especially loud and sharp echo bounces straight back at the bat. This is called an acoustic glint. It's a dead giveaway that an insect is flying nearby. FM bats can theoretically detect these glints, but they're unlikely to. Their brief sonar pulses are separated by long gaps, so an FM bat has to get very lucky to hit an insect's wing at exactly the right moment to return a glint. By contrast, the pulses of CF bats are long enough to cover an entire wingbeat. They catch glints galore. And since leaves and other background objects don't flap in the same rhythmic way as wings, a CF bat can use glints to distinguish fluttering insects against cluttering foliage. They must be the auditory equivalent of flashes of light.

Hans-Ulrich Schnitzler, who has been studying CF bats since the 1960s, has shown that they can recognize different species of insects from the rhythms of their wingbeats. They can tell if the insect is flying toward them or away from them. And they can absolutely tell liv-

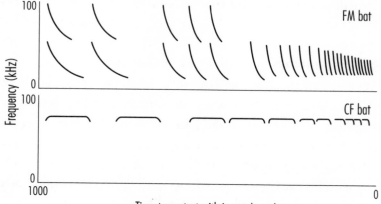

These spectrograms show the echolocation calls of two bats as they approach an insect. Note that the FM bat's calls cover a wide range of frequencies, whereas the CF bat mostly holds the same note. But both bats produce shorter and more rapid calls as they approach their prey.

ing targets from inanimate ones: Unlike big brown bats, CF bats won't go after airborne pebbles.*

The ears of CF bats are as specialized as their calls. The greater horseshoe bat, for example, makes a call with a constant frequency of around 83 kHz, and has a disproportionate number of auditory neurons devoted to exactly this pitch.† It hears the sounds of its own echoes more sensitively than anything else. Other species have their own signature frequencies, as if each CF bat has shaved off a thin slice of the full auditory world and claimed that slice for itself. But this strategy also creates a major problem—an 11th challenge that FM bats do not face.

Sounds seem to rise in pitch as you get closer to their sources— think about what a siren sounds like when an ambulance drives toward you. This is called the Doppler effect. It means that when a CF bat flies at an insect, the echoes it hears get higher in frequency and should eventually overshoot the bat's zone of best hearing. But as Schnitzler discovered in 1967, CF bats can compensate for Doppler shifts. When closing in on a target, they produce calls that are lower than their normal resting frequency, so the upshifted echoes hit their ears at exactly the right pitch. And they do this (quite literally) on the fly, constantly tweaking their calls so that the echoes from targets ahead stay within 0.2 percent of the ideal frequency. This is a staggering feat of motor control that's almost unmatched in the animal kingdom.

Imagine that you have a mistuned piano that always produces notes three tones higher than what you're actually trying to play. If you want middle C, you'll have to press the A on its left. You'd soon get the hang of it—but imagine now that the piano's mistakes aren't system-

---

* In practice, many bats use a mix of CF and FM calls. When FM bats like big browns are searching in open air, they produce CF-like pulses. Meanwhile, CF bats will add a brief frequency sweep at the end of their pulses to better judge the distance to their prey.

† Researchers have called this sensitive band the acoustic fovea, after the part of the retina where visual acuity is sharpest. It's a decent analogy, but also a little off. The fovea is a region of physical space where vision is sharpest, but the acoustic fovea describes a region of informational space where the bat's hearing is most acute. It's more like walking around with eyes that are inordinately good at seeing a particular shade of green.

atic, and the gap between the pressed notes and desired notes changes all the time. Now you must constantly judge the size of the gap by listening to the music coming out of the janky instrument, and adjust your fingers as you play. That is what CF bats are doing—many times a second, with almost no errors. They can even do this for several targets simultaneously. A horseshoe bat can throw its attention between different obstacles at varying distances and perform the right Doppler compensation for each one.[*]

For a nocturnal insect, no environment is safe from bats. If they fly in open air, big brown bats can grab them. If they head for thick foliage, greater horseshoes can track them. If they land on a surface and stay still, common big-eared bats can still find them. Sonar seems like an unbeatable weapon that can be tailored to any possible habitat. But while it is certainly versatile, it isn't invincible. In evolving an incredible sense, bats opened themselves up to equally incredible illusions.

IT IS GENTLY SNOWING inside Jesse Barber's lab, or so it seems. The team members have been carrying moths into the flight room where Zipper and other bats are swooping around, and the insects have left a cloud of white scales hanging in the air. The scales are so pervasive that both Barber and Juliette Rubin have become horribly allergic to them and are now wearing face masks. This, they tell me, is a common occupational hazard among lepidopterists—people who study moths and butterflies. In some circles, it's called lep lung.

When not inflaming the airways of scientists, the scales protect the bodies of moths, by absorbing the sound of a bat's calls and muffling the resulting echoes. This acoustic armor is just one of several anti-bat defenses. As we saw in the previous chapter, more than half of moth species have ears that can hear bat sonar. Such ears offer a considerable

---

[*] In this way, CF bats use the potential problem of the Doppler effect to their advantage. FM bats must keep their calls short to avoid overlaps with the returning echo. But CF bats separate their calls and echoes in frequency rather than time. Thanks to the Doppler effect, the echoes are usually higher in pitch than the calls, and more obvious to the bat's finely tuned ears. That's why their calls can afford to be long—long enough to return an acoustic glint and reveal the presence of fluttering prey.

advantage. Bats are listening for sounds that have traveled to a moth and back again, but moths only have to detect the same sounds after their initial outward journey, when they're much stronger. So while bats can hear small moths from no more than 9 yards away, moths can hear bats from 15 to 33 yards away. Many of them exploit this lead by executing dodges, loops, and power dives whenever they hear bat voices. Others talk back.

Tiger moths, a diverse group of 11,000 species, have a pair of drum-like organs on their flanks. These vibrate to produce ultrasonic clicks that seem to baffle bats, causing them to miss the moths.[*] Sometimes these clicks are acoustic versions of warning colors: Many tiger moths are full of foul-tasting chemicals, and they click to tell bats that they aren't worth eating. The clicks can also jam a bat's sonar. In 2009, Aaron Corcoran and Jesse Barber found clear evidence that this happens by pitting big brown bats against *Bertholdia trigona*—a stunning American tiger moth that's clad in the colors of a burning log. These moths have no chemical defenses and bats will eat them if they can. But the big browns frequently flubbed their attacks when they approached a clicking *Bertholdia,* even when the moths were tethered in place. The clicks overlapped with the bats' echoes and messed with their ability to gauge distance. From their perspective, a target that was once sharply defined and precisely pinpointed would have suddenly blurred into a nebulous cloud with ambiguous position.[†]

Other moths can cast illusions without incantations. Barber and Rubin have been breeding luna moths—unmistakable, palm-sized in-

---

[*] Dorothy Dunning and Kenneth Roeder first demonstrated this in 1965, showing that the clicks can stop little brown bats from successfully catching their prey. The duo had trained the bats to catch mealworms that were shot into the air—a task they did almost perfectly. But when they heard recordings of tiger moth clicks, they usually missed.

[†] Around half of hawkmoths—another major group of around 1,500 species—can also jam bats. But unlike the tiger moths, hawkmoths produce their confounding clicks by rubbing their genitals together. They seem to have evolved this ability on three separate occasions, with each group repurposing a different section of their sex organs into bat-befuddling instruments. But bats, in turn, have evolved counters to moth defenses. At least two species—the barbastelle of Europe and the Townsend's big-eared bat of North America—make very quiet calls that allow them to sneak up on moths unnoticed. With their stealthy whispers, they can get so close that their prey don't have time to either dodge or jam.

sects with a white body, blood-red legs, yellow antennae, and lime-green wings that end in a pair of long, streaming tails. When I open a cupboard in their lab, a few of these moths are just hanging calmly on the door, their empty chrysalises strewn across the shelves. In their adult form, they have no mouths and little time. In a week, they'll be dead. Until then, "all they do is mate and evade bats," Barber says. They have no noxious chemicals. They can't make jamming clicks. They can't even hear bats coming because they have no ears. But those long tails that grow from their hindwings flap and spin behind them as they fly, producing echoes that distract echolocating bats into attacking an inessential body part. On average, a luna moth without tails is nine times more likely to be eaten than one whose tails are intact. "When I discovered that, I thought: This can't be real," Barber says. "Echolocation is such a remarkable sense. How can a spinning piece of membrane fool the bats? But we see it, and consistently."

I see it, too, on Barber's monitor. When a luna moth is released into the flight room, Zipper the bat attacks it, and misses. She turns, attacks again, tears off a mouthful of tail, and spits it out. As the unappetizing fragment drifts to the floor, Barber looks at me, grinning, and says, "I told you." The handlers bring out the moth: It's missing the left tail but is otherwise unharmed. They take a second luna inside, this time with its tails already removed. Zipper catches it almost immediately.*

When I first looked at the luna moths, I thought their tails were like those of a peacock. But that was my visual bias leading me astray again. These moths find their mates through smell, and there's no evidence that the tails make them more attractive. They are meant not to delight the eyes of prospective mates but to fool the ears of prospective predators.

Donald Griffin once described bat echolocation as a "magic well" that, when uncovered, became an endless source of surprising discov-

---

* It's still unclear how the tails work. The echoes they produce might fuse with those from the moth's body, tricking the bat into thinking that it's hunting a much larger animal that's closer to its jaws. Alternatively, they might sound like entirely separate targets, or more conspicuous ones. Whatever the case, they work. Moths have evolved long tails on at least four separate occasions, and some of these can be twice as long as the rest of the insects' wings.

eries. By understanding what bats can do, we can appreciate them for the biological marvels that they are instead of the unsavory creatures they are reputed to be. We can better understand the creatures they hunt. And, as many scientists did after Griffin's work, we can look for other creatures that perceive the world through echoes.

BATS AND DOLPHINS ARE about as different as two groups of mammals can be. Bats' front legs have stretched into wings, while dolphins' have flattened into flippers. Bat bodies are svelte and lightweight; dolphin frames are streamlined and blubbery. Bats cut paths through the open air; dolphins, the open seas. But both groups must move and forage through three-dimensional and often dark spaces. Both groups did so by evolving echolocation. And both groups surrendered their secrets to science in roughly the same way: Researchers first noticed that dolphins could avoid obstacles in the dark even when blindfolded, and then that they made and heard ultrasonic clicks.* These observations were easier to interpret because, thanks to the pioneering work of Griffin and others, people already knew that echolocation existed. Researchers working with dolphins could test for a skill that just two decades earlier had seemed inconceivable.

Despite that advantage, research on dolphin sonar has progressed rather slowly, because the animals are not easy to work with. Their size alone is a problem. The smallest dolphin is around 40 times heavier than the biggest bat and requires a large saltwater tank instead of a small room. Dolphins are also smarter, harder to train, and more willful than bats: Kathy, a bottlenose dolphin who took part in a seminal early experiment, would agree to wear eye cups, but absolutely refused

---

* In the 1950s, Arthur McBride wondered if dolphins, porpoises, and other toothed whales might share the same ability. After watching porpoises evading fishing nets in the dark, he was reminded of bats. Ken Norris carried out a particularly illuminating experiment in 1959 when he trained a bottlenose dolphin named Kathy to wear latex suction cups over her eyes. Without vision, Kathy could still find floating pieces of fish by releasing volleys of rapid clicks, or swim through a maze of vertical pipes just like the bats flew through curtains of wires. If anything, she was more agile. While Griffin's bats would often brush the wires with their wingtips, Kathy only ever once bumped a pipe in two months of testing—and even then, she seemed to do it on purpose.

to don a sound-blocking mask that covered her jaw and forehead. And while bats can be easily found in buildings and woods, dolphins live in a habitat so inaccessible that most humans only skim across its surface. So researchers who study dolphins have been mostly forced to work with animals that live either in aquariums or in naval facilities.

The U.S. Navy started training dolphins in the 1960s to rescue lost divers, find sunken equipment, and detect buried mines. In the 1970s, it invested heavily in echolocation research, not to understand how the dolphins themselves perceived the world but to improve military sonar by reverse-engineering the animals' superior capabilities. A field station in Hawaii's Kāneʻohe Bay became a hub of important research, led by psychologist Paul Nachtigall and electrical engineer Whitlow Au. "The dolphin was a black box, and my interest lay in defining the parameters of that box," Au tells me. "I used to get my kids very upset because they just wanted to hug the animals, and I would say that they were just test subjects." (I ask him if he still regards them that way after working with them for decades. He pauses, then says, "I see them as more complex test subjects.")

At Kāneʻohe Bay, where bottlenose dolphins like Heptuna, Sven, Ehiku, and Ekahi could swim in large, open-water pens, Au and his colleagues realized that dolphin sonar was even more impressive than anyone had guessed. Dolphins could discriminate between different objects based on shape, size, and material. They could distinguish between cylinders filled with water, alcohol, and glycerine. They could identify distant targets from the information in a single sonar pulse. They could reliably find items buried under several feet of sediment, and they could tell if those objects were made of brass or steel—feats that no technological sonar can yet match. To date, "the only sonar that the Navy has that can detect buried mines in harbors is a dolphin," Au says.

Dolphins belong to the group of whales known as odontocetes, or toothed whales.* The other members of this group—porpoises, belu-

---

* A brief note on terminology: Dolphins, whales, and their relatives all belong to the group known as the cetaceans, which are colloquially just known as whales. There are two main groups: baleen whales (mysticetes) and toothed whales (odontocetes). Dolphins are one group

gas, narwhals, sperm whales, and orcas—also echolocate, and many do so just as well as the familiar bottlenose. In 1987, Nachtigall's team started working with a false killer whale—an 18-foot-long, black-skinned dolphin species known for being smart and sociable. The animal, Kina, could use her sonar to tell the difference between hollow metal cylinders that looked identical to the human eye and that differed in thickness by the width of a hair. On one memorable occasion, the team tested Kina using two cylinders that had been manufactured to the same specifications. To everyone's confusion, Kina repeatedly indicated that the objects were different. When the team had the cylinders remeasured, they realized that one had a minuscule taper and was 0.6 millimeters wider at one end than the other. "It was incredible," Nachtigall recalls. "We ordered them to be the same, the machinists said they were the same, and the animal said, 'No, they're different.' And she was right."

Dolphins can also echolocate on a concealed object and then recognize the same object visually—even on a television screen. This might seem like an obvious feat, but stop to consider what it involves. The animal isn't just working out the object's position but constructing a mental representation of that object, which can be translated to its other senses. And it's doing that with *sound*—a stimulus that doesn't naturally carry rich, three-dimensional information. If you heard a saxophone, you might recognize the instrument and work out where its music is coming from, but good luck predicting its shape from sound alone. You could, however, *touch* a saxophone and get a solid impression of what it should look like. So it is with echolocation. This sense is often described as "seeing with sound," but you could just as easily think of it as "touching with sound." It's as if a dolphin is reaching out and squeezing its surroundings with phantasmal hands.

I'm not used to thinking about sound in this way. Outside my win-

---

within the toothed whales, and they include killer whales and pilot whales. Dolphins and porpoises are different kinds of toothed whales, but the two terms have sometimes been used interchangeably; some early echolocation papers refer to "bottlenose porpoises." So, to recap, dolphins are whales, killer whales are dolphins, and porpoises are not dolphins, except when they are.

dow, I can hear barking dogs, singing starlings, and chirping cicadas, all using sound to convey information to their audiences. But the air and water of this planet also abound with sounds that animals use to convey information to *themselves*—sound produced not for communication but for exploration. Other senses *can* be used in this way to explore, but echolocation is inherently exploratory. And it absolutely feels that way when deployed by an animal as inquisitive as a dolphin. "The animals aren't echolocating all the time, but any time you put a new object in with them, they'll buzz the crap out of it," Brian Branstetter, who started working with dolphins in Oahu in the 1990s, tells me. "And when I'm swimming with them, I can hear and feel their clicks: This animal is checking me out right now!"

MUCH ABOUT DOLPHIN SONAR is counterintuitive, including the way they produce it. At the top of the dolphin's head is the blowhole, which is equivalent to your nostrils. Just below the blowhole, in the animal's nasal passages, are two pairs of organs called phonic lips. The dolphin clicks by forcing air through those lips and making them vibrate. The sound then travels forward and is focused by a fatty organ called the melon, which is what gives the dolphin its bulging brow. So while a bat's call begins in its throat and goes out through its mouth or nose, a dolphin's click begins in its nose and goes out through its forehead.

The sperm whale—the biggest odontocete of all—does something even stranger. Its titanic barrel of a nose can make up a third of its 52-foot body, and the phonic lips lie at the very front. When they vibrate, most of their sound goes *backward* through the whale's head. It passes through a fat-filled organ called the spermaceti (the contents of which whalers once prized), bounces off an air sac at the back of the head, and then moves forward through another fatty organ called the junk (which whalers deemed worthless). The sound that emerges from this absurd detour is the loudest in the animal world. At 236 decibels, it's basically an explosion. When scientists want to calibrate hydrophones to record sperm whale clicks, they throw cherry bombs into the water. The clicks are also focused into an extremely thin beam that's around 4 degrees wide. If a bottlenose dolphin perceives the ocean with a sonar flashlight, then a sperm whale fires a laser.*

Odontocetes also intercept their own echoes in a bizarre way. In the 1960s, Ken Norris found a dolphin skeleton on a Mexican beach, and noticed that part of its lower jaw was so thin that it was almost translucent. This hollow stretch of bone is filled with the same fats that make up the melon. These "acoustic fats" are never burned for energy, no matter how starved a dolphin gets. Their purpose is to channel sound toward the inner ear. A dolphin is an echolocator that clicks with its nose and listens with its jaw.

Despite these weird traits, odontocetes use many of the same echolocating tricks as bats. When they need more information, they can speed up the pace of their clicks (as in the terminal buzz) or group those clicks into packets (as in the strobe groups). They can adjust the sensitivity of their ears to dampen their own booming noises and to perceive the returning echoes at the same steady loudness. But odontocetes can also pull off feats of sonar that bats cannot. Sound behaves differently in water than in air. It travels faster and farther, so dolphin sonar operates over ranges no bat can manage.† In an early experiment, Au showed that blindfolded dolphins could detect steel spheres at a distance of 110 yards, far enough that the team had to use binoculars to check that the targets were correctly positioned. The dolphins didn't need the help—and it later transpired that they were working under difficult conditions. Unbeknownst to anyone at the time, Kāne'ohe Bay was full of snapping shrimps, whose large claws fill the water with cacophonous pops. The dolphins were using sound to spot tennis balls

---

* Why are sperm whale calls so ridiculously loud? It might be so that they can detect the ocean floor when they dive after prey. With a top speed of 9 miles per hour and bodies that can weigh 40 tons, it takes some time for them to stop. It might also be that they mainly feed on squid, whose soft bodies are harder to detect through sonar.

† It helps that dolphin sonar pulses tend to be shorter, louder, and more focused than those of bats. A bottlenose's click can contain 40,000 times more energy than a big brown's call.

across the length of a football field, in the underwater equivalent of a rock concert. Later studies showed that echolocating dolphins can detect targets from over 750 yards away.

Sound also interacts differently with objects underwater. Generally, sound waves reflect when they encounter a change in density. In the air, they ricochet off solid surfaces. But in water, they'll penetrate flesh (which mostly has a density similar to water's) and bounce off internal structures like bones and air pockets. While bats can only sense the outer shapes and textures of their targets, dolphins can peer *inside* theirs. If a dolphin echolocates on you, it will perceive your lungs and your skeleton. It can likely sense shrapnel in war veterans and fetuses in pregnant women. It can pick out the air-filled swim bladders that allow fish, their main prey, to control their buoyancy.* It can almost certainly tell different species apart based on the shape of those air bladders. And it can tell if a fish has something weird inside it, like a metal hook. In Hawaii, false killer whales often pluck tuna off fishing lines, and "they'll know where the hook is inside that fish," Aude Pacini, who studies these animals, tells me. "They can 'see' things that you and I would never consider unless we had an X-ray machine or an MRI scanner."

This penetrating perception is so unusual that scientists have barely begun to consider its implications. The beaked whales, for example, are odontocetes that look dolphin-esque on the outside—but on the inside, their skulls bear a strange assortment of crests, ridges, and bumps, many of which are only found in males. Pavel Gol'din has suggested that these structures might be the equivalent of deer antlers— showy ornaments that are used to attract mates. Such ornaments would normally protrude from the body in a visible and conspicuous way, but that's unnecessary for animals that are living medical scanners. With "internal antlers," beaked whales could conceivably advertise to mates without needing to disrupt their sleek silhouettes.

This idea is hard to test because beaked whales are so elusive.

---

* Most fish cannot hear very high frequencies, but there are exceptions. The American shad, the Gulf menhaden, and a few other species have evolved ears that can hear dolphin sonar, just as some moths can hear the cries of bats.

They've never been kept in captivity, and, since they can dive for several hours on a single breath, many species are rarely seen. But despite their rarity, they have unexpectedly helped to address one of the biggest mysteries of odontocete sonar: how the animals use it in the wild. They certainly don't care about the distances to steel spheres, or the width of brass cylinders—but what *do* they care about? How do they use their sonar to orient, hunt, or solve problems? Do diving sperm whales echolocate on the ocean floor to avoid literally hitting rock bottom? Do belugas and narwhals scan for distant breathing holes among Arctic ice? When dolphins swim into a school of sardines, do they focus their perception on one fish, or all of them? Have any of them developed specialized strategies akin to CF bats detecting the fluttering wings of insects?

One way to find out is to use an acoustic tag—an underwater microphone on suction cups. When an odontocete surfaces for air, scientists can sidle over in a small boat, lean across with a long pole, and plonk the tag on the animal's flank. When it dives out of view, the device records both its clicks and the returning echoes. It captures a detailed journal of the animal's dive—everything it hears and everything it's *trying* to hear. Since 2003, one team of researchers has deployed acoustic tags on dense-beaked whales near the Canary Islands. These animals are silent when they first start to dive, perhaps to avoid attracting eavesdropping predators like orcas. Once they hit 400 meters, they start to click, and they'll typically find something to eat within minutes. These dark depths are apparently so rich in fish, crustaceans, and squid that the dense-beaked whale can afford to be picky. It might ensonify thousands of creatures but chase just a few dozen, selecting only the best morsels using the fine discriminatory abilities that Au and Nachtigall saw in their captive animals. The whales are so efficient that they only need about four hours of daily foraging to sustain their large bodies.

The dense-beaked whale's foraging style is only possible because underwater sonar has such a long range. A flying bat has less than a second to decide what to do about an insect-sized target that enters its sonar field, but a swimming odontocete has around 10 seconds to make its decision. A bat must always react. A whale can *plan*. In the introduction, I wrote about Malcolm MacIver's hypothesis that when animals moved from the water to the land, the extra range of their vision enabled the evolution of more sophisticated minds, capable of planning. I wonder if the same hypothesis might work in reverse for echolocation.

Underwater sonar not only gives odontocetes a chance to deliberate but also allows them to coordinate. At night, spinner dolphins— a small and especially acrobatic species—capture prey by working together in teams of up to 28 individuals. Kelly Benoit-Bird and Whitlow Au showed that these hunts go through several distinct phases. First, the spinners patrol in a widely spaced line. Then, once they've found a group of fish or squid, they cluster together into a tight row and bulldoze their prey. The victims pile on top of each other, and the spinners encircle them to cut off any escapees. Pairs of dolphins then take turns darting into the circle from opposite ends, picking off the trapped animals. Throughout this sequence, the spinners switch formations seamlessly and simultaneously, and at those transition points they're especially likely to click. Are they shouting commands at each other? Are they echolocating on their teammates to track their positions? Could they be using each other's echoes to extend their own perceptions? Whatever the case, their coordinated, intelligent behavior is made possible by sonar—a sense that works over distances longer than a single dolphin. The pod might be spaced over 40 meters of water, but they're connected by sound and can act as one.

Daniel Kish envies them. "Waterborne sonar is sort of cheating," he tells me. "It gives you enormous advantages, having a medium like that. Air is not conducive to sonar, and yet it still works." And he should know. Kish isn't a bat researcher or a dolphin researcher. He doesn't study animal echolocation.

He echolocates.

WHEN I TRY TO click with my tongue, the sound has a muffled wetness to it, like a stone being thrown into a pond. When Daniel Kish clicks, the sound is sharper, crisper, and *much* louder. It is the sound of someone snapping their fingers, a sound that will make you snap to attention. It's a sound that Kish has been practicing for almost all of his life.

Born in 1966 with an aggressive form of eye cancer, Kish had his right eye removed at 7 months, and his left at 13 months. Shortly after he lost his second eye, he started clicking. At the age of two, he would routinely climb out of his crib and explore his house. One night, he crawled out of his bedroom window, dropped into a flower bed, and toddled around the backyard, clicking as he went. He remembers sensing the acoustically transparent chain-link fence, and the large house on the other side. He remembers climbing the fence, and then others like it, until a neighbor finally called the police, who brought him home. It wasn't till much later that Kish learned what echolocation was, or that he'd been doing it for about as long as he'd been walking.

Now in his 50s, Kish is still clicking and still using the rebounding echoes to perceive the world. I meet him at his house in Long Beach, California, where he lives by himself. Inside, he doesn't need to echolocate; he knows exactly where everything is. But when we go for a walk, the clicks come into play. Kish walks briskly and confidently, using a long cane to sense obstacles at ground level and echolocation to sense everything else. As we head down a residential street, he accurately narrates everything that we pass. He can tell where each house begins and ends. He can locate porches and shrubbery. He knows where cars are parked along the road. An overgrown tree stretches a large branch across the sidewalk, and although my natural inclination is to warn Kish about it, I don't need to. He ducks, effortlessly. "If I wasn't echolocating, I'd have definitely bumped into that," he tells me.

Besides bats and odontocetes, several animals use a simpler form of echolocation. Small mammals might make ultrasonic clicks to find their way around, including various shrews, the solenodons of the Caribbean (which look like shrews), and the tenrecs of Madagascar (which look like hedgehogs). Certain fruit bats, which supposedly don't echolocate, create clicking noises with their wings and can use these to distinguish different textures. The oilbird, a large South American fruit-eater, makes audible clicks, perhaps to navigate the caves in which it roosts. Swiftlets, small insect-eating birds, might click for the same reason. And as Kish and many other people demonstrate, humans can navigate with echoes, too.*

Human echolocation isn't as sophisticated as that of a bat or a dolphin, but as Kish likes to point out, those species have a several-million-year head start. And Kish does have a skill that Zipper the bat and Kina the false killer whale lack—language. He can give words to his experience. This should neatly solve Nagel's philosophical dilemma: We might never know what it's like to be a bat, but Kish can explain what it's like to be Kish. And yet he mostly describes his decidedly non-visual experiences in visual terms, even though he has no memory of what it was like to see. Glass panes and stone walls, which return sharp echoes, are "bright." Foliage and rough stones, which produce coarser echoes, are "dark." When Kish clicks, he gets a series of "flashes," like matches being repeatedly struck in the dark, each one briefly illuminating the space around him. "I live on a planet of seven and a half billion sighted people, so you tend to absorb the way people language their experience," he tells me. And since he doesn't know what it's like to see, and I can't fully appreciate his experience of sonar, there's still a barrier between us that words can't fully bridge. We're both guessing at each other's Umwelt, trying to use a vocabulary we share to describe experiences we don't.

When fictional characters echolocate—think Toph Beifong from *Avatar: The Last Airbender* or Daredevil from Marvel comics† —their abilities are usually portrayed as white concentric lines, spreading over

---

* Griffin predicted that owls might echolocate—and they don't. After the discovery of echolocation in dolphins, some scientists suspected that seals might share the same skill—and they don't. Why don't seals echolocate? One reason might be that they are amphibious. A dolphin is completely tied to the water, but seals and sea lions must venture out onto land, and it is very hard to develop a sonar system that works in both worlds. Instead of sonar, they rely on their eyes, their ears, and the incredible wake-sensing whiskers that we met in Chapter 6. Notably, every species that's known to echolocate is warm-blooded, and none of the countless invertebrates are known to use this ability. Is there some reason for that, or have scientists just not looked hard enough?

† Toph's skill is more like the seismic senses of treehoppers, while Daredevil doesn't have to make sound to use his "radar sense," so neither is true echolocation. Also, Kish and other

a black background and delineating the edges of objects. Some of this is correct in spirit: Kish does get a sense of the three-dimensional space around him. But without the ultrasonic frequencies available to bats, the resolution of his sonar is lower. Edges aren't clear-cut. Objects are defined less by their borders and more by their densities and textures. Those qualities "are like the color of echolocation," Kish tells me. When I think about his sensory world, I imagine a watercolor sculpture popping into awareness with every click. Objects are represented by splotches whose outlines are indistinct, and whose "hues" represent different textures and densities.* A tree, Kish tells me on our walk, sounds like a solid vertical post that is topped by a larger, softer blob. A wooden fence will sound softer than a wrought iron one, and both will sound more solid than a chain-link fence. On his street the crisp sound of the hardwood door sandwiched between the fuzzier sounds of the surrounding bushes tells him when he's back at his house. Occasionally, unexpected combinations of texture confuse him. We pass a car that's parked in an incompletely paved driveway, with concrete beneath its tires but turf beneath its undercarriage. Kish pauses as we pass it, and asks me if someone has parked on their lawn.

For Kish, echolocation is freedom. He walks around the city, rides his bike, and goes on solo hikes. And he's not unusual in that: Since at least 1749, there have been anecdotes about blind people who could walk unassisted through crowded streets, or (in later centuries) cycle around obstacles and skate in busy rinks. Humans had been echolocating for hundreds of years before anyone had even defined echolocation as a concept. The ability was historically described as "facial vision" or an "obstacle sense." As with bats, researchers believed that practitioners were sensing subtle changes in airflow over their skin. The practi-

human echolocators are often described as "real-life Batmen," which is an appropriate comparison since bats echolocate, but also an inappropriate one, since Batman does not.

* In the *Daredevil* series that appeared on Netflix, the character's radar sense is portrayed differently than in the comics. He describes it as a "world on fire," with one character appearing as a red smudge against a cooler backdrop. This, to me, comes a little closer to capturing the textural detail of actual human echolocation.

tioners, meanwhile, were mostly mystified about the nature of their perceptions.*

Take Michael Supa. A psychology student, Supa had been blind since childhood. He would regularly detect distant obstacles in his daily life but couldn't explain how he did it. He suspected that hearing was involved, since he'd often snap his fingers or click his heels to find his way around. In the 1940s, he tested that idea. In a large hall, Supa showed that he and other students—one also blind, and two sighted but blindfolded—could use their hearing to detect a large Masonite screen. This worked best if they wore shoes on a hardwood floor, less well if they wore socks on carpet, and not at all if their ears were plugged. In an even more dramatic demonstration, Supa asked a blind-folded experimenter to carry a microphone and walk toward the screen. Sitting in a nearby soundproofed room and listening through earphones, Supa could work out where the screen was and tell his colleague when to stop.

By coincidence, these experiments were taking place at roughly the same time Griffin and Galambos were working with bats. Supa referenced the bat studies when he published his results in early 1944, and when Griffin coined the term *echolocation* later that year, he was describing the skills of both bats *and blind people,* citing Supa. But while bat sonar became a common part of popular knowledge, human echolocation did not. To this day, Kish will meet echolocation researchers "who have no idea that humans can echolocate," he says. "Human biosonar has been dismissed as too crude to be worthy of study." I suspect that's because blindness still carries so much stigma. To be blind to something is to be oblivious to it. To have a blind spot is to have a zone of ignorance. To lack vision is to lack creativity. These ableist phrases equate lack of sight with lack of awareness. And yet blind people are profoundly aware of their surroundings.†

---

* Kish tells me that it took him a long time to articulate how his clicks were working; he just knew they worked.

† Kish says that most blind people use at least a rudimentary version of echolocation that's enough for them to avoid walls or walk down corridors. He describes this as "monochromatic"—a basic awareness of what's around. Even sighted people can quickly learn

With echolocation, Kish can do things that sighted people cannot, like perceive objects behind him, around corners, or through walls. But some tasks that are easy with vision are very hard through sonar. Large objects in the background will mask the echoes of smaller objects in the foreground. Just as bats struggle to detect insects on leaves, Kish and other echolocators struggle to locate objects on tabletops—a task that, somewhat annoyingly, they're often asked to try. "You're trying to discriminate a Kleenex box, a stapler, or some other piece of rubbish off of this massive target," he says. "It's like reading white text on white." Similarly, if a person is standing right up against a wall, Kish will sometimes miss them entirely if he's clicking from the wrong angle. Surfaces that slope up away from him are easier to detect than those sloping down. Angled objects are easier than curved ones. Harder objects are easier than soft ones. In one memorable test involving a German TV show, Kish realized that his echolocation couldn't distinguish between a champagne bottle and a stuffed toy. The curved and tapered bottle reflected his clicks in too many directions, while the fluffy toy absorbed them. Ultimately, neither reflected enough energy to produce a clear sense of shape or texture, "and so my brain equated the two," Kish says. "I just couldn't tell them apart."

In practice, these challenges aren't actually that challenging because Kish almost never relies on echolocation alone. When moving around his house, he remembers where he placed his stuff. When he's walking around his neighborhood, he remembers the layout of streets. He'll use other senses, including passive hearing and touch. If he's walking down a road, he can hear oncoming vehicles before he can echolocate them. If he's standing on a sidewalk, his sonar can't tell him where the edge of the curb is, but his cane easily can. Years ago, when he was a little

---

to do this. What distinguishes the most proficient echolocators is their ability to make out finer details at greater distances with less effort. Our sense of hearing, like all our senses, is built to extract the signal from the noise—speech over background noises, our names at a cocktail party, a siren across a street. In the process, we downplay ambient sounds, including echoes. "If you're echolocating, you almost have to invert that filter because those ambiances and reverberations—sounds that we would normally dismiss as background—are now actually the elements needing to be discriminated," Kish tells me. For him, signals are embedded in what most other ears would hear as noise. That's why it takes so much practice.

younger and bolder, he and other blind friends would go mountain biking. A sighted friend would lead the way, and the group would follow. They fixed zip ties to the backs of their bikes so that the rattling of plastic against metal would tell them where their fellow cyclists were. They chose bikes with hard suspensions to better feel the terrain. "And then, yeah, a heck of a lot of clicking," Kish says.

In 2000, Kish founded a nonprofit called World Access for the Blind to teach other blind people to echolocate. He and his fellow instructors, who are also blind, have trained thousands of students in dozens of countries. Echolocation is still a niche skill and one that's frowned upon by some parts of the blind community for being socially inappropriate, counter to tradition, or too hard for all but a prodigious few. Kish disagrees. Echolocation could be more common if only more echolocators were allowed to teach. Kish himself was the first fully blind person in the United States to be certified as an orientation and mobility specialist—someone who helps blind people learn to get around. "There is active resistance to blind people teaching other blind people how to be blind," he tells me. "It's a sort of reinforced custodialism." Kish says that many blind children will naturally try to explore through noise. If they're not using their tongues, they might snap their fingers or stomp their feet. But parents often see these behaviors as weird or antisocial, and put a stop to them before they can bloom into a sophisticated sonar sense. Kish's parents never did that. They allowed him to click. They bought him a bicycle. "They regarded my blindness as very much incidental and supported my freedom to move, to discover, to learn how to relate to my environment," he says. That freedom eventually changed the nature of his brain.

Neuroscientist Lore Thaler has worked with Kish since 2009. Using brain scanners, she has shown that when he and other echolocators hear echoes, parts of their visual cortex—the region that normally deals with vision—are highly active. When sighted people hear the same stimuli, those same brain regions lie dormant. This doesn't mean Kish is "seeing" echoes. It's more that he's organizing the information from those echoes to build a spatial map of his surroundings—a task that vision naturally excels at. Without vision, the brain can still construct similar maps by repurposing the so-called visual cortex into an echo-processing cortex.* So Kish can hear where things are relative to him, but he also knows where they are relative to each other. This ability likely explains many of the more impressive things that he does, from hiking to biking. His memory, his cane, and his other senses give him information, but his clicks ground that information in space. "His ability to understand space is fundamentally better than most people who have no vision from an early age," Thaler tells me. And that ability comes from a lifetime of practice and active exploration.

Earlier in this chapter, when talking about dolphins, I wrote that echolocation could just as easily be described as "touching with sound." That's also roughly how Kish thinks about it. "It feels like an extension of my sense of touch," he says. It's purposeful and probing. Like a bat, Kish is forcing the world to reveal itself. In some ways, all senses can be like this. A raptor can look around with its eyes, a snake can flick its tongue to collect scent, a star-nosed mole can press its starry nose upon the walls of its burrow, a rat can whisk with its whiskers, and a fire-seeking beetle can sensitize its infrared detectors by flapping its wings. But an echolocating bat, dolphin, or human is *always* exploring, by default. So far, echolocation is the only sense we've met that works in this permanently active way.

There is another.

---

* One might ask whether the "visual cortex" is accurately named, and whether it's really a "spatial mapping cortex that's usually but not always connected to the eyes."

10.

Living Batteries

Electric Fields

I AM IN ERIC FORTUNE'S LAB IN NEWARK, NEW JERSEY, STARING into an aquarium tank that houses an electric catfish, one of many fish that can generate electricity. Stout and russet brown, it looks like a sweet potato with fins. Fortune has named it Blubby. Its shock, he assures me, is punchy but no worse than licking a battery. "If you want to be electrocuted, you can be," he says. Despite a niggling concern that he does this to haze visiting journalists, I stick my hand in the tank. Blubby doesn't flinch. I quickly do. As the fish's discharge forces my muscles to contract, I reflexively yank my arm out of the tank, splashing water over my notepad. My fingers tingle for an hour afterward. "That's about 90 volts," Fortune says. "I'm glad you had that experience."

Around 350 species of fish can produce their own electricity, and humans have known about their ability since long before anyone knew what electricity was. Around 5,000 years ago, the Egyptians carved depictions of Blubby's ancestors onto tombs. The Greeks and Romans wrote about the torpedo ray's "benumbing" power—a strange force that could kill small fish, run up a spear into the arm of the person fishing, and treat everything from headaches to hemorrhoids.* The true

---

* The Greeks referred to the torpedo ray as *nárkē*, from which the modern word *narcotic* derives. The history of electric fish and their contributions to science is fascinating, and far richer

nature of these discharges only became clearer in the seventeenth and eighteenth centuries, when scientists defined electricity as a physical entity and realized that animals can produce it.

The study of electric fish then became entwined with the study of electricity itself. These animals inspired the design of the first synthetic battery. They fueled the discovery that muscles and nerves in all animals run on minute currents. Indeed, electric fish evolved their unique powers by modifying their own muscles or nerves into special electric organs. These organs consist of cells called electrocytes, stacked together like towers of pancakes flipped on their sides. By controlling the flow of charged particles called ions through an electrocyte, a fish can create a small voltage across it. And by lining these cells up and triggering them together, it can combine the minuscule voltages into substantial ones.

None do this better than electric eels. Their electric organs take up most of their 7-foot-long bodies, and contain around 100 stacks of between 5,000 and 10,000 electrocytes. The most powerful of the three electric eel species can discharge 860 volts—enough to incapacitate a horse.* It uses its brutal powers with sinister finesse. When hunting small fish and invertebrates, it delivers pulses that force the muscles of its prey to twitch, giving away its position. Stronger pulses then cause those same muscles to lock, paralyzing the victim. The electric organ is both remote control and Taser, allowing the eel to commandeer the bodies of other animals from afar.†

Most electric fish are more benign. Their discharges are so faint that

---

than the meager paragraph I've allotted to it. For a fuller account, try *The Shocking History of Electric Fishes* by Stanley Finger and Marco Piccolino.

* This is not apocryphal hyperbole. In 1800, Chayma fishers in South America helped the naturalist Alexander von Humboldt collect electric eels by driving 30 horses and mules into a pool filled with the fish. The eels leapt out of the water, pressed themselves against the horses, and electrocuted them. After the chaos died down, the exhausted fish could be easily scooped up. Two horses died in the process.

† Though electric eels have been known for centuries, much of what we know about them was only discovered recently. Ken Catania, that eclectic aficionado of star-nosed moles, earthworms, and crocodiles, showed that they can remote-control their prey. And a team led by Carlos David de Santana showed that the iconic animal is actually three separate species, one of which packs a much stronger voltage than anyone had previously measured.

they can barely be felt by humans. Known as weakly electric fish, they belong to two main groups—the elephantfishes (mormyroids) of Africa, and the knifefishes (gymnotiforms) of South America. (The electric eel, despite its name, is actually a knifefish—and the only member of the order that produces strong discharges.) Weakly electric fish perplexed nineteenth-century scientists, including Charles Darwin. He correctly theorized that the strong electric organs of electric eels and torpedo rays must have evolved from normal muscle via a weaker, intermediate stage. But weaker electric organs wouldn't have evolved at all if they weren't of some use. And if they are too feeble for offense or defense, what were they for? "It is impossible to conceive by what steps these wondrous organs have been produced," Darwin wrote in 1859, in his landmark opus *The Origin of Species*. "But this is not surprising, for we do not even know of what use they are."

Darwin can rest easy. After 160 years of research, it is clear that the knifefishes and elephantfishes use their electric fields to sense their surroundings, and even to communicate with each other. Electricity is to them what echoes are to bats, smells are to dogs, and light is to humans—the core of their Umwelt.

MALCOLM MACIVER TELLS ME to listen and then dips an electrode into a small tank. The device detects an electric field oscillating 900 times a second. It converts that field into a sound that emerges from a nearby speaker as a haunting soprano note, roughly two octaves above middle C. This is how we hear the tank's silent resident—a black ghost knifefish.*

The black ghost is as long as my hand. Its skin is the color of dark chocolate, and its blade-like body tapers from a broad head toward a pointed tail. A single ribbon-like fin runs along its underside, undulat-

---

* MacIver once created a musical installation consisting of 12 electric fish of different species, each housed in a separate tank. The fish all produced electric fields at different frequencies, and electrodes in their tanks converted those fields into musical tones. Visitors could stand at a mixing board and turn the volume on each tank up or down, conducting the electric orchestra. "I was getting a little tired of people not appreciating electric fish, and I wanted to highlight that these are amazing animals that can give you a sense of wonder," MacIver says.

ing constantly. This fin propels the fish in every possible direction with uncanny agility. At first, the fish hovers in the middle of a cylinder at the bottom of the tank. It darts out and, with equal ease, reverses. It turns upside down. It zips backward, and just before it collides with the back wall of the tank it curves its body and slides up the wall tail-first. "That's how Hans Lissmann worked out what was happening," MacIver tells me.

Hans Lissmann was a Ukrainian-born zoologist who studied with Jakob von Uexküll, the man who coined the Umwelt concept. After surviving two world wars, Lissmann ended up in Britain. During a fateful visit to the London Zoo, he watched an African knifefish deftly avoiding obstacles while reversing around its tank. In a neighboring display, he saw an electric eel perform the same feat, and wondered if both fish were somehow using electricity to sense the objects around them. He soon got the chance to test this idea when a friend gave him a knifefish as a wedding present.*

In 1951, Lissmann used electrodes to confirm that the animal produced a continuous electric field from the organ in its tail. He realized that objects would distort this field if they were either more or less electrically conductive than water. And by sensing those distortions, a knifefish could conceivably detect whatever produced them. Lissmann and his colleague Ken Machin probed the limits of this ability and were astonished. After some training, a knifefish could distinguish between a clay pot that contained an insulating glass rod and an identical pot that was empty. It could even discriminate between different blends of water that varied only in their purity. It clearly had an electric sense unlike anything that humans possess. Lissmann and Machin published their results in 1958, marking the second time in as many decades that a strange new sense had been formally documented. Just 14 years earlier, Donald Griffin had coined the term *echolocation* to describe the sonar of bats. Fittingly, the equally strange ability of electric fish be-

---

* Confusingly, the species that Lissmann studied is called the African knifefish, but is more closely related to the elephantfishes than the actual knifefishes (which are all South American). The black ghost knifefish, you'll be glad to know, is actually a knifefish, certainly blackish, and rather ghostly.

came known as *active electrolocation*. (Why the "active" qualifier? We'll find out later.)

The electric organ in the fish's tail is like a small battery. When it switches on, it creates an electric field that envelops the animal. Current flows through the water from one end of the electric organ to the other. Nearby conductors, like animals (whose cells are essentially bags of salty liquid), increase the flow of that current. Insulators, like rocks, reduce it. These changes affect the voltage on different parts of the fish's skin. The fish can detect these differences using sensory cells called electroreceptors. The black ghost knifefish has 14,000 of these scattered over its body, and it uses them to work out the position, size, shape, and distance of the objects around it. Just as sighted people create images of the world from patterns of light shining onto their retinas, an electric fish creates electric images of its surroundings from patterns of voltage dancing across its skin. Conductors shine brightly upon it. Insulators cast electric shadows.

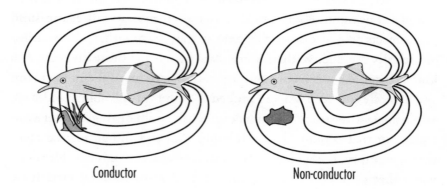

Conductor   Non-conductor

An elephantfish produces its own electric field, which is distorted
by conducting and nonconducting objects in its environment.

Visual terms like *image* and *shadow* are useful when describing such an alien and unfamiliar sense. But electrolocation is very different from vision. The fish that possess this sense care about physical qualities that many other creatures never notice, while ignoring traits that seem (quite literally) blindingly obvious. When Eric Fortune collects electric fish from the wild, he can shine a flashlight upon them to no effect. But once he reaches into the water with a net, "if there is any exposed metal, you can't catch them," he tells me. Conductive metal is more of a beacon to them than actual light.

They are also sensitive to salinity. In the Amazon basin, where many knifefishes live, heavy rainfall regularly flushes ions out of the water. Against this desalinated background, the conductive, salt-filled bodies of other animals pop out to fish that can electrolocate. But in North American tap water, which is relatively laden with ions, those same animals would blend into the background. MacIver's lab is based in Evanston, Illinois; he tells me that if he were to release his captive black ghost knifefish into a local river, it would probably struggle to detect any food, and die. As it is, he adjusts the ion levels in the fish's tank to mimic its natural environment, using a recipe handed down between generations of electric fish researchers.[*] The black ghost is far from the Amazon, but perhaps its water might at least feel like home.[†]

Active electrolocation is similar to echolocation in that it always involves effort. In other senses, activity is optional—noses can sniff, eyes can dart, and hands can stroke, but these organs can also wait for stimuli to come to them. Echolocating bats and electrolocating fish cannot wait. Both must create the stimuli that they then detect. But there is one key difference between these senses: *Electric fields do not travel.* Almost all other senses depend on stimuli that move. Odor molecules, sound waves, surface vibrations, and even light must all make journeys from sources to receivers. But whenever a knifefish fires its electric organ, electric fields immediately materialize around it. It doesn't have to wait, as a bat must, for a returning echo. Electrolocation is an instantaneous sense.

It is also omnidirectional. Since an electric fish's field extends in every direction, so does its awareness. That's why the black ghost

---

[*] The recipe is called Maler's muck, after pioneering researcher Leonard Maler.

[†] Some species of electric fish seem to have evolved electric senses that work best within narrow ranges of salinity. "A most interesting consequence may be that these fish might confront invisible barriers when they attempt to disperse into river systems that differ in water conductivity," wrote Carl Hopkins in 2009.

knifefish that I saw, and the African knifefish that entranced Hans Lissmann, could avoid obstacles behind them. These fish have been filmed swimming backward for meters at a time. "Imagine walking backward for five meters—you just wouldn't," Fortune tells me. "Electric fish can."

Their wraparound sensing comes with a significant catch. Electric fields rapidly weaken the farther they are from their source, so electrolocation only works at very short ranges. The black ghost knifefish eats water fleas that are just a few millimeters long, and can sense these tiny morsels as long as they're within roughly an inch of its body. Beyond that, the water fleas are undetectable, and even larger objects are indistinct. "I think of the fish as being in heavy fog all the time," MacIver tells me. The black ghost can extend the range of its awareness by generating a stronger electric field, and it does this every night when it starts to forage. But extra effort can only go so far. To double the range of its electric sense, it would have to expend eight times more energy—and it already spends a quarter of its total calories on generating its fields.*

These limitations help to explain why many of these fish are so agile. With their awareness mostly confined to a small sensory bubble, they must quickly react to whatever they detect. By the time they sense an obstacle, they have to brake suddenly or swerve quickly. When they detect something edible, they might have already passed it, and must backtrack. MacIver shows me a video in which a black ghost does exactly that. It initially swims past a water flea, but then reverses until its head is close enough to grab the morsel. If it did a U-turn, the flea would have left the range of its electric sense and been lost. Instead, it pulled off a parallel-parking maneuver and kept its prey inside its sensory bubble. This is another example of the intimate connections

---

* Of course, electric fish have other senses at their disposal, including those, like vision, that work over longer ranges. Elephantfish eyes seem to be tuned to large, fast-moving objects at a distance, which might theoretically help them detect predators before they come within range of the electric sense. Then again, many of these animals live in murky water, where long-range vision is impossible. And in the wild, many knifefish live perfectly well with parasitic worms in their eyes—a grisly sign that they can survive without vision.

between an animal's body and its sensory systems. The black ghost knifefish's agility wouldn't be much use without its wraparound electric sense, and its sense would be of little use if the fish weren't so agile.

The omnidirectional nature of electrolocation means that of all the senses we have encountered so far, it is perhaps most similar to touch. "We don't find it weird that we can sense touch all over our body," MacIver says. "Now imagine that's extended out a little bit. That's what the electric sense is like, I think. But who knows what it's like for the fish?" Bruce Carlson, who also studies electric fish, imagines that the fish might feel a kind of pressure on its skin. Conductors and insulators might feel different, just as hot and cold objects or rough and smooth ones do to our fingers. "I can imagine that if I swam past a metal ball, I'd get a small cool sensation like a piece of ice rolling down one side of my body," he tells me. This is speculative, of course, but electric fish really do behave as if they're touching their surroundings from a distance. They'll investigate objects by shimmying back and forth next to them, just like humans running their fingertips over a surface. They'll wrap their bodies around mystery items to get clues about their shape, just as we might grasp unfamiliar things in our hands. Daniel Kish said that he thought about echolocation as a tactile sense: He uses sound to extend his sense of touch and to purposefully probe his world. Electric fish use electric fields in the same way.[*]

If all this sounds eerily familiar, think back to how swimming fish create fields of flowing water around their bodies. Objects around them distort those flow fields, and fish can use their lateral lines to sense those distortions. Sven Dijkgraaf called this "touch at a distance," which is exactly what electrolocating fish are doing, only using electric currents instead of water currents. This resemblance isn't a coincidence. The electric sense *evolved from the lateral line*. Electroreceptors grow from the same embryonic tissues that create the lateral line, and both sense organs contain the same kinds of sensory hair cells (which

---

[*] Angel Caputi has argued that for electric fish, the electric sense likely combines with the lateral line and proprioception—an animal's awareness of its own body—to form a single integrated sense of touch.

are also found in your inner ear).* The electric sense really is a modified form of touch, repurposed for sensing electric fields instead of flowing water.†

But if the lateral line already existed, why evolve electrolocation on top of it? It might be that electric fields are more reliable than almost any other stimulus. They aren't distorted by turbulence, so electric fish can thrive in fast-flowing rivers, where torrents and eddies befuddle the lateral line. Electric fields aren't obscured by darkness or murkiness, so electric fish can stay active in turbid waters and nighttime hours. Electric fields aren't blocked by barriers as light and smells are, so electric fish can sense through solid objects to detect hidden treasures. Indeed, it's very hard to hide from these animals. They are sensitive not only to conductance, which is an object's ability to carry a current, but also to capacitance, which is its ability to store a charge. And in natural environments, "capacitance is a mark of the living," MacIver says. Prey animals can freeze, hide, and hush to fool predators that rely on vision and hearing. But stillness, concealment, and silence don't work against electrolocation. To an electric fish, all that's alive stands out against all that isn't. And other electric fish stand out most of all.

SHORTLY AFTER THE 9/11 ATTACKS, Eric Fortune got a call from the dean of his university. One of Fortune's colleagues was part of the Air Force Reserve and had been called for duty. The man had been scheduled to go to Ecuador on a field trip, and his spot was now open. It was Fortune's if he wanted it—and he did.

Fortune ended up in the middle of the Amazon rainforest, in a lodge overlooking an oxbow lake. One evening, while bats gleaned insects off the lake's surface and huge spiders fished by its edge, Fortune walked onto a pier, connected an electrode to an amplifier, and

---

* It is frankly incredible that the same basic sensor—the hair cell—has been adapted for sensing sound, water flow, and electrical fields.

† This isn't as much of a stretch as it might seem. The neuromasts of the lateral line are already electrically sensitive, but a hundred to a thousand times less so than the electroreceptors of electric fish.

lowered it into the water. Immediately, he heard a familiar sound—the distinctive hum of *Eigenmannia,* the glass knifefish. These are among the most widely studied electric fish, and Fortune had worked with them before. But he had only ever listened to a few dozen in his laboratory. Standing on that pier, he heard what must have been hundreds. He couldn't see any of them, but he knew there was a bustling electric world below his feet. "It was a moment I can still close my eyes and go back to," he tells me. "It was the most amazing experience I've ever had, and I'm so sad I'm not there right now."

For decades, scientists have studied electric fish in laboratories. It's so easy to record, tweak, and play back the discharges of these animals that they have become mainstays of research in neuroscience and animal behavior. Researchers can, for example, play signals that mimic something moving against a fish's body, and watch how it responds. They've been doing this since the 1960s, creating virtual-reality worlds for electric fish. But the animals' actual worlds are still mysterious because they are very hard to study in the wild. Both the African elephantfishes and the South American knifefishes tend to live within dense rainforests, in murky rivers, and amid tangled underwater vegetation. In some places, they are easily the most common fish around. But you'd never be able to tell unless, as Fortune did, you dropped an electrode into the water and converted their electric chorus into audible sounds.

Such electrodes have improved over time, from simple ones you can buy at a local store* to complex grids that can determine the position of every individual within a shoal. These devices have revealed that fish use electric fields not just to sense their environment but also to communicate. They court mates, claim territory, and settle fights with electric signals in the same way other animals might use colors or songs.

Electric fields are great for communication because they don't get distorted in the way that sounds do. They aren't absorbed by obstacles. They don't echo. They don't even travel; instead, they instantly appear

---

* "One of the worst things that happened to our field was when RadioShack went out of business," Fortune tells me.

in the space between the fish that generates them and the one that detects them.* This means that electric fish can encode information
within fine-grained features of their discharges, without any risk that
their messages will be corrupted. In the chapter on hearing, we saw
that zebra finches pay attention to the temporal fine structure of their
songs—that is, how notes change from one thousandth of a second to
the next. Electric fish do the same with their electric discharges, but
they're sensitive to *millionths* of a second. They can cram information
into even simple signals.

Some species of electric fish turn their fields on and off to produce
strong staccato pulses, like drumbeats. The shape of these pulses—
their duration and how their voltage changes over time—contains information about the animal's species, sex, status, and sometimes
identity. Over short timescales, every individual produces the same
pulses again and again: "I like to think of it as the sound of your voice,"
says Bruce Carlson. The timing of the pulses, however, can vary considerably. If the shape of the pulse conveys identity, the *timing* of the
pulses conveys meaning. One rhythm might be as attractive as birdsong; another could be as threatening as a snarl.

Other species, like the black ghost and glass knifefishes, produce
pulses in such quick succession that they blend into a single, continuous wave, like an endless violin note. The frequencies of these waves
differ between species (and sometimes sexes), and the fish control their
timing with unbelievable precision. The neuroscientist Ted Bullock
once showed that the black ghost's electric field usually oscillates once
every 0.001 seconds, with an error of just 0.00000014 seconds. It's one
of the most accurate clocks in the natural world, and was almost too
precise for Bullock's instruments to measure.† By minutely changing
the frequencies of these carefully controlled signals, wave-type electric
fish can send messages. By briefly and sharply increasing the frequency

---

* They aren't much troubled by ambient noise, either, with one exception—distant lightning storms create electromagnetic waves that travel for thousands of miles. These create clicks
that electrodes can certainly detect, and that electric fish possibly could.

† In *Sensory Exotica*, Howard Hughes wrote that if you set a clock by a black ghost knifefish's
electric field, the device would only lose an hour every year.

of their signals, they can produce "chirps," which are "short and abrupt during aggressive encounters but assume a softer and more raspy quality during courtship," Mary Hagedorn and Walter Heiligenberg once wrote.[*]

Such messages don't carry far, but electrocommunication is less limited by range than active electrolocation. When electrolocating, a fish can only extend the range of its sense by producing a stronger electric field, which, at some point, just takes too much energy. But when "listening" to another fish's electric signals, it doesn't need to generate a field at all. It only needs more sensitive electroreceptors, and those are easier to evolve. A fish might only be able to sense prey an inch around its body, but it can detect the signals of other electric fish from a few feet or more away. Its own kind shine out in the perceptual fog that Malcolm MacIver imagined.

Electrocommunication is especially important for one group of elephantfishes called the mormyrins, which have taken the skill to extreme heights. All elephantfishes have a unique type of supersensitive electroreceptor called the knollenorgan, which is not used for electrolocation and is tuned only to the electric signals from other fish. The mormyrins have altered these special receptors even further, retooling them to detect subtle features of electric signals that other elephantfishes can't spot. According to Bruce Carlson, who discovered these differences, it's as if the mormyrins have the electric version of color vision, while other elephantfishes are stuck with monochrome.

Carlson suspects that these changes were triggered by a shift in the fishes' social lives. The elephantfishes with simpler knollenorgans live in large schools and open water. They only need to know if others are around and where they are. The mormyrins, however, are mostly solitary, territorial, and found at the bottom of dark rivers. "If they detect another fish, they want to know exactly where that fish is and *who* it is," Carlson says. "A potential rival? A mate? Another species they don't care about?" This need to know about others has changed their

---

[*] If two *Eigenmannia* meet and their electric discharges are close in frequency, they'll shift their signals away from each other. This is called the jamming avoidance response, and it's one of the most thoroughly studied behaviors in any vertebrate.

electric sense. It has also altered the course of their evolution in at least two important ways.

First, mormyrins are very diverse. Since they can sense tiny variations in each other's electric signals, they can also develop sexual preferences for those minute quirks. These predilections can quickly split a single population of fish into two, each with its own electric penchants and the signals to match. This process is called sexual selection, and it runs at high gear within the mormyrins. These fish have diversified their electric signals 10 times faster than other elephantfishes and given rise to new species at three to five times the rate seen elsewhere. Today, there are at least 175 species of mormyrins, compared to just 30 or so species of other elephantfishes. From precision in their senses came variety in their forms.

Second, mormyrins have evolved more complex brains, perhaps in part to process the information that their souped-up knollenorgans detect. One species, the Ubangi elephantfish (or Peters's elephantnose), has a brain that makes up 3 percent of its body weight and consumes 60 percent of its oxygen.* "With such a brain, you'd imagine that they're building castles or composing symphonies," Nate Sawtell, who studies these fish, tells me. "We haven't seen that, but when you look at them, you can tell they're not goldfish. They're canny and aware."

He illustrates this by taking me to see a group of Ubangi elephantfish that live in his New York lab. Their bodies are long, brown, and flattened; their tails are forked; and their faces end in a mobile appendage called the schnauzenorgan. This is why they're called elephantfish, but the appendages are chins, not noses—think pharaoh, not Pinocchio. While the other electric fish that I've met were placid and ethereal, these seem frenetic and high-strung.† They *explore* the electrode

---

* For comparison, human brains make up around 2 to 2.5 percent of our body weight and soak up 20 percent of our oxygen. One can't directly compare these proportions between animals of different sizes and that are variously warm-blooded and cold-blooded. Also, intelligence can't be measured by brain size alone. Still, the point remains: The elephantfish is unusually big-brained.

† Carlson has shown that one mormyrid—the Cornish jack—hunts in packs. "In the lab, if we were to put two of them in the same tank, at least one of them would die, and quite possibly both," he tells me, because they would fight to the death. But in Lake Malawi (one of the

that Sawtell dips in the water. They probe the sandy floor of their tanks with their schnauzenorgans, which are especially rich in electro-receptors. Sometimes two individuals line up so that the electric or-gans in their tails are right next to the glut of electroreceptors in their partner's head. Then they buzz frantically, like two people shouting a duet into each other's ears. They chase each other. They seem to play.*

As I watch these fish, I wonder what a social life governed by elec-tric signals must be like. These animals can't hide from each other. In setting off their electric discharges to sense their environment, they unavoidably announce their presence and identities to any other elec-tric fish within range. A river full of electric fish must be like a cocktail party where no one *ever* shuts up, even when their mouths are full.

And here's the part that really baffles me: The fish use the same dis-charges for navigation and communication. The electric fields they generate to send signals to other fish are the very ones they use when electrolocating. This simple fact means that when the fish alter their fields to convey messages, they must also change their own ability to navigate or forage. For example, electric fish that lose fights will often briefly pause their pulses as a sign of submission—but this also tempo-rarily shuts down their awareness of their surroundings. For them, communication alters perception. When you listen to a bird's song, you might not be able to hear everything the creature is saying, but you can be sure it's saying *something*. But if you hear one electric fish buzzing near another, is it trying to send a message or work out where the other animal is, or some combination of the two? Does the distinc-tion between navigation and communication even matter to the fish?

"We don't know much about the richer aspects of their lives, the cognitive aspects, what you know about your pet cat or dog," Sawtell tells me. After decades of work, scientists know more about an electric

---

few electric fish habitats with water clear enough to see through), the jacks would come out at night, gang up with the same group of peers, and chase after smaller fish. They often produce bursts of electric pulses when reuniting, which might act as a mutual acknowledgment—a signal that keeps the pack together.

* Bruce Carlson tells me he has seen large elephantfish playing with the tubes in their tanks. "They'll swim into one, lift it up to the surface, and try to balance it there as long as they can until it falls," he says. "Then they go and do it again."

fish's nervous system than that of most other animals. They can draw detailed maps of the neural circuits that drive the electric sense, but the sense still seems otherworldly. And yet, it is surprisingly common.

IN 1678, ITALIAN PHYSICIAN Stefano Lorenzini noticed that the face of the electric torpedo ray was freckled with small pores—thousands of them, each opening into a jelly-filled tube. Other rays had similar pores and tubes, as did their close relatives, the sharks. These structures eventually became known as the ampullae of Lorenzini, but neither he nor any of his contemporaries knew what they were for. Clues slowly trickled in over several centuries. Better microscopes revealed that each tube ended in a bulbous chamber (or ampulla) that was connected to a single nerve—imagine a butternut squash with a string coming from its bottom. They must be sense organs. But what did they sense? In 1960, biologist R. W. Murray finally showed that the ampullae responded to electric fields. A few years later, Sven Dijkgraaf and his student Adrianus Kalmijn confirmed his idea. The duo showed that sharks will reflexively blink when exposed to electric fields, but not if the nerves in their ampullae of Lorenzini have been cut. These squash-shaped structures were electroreceptors.[*]

The answer to this three-century mystery only raised more questions. By the 1960s, Hans Lissmann had already shown that weakly electric fish could navigate by sensing their own electric fields. But sharks and rays couldn't possibly be electrolocating because, aside from the torpedo ray, they didn't produce their own electricity. Why, then, did they have electroreceptors?

It turns out that *all* living things produce electric fields when submerged in water. Remember that animal cells are bags of salty liquid. The concentration of those salts differs from that of the surrounding

---

[*] The jelly inside the ampullae of Lorenzini is extremely conductive. It acts like a cable, transferring the electric field of the surrounding water into the bottom of the ampullae, where it is detected by a layer of sensory cells. The cells compare those properties to those of the animal's own body, and relay that information to the brain. By combining the signals from these cells across thousands of ampullae, the shark can build up a sense of the electric field around it.

water, setting up a voltage across the cells' membranes. When charged ions move across the membranes, they create a current. This is the same basic setup as a battery—charged particles create currents when they move between two salt solutions separated by a barrier. Animal bodies, then, are living batteries, producing bioelectric fields through the mere act of existing. These fields are thousands of times fainter than those produced by even weakly electric fish, and they're damped further by insulating coverings like skin and shells. But at certain exposed body sites like mouths, gills, anuses, and (important for sharks) wounds, they're strong enough to be detected. Sharks and rays can home in on these fields to find their prey, even when their other senses fail them.*

Kalmijn proved as much in 1971. He showed that the small-spotted catshark could always detect tasty flounders, even when the fish were buried in sand, and even if they were first put in an agar chamber that blocked smells and mechanical cues. The sharks only failed when the flounders were covered by an electrically insulating plastic sheet. When Kalmijn removed the flounders altogether, and instead duplicated the fish's weak electric fields using buried electrodes, the sharks "dug tenaciously at the source of the field, responding again and again when coming across the electrodes," he wrote. Wild sharks will also bite at buried electrodes. Some do so from birth.

The shark's electric sense is known as passive electroreception, and it's different from what we've seen so far. Sharks and rays aren't actively producing their own electric fields to locate objects around them, but passively detecting the electric fields of other animals—and mostly prey.† They are exceptionally good at that, perhaps more so than any other group of animals.‡ Stephen Kajiura showed that a small species of

---

* It's sometimes said that sharks and rays detect electric fields produced by moving muscles. But while such movements do produce electric fields, they are typically below the detection range of electroreceptors.

† Not always, though. Some stingrays use electric fields to find buried mates. And some embryonic sharks freeze when they detect the electric fields of passing predators—a feat that reminds me of Karen Warkentin's tree frogs.

‡ Technically, even humans can sense electricity if it is strong enough. We just don't have any sense organs dedicated to the task. Instead, strong currents indiscriminately stimulate our nerves, producing tingling, pain, and twitching. Even then, we can only feel electric fields of

hammerhead can detect an electric field of just one nanovolt—a *billionth* of a volt—across a centimeter of water.* A shark's electric sense only works at short range, however. It can't sense a buried fish (or electrode) from across an ocean, or even from across a pool. It has to be within an arm's length of its target. Over mile distances, a shark sniffs out its food. As it draws near, vision takes over. Nearer still, the lateral line chips in. Its electric sense only enters the fray at the close of the hunt, to pinpoint the exact position of its prey and guide its strike. That's why the ampullae of Lorenzini are usually concentrated around the mouth.†

Passive electroreception is especially useful for finding hidden prey. Animals, after all, can't turn off their natural electric fields.‡ But if a shark can't rely on other senses—say, when its prey are buried, as in Kalmijn's experiment—it has to swim around until its ampullae of Lorenzini are close enough to a target. Some species have expedited that search by enlarging their heads. Instead of conical snouts, hammerhead sharks have broad, flattened heads that look like car spoilers. The undersides of their "hammers" are loaded with ampullae, and the sharks use these as one might use metal detectors, sweeping them over the seafloor in search of buried (edible) riches. They're not more electrically sensitive than other sharks, but their heads allow them to scan a wider area in a given time.

Sawfish can do this, too. These animals are actually rays, but their bodies look more like sharks and their heads look more like medieval

---

0.1 to 1 volt per centimeter. Sharks are around a billion times more sensitive, and the experience for them doesn't suck.

* It's often said that to set up a field that faint with a normal AA battery, you'd have to connect its ends to electrodes dipped into opposite sides of the Atlantic. This metaphor, though evocative, conjures up an entirely inappropriate sense of scale. In reality, sharks are after electric fields considerably fainter than those of a battery, and said fields weaken with distance, which is why a shark's electric sense only works at short range.

† It's also why electric fields trigger the blinking reflex that Dijkgraaf and Kalmijn saw: Sharks protect their eyes in anticipation of a lunge.

‡ They can reduce their fields, though. When cuttlefish see the looming shapes of sharks, they'll stop moving, hold their breath, and cover their gill cavities. Christine Bedore showed that these acts reduce the voltage of their electric fields by almost 90 percent, and halve their risk of being bitten. Cuttlefish don't behave in this way when menaced by crabs, which can't sense electric fields.

weaponry. Their snouts end in long, flattened blades with fiendish teeth protruding from both sides. This "saw" can make up a third of its owner's body length, and it is packed with ampullae, top and bottom. It greatly extends the sawfish's electrical awareness into the space ahead of it—a useful trait in turbid water. "We find them in rivers where we can't even see our boat's propeller," says Barbara Wueringer, who studies these animals. She showed that the saw doubles as both sensor and weapon. When fish swim above the saw, the sawfish slashes at them, using its sideways teeth to impale, stun, and bisect. When the wounded fish fall to the bottom, the sawfish uses the underside of its saw to find and pin them. "Whenever I see them, I think: How is this a thing?" Wueringer tells me.*

THE ABILITY TO DETECT electric fields is not unique to sharks and rays. Among vertebrates, around one in six species shares this sense. The list includes lampreys, sinuous fish with toothy suckers instead of jaws; coelacanths, ancient fish that were thought to have gone extinct until they were found alive in the 1930s; other groups of ancient fish including paddlefishes, which use their long, electroreceptor-rich snouts to find prey much the way sawfishes use their saws; the knifefishes and elephantfishes, which can sense the electric fields of other creatures as well as their own; the thousands of species of catfish, many of which hunt electric fish; and some amphibians like salamanders and the worm-like caecilians.

There are even mammals with electric senses.† At least one species of dolphin—the Guiana dolphin of South America—has this skill, although it's hard to imagine what benefit it could get from just 8 to 14 electroreceptors, when it already has echolocation at its disposal. Simi-

---

* Wueringer founded an organization called Sharks and Rays Australia to save sawfish and their relatives. The same saws that make them masters of electroreception also make dramatic trophies, and get easily caught in nets. The five species are all endangered, three of them critically so.

† There's a paper claiming that star-nosed moles have an electric sense, but Ken Catania, who looked for such a sense when he first studied the animal, tells me he found no evidence for it.

larly, it's unclear how the echidnas—egg-laying mammals from Australia that resemble bulky hedgehogs—use the electroreceptors on the tips of their snouts. Perhaps they sense small insects moving about within moist soil. Their close relative, the platypus, also has over 50,000 electroreceptors on its famous duck-like bill. As it dives for food, it frenetically sweeps the bill from side to side like a hammerhead shark. Underwater, its eyes, ears, and nostrils are closed; it relies on touch and its electric sense alone.

This extensive cabal of electroreceptive critters tells us three important things. First, this is an ancient sense. Electroreceptors first evolved from the lateral line a long time ago, and the common ancestor of all living vertebrates might well have sensed electric fields. You do not have an electric sense, but if you traced your family tree back 600 million years, your ancestors almost certainly did. Second, vertebrates have *lost* the electric sense on at least four occasions during their evolutionary history, which is why hagfish, frogs, reptiles, birds, almost all mammals, and the majority of fish don't have it.* Third, having lost the sense, several vertebrate groups, including the platypuses and echidnas, Guiana dolphins, and electric fish, then regained the ability that their ancestors had but their relatives don't.† The knifefishes and elephant-fishes are special cases. On opposite sides of the world, they independently and successively evolved three kinds of electroreceptors: first, for passively detecting the electric fields of other fish; then, for actively sensing their own self-made fields; and finally, for detecting the fields of other electric fish.‡ The history of these two groups is a spectacular

---

* No one really knows why so many creatures have lost electroreception, especially since the sense is so useful for finding hidden prey underwater. Bruce Carlson tells me he hasn't even heard any good hypotheses. "It's kind of a mystery," he says.

† Each of these groups ended up with its own distinctive electroreceptors (and only those in sharks and rays bear Lorenzini's name). But despite their variety, these organs share the same basic structure. There's almost always a pore that leads from the surface into a jelly-filled chamber, with sensory cells at its base. In many cases, these structures are derived from the lateral line. But the Guiana dolphin evolved its electroreceptors by modifying whisker pits, which are now devoid of hairs and full of conductive jelly.

‡ These events happened at roughly the same times, too. Both groups of fish evolved passive electroreceptors between 110 and 120 million years ago, before evolving active electroreceptors after another 15 to 20 million years.

example of convergent evolution, where two different groups of organisms accidentally show up at life's party in the same outfits.

The convoluted history of the electric sense also hints at something special about electroreceptors. The language of the brain is electricity, and as we've seen, animals have had to evolve weird ways of converting light, sound, odorants, and other stimuli into electrical signals. But electroreceptors are just translating electricity into electricity. They're the only sense organs that detect the very entity that powers our thoughts. Perhaps it's not that difficult to evolve an electroreceptor, and that's why they repeatedly blink in and out of the vertebrate evolutionary tree.

Electroreceptors do seem to have one important limitation: They only work when immersed in a conductive medium. Water certainly counts, and it's no coincidence that almost every electroreceptive animal we've met so far is aquatic.* Air, by contrast, is an insulator, with a resistivity 20 billion times higher than water. For good reason, scientists have long assumed that an electric sense simply couldn't work on land.

And then Daniel Robert did an incredible experiment with bees.

EVERY DAY, AROUND 40,000 thunderstorms crackle around the world. Collectively, they turn Earth's atmosphere into a giant electric circuit. Whenever lightning strikes the ground, electric charge moves upward, so the upper atmosphere ends up with a positive charge and the planet's surface with a negative one. This is the atmospheric potential gradient—a strong electric field that stretches from sky to ground. Even on calm, sunny days, the air carries a voltage of around 100 volts for every meter off the ground. Whenever I write about this, someone inevitably tells me that I must have made a misprint, and I assure you that I have not: There really is a gradient of *at least* 100 volts per meter outside your door.

---

* Echidnas are the exception, but they still probably have to dip their electroreceptors into wet soil.

Life exists within that planetary electric field and is affected by it. Flowers, being full of water, are electrically grounded, and bear the same negative charge as the soil from which they sprout. Bees, meanwhile, build up positive charges as they fly, possibly because electrons are torn from their surface when they collide with dust and other small particles. When positively charged bees arrive at negatively charged flowers, sparks don't fly, but pollen does. Attracted by their opposing charges, pollen grains will leap from a flower onto a bee, even before the insect lands. This phenomenon was described decades ago. But when Daniel Robert read about it, he realized there must be more to the electric world of bees and flowers. (We met Robert in the chapter on hearing because of his work on the *Ormia* fly.)

Although flowers are negatively charged, they grow into the positively charged air. Their very presence greatly strengthens the electric fields around them, and this effect is especially pronounced at points and edges, like leaf tips, petal rims, stigmas, and anthers. Based on its shape and size, every flower is surrounded by its own distinctive electric field. As Robert pondered these fields, "suddenly the question came: Do bees know about this?" he recalls. "And the answer was yes."

In 2013, Robert and his colleagues tested bumblebees with artificial "e-flowers," whose electric fields they could control. They baited a charged e-flower with sweet nectar, and a charge-less one with bitter liquid. The fake blooms were otherwise identical, but the bees quickly learned to tell them apart using electric cues alone. They could even distinguish between e-flowers with differently shaped electric fields— one with voltage spread evenly over its petals, and another with a field shaped like a bullseye.* These patterns are artificial, of course, but real flowers have similar ones. Robert's team visualized these by spraying foxgloves, petunias, and gerberas with charged colored powder. The powder settled around the edges of petals, demarcating patterns that would be otherwise invisible. Alongside the bright colors that we can see (and the ultraviolet ones we cannot), flowers are also surrounded

---

* The bees also learned to more quickly distinguish between flowers of similar colors if electric cues were present as well.

by invisible electric halos. And bumblebees can sense these. "We just jumped to the ceiling when we saw what the bees were telling us," Robert tells me.*

Bumblebees don't have ampullae of Lorenzini. Instead, their electroreceptors are the tiny hairs that make them so endearingly fuzzy. These hairs are sensitive to air currents and trigger nervous signals when they are deflected. But the electric fields around flowers are also strong enough to move them. Bees, though very different from electric fish or sharks, also seem to detect electric fields with an extended sense of touch. And they are almost certainly not the only land-based animals to do so. As we saw in Chapter 6, many insects, spiders, and other arthropods are covered in touch-sensitive hairs. If these hairs can also be deflected by electric fields, and Robert suspects they can, then electric senses might be even more common on land than in the water.

The mere possibility of widespread aerial electroreception has staggering implications. Just think about pollination. Have flowers evolved shapes that produce especially attractive electric patterns? Honeybees tell each other about food sources through their famous waggle dances, and they can sense the electric fields produced by waggling hive-mates; do these fields add another layer of meaning to the dance? A visiting bee temporarily changes a flower's electric field; could this tell other bees that a flower has been recently visited and might be out of nectar? Could flowers lie to bees by quickly resetting their fields to signify that they're open for business? Do flowers feel different in rain and fog, when the atmospheric potential gradient can be 10 times stronger than on clear days? "We don't feel it," Robert says, "but do they?"

What about other arthropods? Atmospheric electric fields are most strongly distorted by the extremities of plants, but many insects that live on plants have spikes, hairs, and strange protrusions. Could these

---

* Although other scientists had already shown that cockroaches, flies, and other insects can react to electric fields, they normally ran experiments with fields that are much stronger than natural ones. That's not very instructive: Even humans can detect extremely strong electric fields, because our hairs stand on end. Robert's study was important because it showed that bumblebees detect electric fields at biologically relevant strengths, that they can use that information to guide actual meaningful behaviors like choosing where to drink, and that they sense subtle cues like the bullseye pattern.

be antennae for detecting the charges of incoming threats? Could they be similar to a luna moth's long tails—decoys that alter the way these insects appear to electrically sensitive predators? The answer to all these questions might well be no, but what if the answer to just a few of them is yes? We've already seen that the insect world must be radically richer than what we imagine, full of subtle air currents, vibrational signals, and other stimuli to which we are oblivious. Now we must add electric fields to the mix. It's telling that just five years after his bumblebee experiments, Robert found evidence of electroreception in another familiar group of arthropods. He found that spiders can sense Earth's electric field and ride it.

Many spiders travel over long distances by "ballooning." They stand on tiptoe, raise their abdomens to the sky, extrude strands of silk, and take off. Carried aloft, they can float for miles. It is commonly said that the silk catches the wind and pulls the spider along, but spiders can still balloon successfully on calm days.* In 2018, Robert's colleague Erica Morley found a better explanation. Spider silk picks up a negative charge as it leaves a spider's body, and is repelled by the negatively charged plants on which they sit. That force, though tiny, is enough to launch the spider into the air. And since the electric fields around plants are strongest at points and edges, spiders can ensure a vigorous takeoff by ballooning from twigs and blades of grass. In her lab, Morley gave them cardboard strips instead of grass. She then exposed them to artificial electric fields that mimicked those outside. When the fields ruffled the sensory hairs on the spiders' legs, the animals adopted the characteristic tiptoe posture and began releasing silk. Even without the slightest breeze around them, some managed to take off. "I could see them levitating," she tells me, "and if I switched the electric field on and off, they would move up and down."

Through these experiments, Morley proved a very old idea. Back in 1828, another scientist had suggested that spiders ride electrostatic

---

* The wind explanation also makes no sense because most spiders don't shoot silk from their abdomens. The silk must be pulled out. Spiders normally do this with their legs, or by first attaching the silk to a surface. But ballooning spiders are doing neither, and it's unlikely that gentle breezes are strong enough to yank out the threads. Electrostatic forces are.

forces, but the idea was dismissed by a rival who favored the wind idea (and wrote a very long-winded letter about it). The rival won, and the electrostatic idea fell out of favor for two centuries. "Wind is tangible," Robert tells me. "People could feel it. Electrics were more elusive."

They still are. Electric senses are still hard to study, though Robert is trying. His work on bumblebees and spiders has changed the way he thinks about the insect and arachnid world. In his own garden, he noticed that young ladybugs will drop to the ground when he brings a charged acrylic rod near them. These larvae have tiny tufts of hair on their backs, and Robert wonders if they can sense the electric charge of an approaching predator. This is what he does now—reimagining his own backyard in a way that reminds me of Rex Cocroft prospecting for new vibrational songs. But while Cocroft can easily convert vibrations into audible noises, Robert can't do the same for electric fields. There are no cameras that can photograph those fields. There's no rich lexicon of words for describing them. *Current, voltage,* and *potential* carry none of the evocative appeal that *sweet, red,* and *soft* possess. "It is very hard to put myself in the skin [of an insect] and imagine what is happening," he tells me. "This is a young science. But I don't think we can ignore it."

The electric sense might stretch his imagination, but at least he knows that some insects have it. He can guess what the others might do with it, and design experiments to test those reactions. And he knows what the likely receptors are, and how they might work. These are all important boons, and they shouldn't be taken for granted. There is another sense whose scholars are not so lucky.

II.

They Know the Way

Magnetic Fields

AFTER SUNSET, WHEN THE HIKERS AND TOURISTS HAVE ALL gone, Eric Warrant and I drive into Kosciuszko National Park, a protected area within Australia's Snowy Mountains. The kangaroos and wombats are out, but we ignore them on our search for much smaller fauna. At 1,600 meters above sea level, we pull over into a quiet spot. I warm my hands on a cup of tea, while Warrant hangs a vertical white sheet between two trees. From below, he illuminates the sheet with a huge light that he calls the Eye of Sauron. From the sheet's corners, he hangs two smaller lamps, whose ultraviolet hues are calibrated to attract insects. We know there are plenty about, because we can hear the echolocation calls of bats hunting above us. Soon, we also hear the loud thud of a large insect hitting the sheet. As it drops to the grass, so does Warrant, who giddily scoops it up. "Yeah, that's definitely a bogong," he tells me, holding up a plastic jar. Inside is an inch-long moth with drab, bark-colored wings. Outwardly, it's not obvious why this creature should so warrant Warrant's delight.

"They really don't look like much," I say.

"No," Warrant says with a chuckle, "and that belies their hidden talents."

Hinting at said talents, the moth in the jar flutters furiously. Many captured insects will sit calmly, but this one seems possessed by a manic energy, some intense compulsion to be elsewhere. "It's flighty as hell," Warrant says. "It's got places to go."

Every spring, billions of bogongs emerge from their pupal stage in the dry plains of southeastern Australia. Anticipating the arrival of the baking summer, they flee toward cooler climes. And somehow, despite never having flown before, let alone migrated, they know which way to go. They fly over 600 miles and arrive at a few select alpine caves. Within these caves, every square meter of wall might be tiled by 17,000 bogongs, their wings overlapping like the scales of a fish. Safe and cool, they ride out the summer in a state of dormancy before making the return trip in autumn. On some nights when Warrant goes out to collect them with the Eye of Sauron he is "literally inundated by thousands of them," he says.

The only other insect known to make such long migrations to such specific destinations is the monarch butterfly of North America. But while monarchs navigate during the day by using the sun as a compass, the bogongs only fly at night. How do they know the right direction? Warrant, who grew up among the Snowy Mountains and has loved the local insects since he was a child, has always wanted to find out. At first, he thought they might be using their sensitive eyes to observe the stars. And while he was right about that, on his first night of observing captive bogongs he noticed that they could still fly in the right direction without being able to see the sky. Warrant realized that they must be able to sense Earth's magnetic field.

Earth's core is a solid iron sphere surrounded by molten iron and nickel. The churning movements of that liquid metal turn the entire planet into a giant bar magnet. Its magnetic field can be depicted in the style of a school textbook: Lines emerge near the south pole, curve around the globe, and reenter near the north pole. This geomagnetic field is always present. It doesn't change across the day or through the seasons. It's not affected by weather or obstacles. Consequently, it is a boon for travelers, who can always use it to establish their bearings. Humans have done so for more than a thousand years, using compasses. Other animals—sea turtles, spiny lobsters, song-

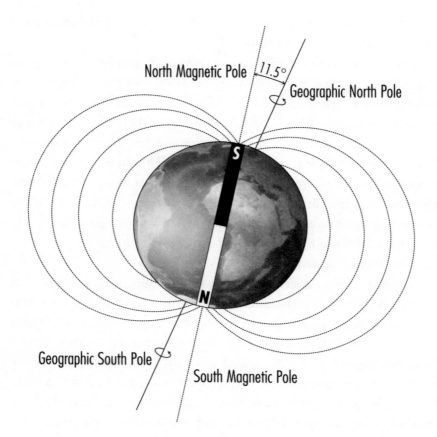

North Magnetic Pole   11.5°   Geographic North Pole

S

N

Geographic South Pole   South Magnetic Pole

birds, and many others—have done so for millions of years, without help.

Their ability, known as magnetoreception, allows them to navigate even when celestial bodies are obscured by clouds or darkness, when large landmarks are wreathed in fog or murk, and when the skies and oceans are devoid of telltale scents. You might think that Warrant, having learned that his precious bogongs are members of the magneto-reception club, would be excited about studying such a fantastical sense. Instead, he jokes, "When I realized that the magnetic sense was important for the bogong, I thought: Oh no."

Magnetoreception research has been polluted by fierce rivalries and confusing errors, and the sense itself is famously difficult both to study and to comprehend. There are open questions about all the senses, but at least with vision, smell, or even electroreception, researchers know roughly how they work and which sense organs are involved. Neither is true for magnetoreception. It remains the sense that we know least about, even though its existence was confirmed decades ago.

THE GEOMAGNETIC FIELD ENVELOPS the entire planet and guides animals over migrations that can span continents. But even the most epic journeys must begin with a few tentative steps, and it was through such steps that magnetoreception was first discovered.

When the time comes for birds to migrate, they become visibly restless. Even in captivity, they'll hop, flit, and flutter. These frantic movements are known as *Zugunruhe*—a German word that means "migration anxiety." The birds know it's time. They long to get going. And as German ornithologist Friedrich Merkel realized in the 1950s, they know the way. Merkel and his students Hans Fromme and Wolfgang Wiltschko captured European robins in the autumn and noticed the birds' migration anxiety wasn't random.* At night, they tended to hop toward the southwest—exactly the direction that, were it not for their cages, would take them to sunny Spain. They did so outdoors when they could see the night sky. But they also kept their bearings in shuttered rooms, where celestial landmarks were hidden from view. This was the same pattern that Warrant would observe in the bogongs half a century later. And in the 1950s, it led Merkel's team to the same epiphany: The birds had to be using another cue, and the geomagnetic field was a possibility.

The idea of magnetoreception wasn't new. In 1859, the zoologist Alexander von Middendorff had suggested that birds, "those sailors of the air," might "possess an inner magnetic feeling." But, for a century, neither he nor anyone else had any data to back up this seemingly outlandish idea. Absent such proof, even Donald Griffin, who was no

---

* The European robin is a completely different bird from the one Americans call a robin. Though both have red breasts, the latter is a medium-sized thrush that was named for the former, which is a small flycatcher.

stranger to unusual animal senses, was skeptical. In 1944, the same year that Griffin coined the word *echolocation,* he wrote that a magnetic sense was "extremely unlikely." The concept was worth taking seriously only because nothing else seemed to adequately explain how migrating birds know where to fly. Magnetoreception was an idea that survived in the absence of better ones. It was a hypothesis in want of evidence.

Merkel and Wiltschko provided that evidence.* First, they recorded the direction of the robins' hops by placing them in an octagonal chamber with a perch on each wall. Every time a bird jumped onto a perch, it triggered a weight-sensitive switch that punched a record of its movements onto paper tape. Later, the team used a simpler but more effective method. They put the birds in a funnel with an ink pad at its base and blotting paper on its sides. Then they counted the inky footprints that the birds left as they tried to jump out.† These were tedious experiments that could only be done in the narrow annual window when birds experience *Zugunruhe.* But they provided clear quantitative evidence that the robins head southwest in the fall. To confirm that the birds rely on a magnetic sense, Wiltschko flipped the magnetic field around them. In the 1960s, he began putting their cages in the middle of Helmholtz coils—pairs of looped wires that can generate artificial magnetic fields between them. When Wiltschko used the coils to rotate the fields around the robins, the birds shifted the direction of their hops accordingly. They had an internal biological compass.

These experiments were still met with skepticism, and for good reason. Earth's magnetic field is extremely weak. It is so faint that the random jiggling movements of an animal's molecules can carry *200 billion times* more energy. No creature should be able to sense such an ab-

---

*   At roughly the same time, other researchers showed that simple animals like flatworms and mud snails can also respond to magnetic fields.

†   This setup is called the Emlen funnel after its creator, Steve Emlen. Cheap and easy to use, it revolutionized the study of bird migration. It is still used today, although the inkpads and blotting paper have been replaced by Tipp-Ex paper or thermal paper that changes color when heated.

surdly weak stimulus. And yet the robins clearly could.* They're not unique, either. Many scientists, including Wiltschko and his wife, Roswitha, have repeated the original robin experiments with several other bird species, including garden warblers and indigo buntings, whitethroats and blackcaps, goldcrests and silvereyes. The "inner magnetic feeling" that Middendorff imagined not only exists but is *common*.

Since Merkel's robins took their pioneering footsteps, scientists have found evidence of magnetoreception throughout the animal kingdom. Yet unlike almost every other sense we've met so far, this one is not used for communication. Animals don't produce magnetic fields, and the only such field that they have evolved to detect is Earth's. They do so mostly to navigate over distances large and small. After a busy night of insect-catching, big brown bats use a compass sense to return to their home roosts. After an early life in the open ocean, baby cardinal fish use a compass sense to swim back to the coral reefs where they were born. Mole-rats use their compass to find their way through their dark underground tunnels. And bogong moths, as Warrant found, use theirs to orient on their trans-Australian flights.

Most of these animals have been tested with some variation of the Wiltschkos' classic experiment: Put the animal in an arena, change the magnetic field around it, and see if it moves in a different direction. That's possible with an animal the size of a robin or a moth. "You can't really do that with a whale," says biophysicist Jesse Granger. "But whales have some of the most insane migrations of any animals on the planet. Some of them almost go from the equator to the poles, and with astounding precision, traveling to the exact same area year after year." It's easy to think that they, too, have a magnetic sense.

To see if they do, Granger looked to the sun itself. The sun periodically throws cosmic tantrums and produces solar storms—streams of radiation and charged particles that affect Earth's magnetic field. Such storms could conceivably mess up the compasses of magnetically

---

* In lab experiments, they can detect a 5-degree shift in the direction of the field they experience. In the wild, where they aren't stressed by confinement, they're probably more precise.

sensitive whales, and if these animals are close to a shoreline, even a small navigational error might send them aground. To test this idea, Granger collated 33 years' worth of records of healthy, uninjured gray whales inexplicably stranding themselves. She compared the timing of these incidents to data on solar activity, wrangled by her astronomer colleague Lucianne Walkowicz. A striking pattern emerged: On days with the most intense solar storms, gray whales were four times more likely to beach themselves.*

This correlation doesn't prove that whales have a compass, but it strongly hints that they do. More than that, it speaks to the awesome nature of magnetoreception. Here is a sense in which the forces produced by a planetary layer of molten metal collide with those unleashed by a tempestuous star, together swaying the mind of a wandering animal and determining whether it finds its way successfully or loses it for good.

FEW MIGRATIONS ARE AS treacherous or as lengthy as those undertaken by sea turtles. Hatching from an egg that was buried in a sandy beach, a baby turtle must run a gauntlet of crab claws and bird beaks on its ungainly crawl toward the ocean. Once in the water, it must flee from the coastal shallows, where it can be easily grabbed from above by seabirds and from below by predatory fish. To find some semblance of safety, it must reach the open ocean as quickly as possible. For a turtle that hatches in Florida, that means swimming due east until it reaches the North Atlantic gyre—a clockwise current that spans the ocean between North America and Europe. The hatchling somehow stays within this loop for 5 to 10 years, hiding out among clumps of floating seaweed and slowly gaining in size. By the time it completes its full (and very slow) lap of the Atlantic and returns to North American waters, it is invulnerable to all but the largest sharks.†

---

* Robins can also be sent off course by artificial magnetic fields that simulate the effects of a solar storm.

† It's estimated that only 1 in every 10,000 hatchlings makes it this far.

By the 1990s, no one had worked out how inexperienced turtles could pull off such grand migrations—a state of ignorance that the late Archie Carr lamented as "an insult to science." At first, Ken Lohmann couldn't understand the fuss. Armed with a newly acquired PhD and the hubris of youth, he thought the answer was obvious: The turtles must use a magnetic compass. It would be a simple matter to build his own magnetic coils and put hatchlings through some version of the then-classic robin experiments. He had signed up for a two-year project, and "my main concern was what I would do for the second year," he tells me. "That was over 30 years ago. The only part I got right was that they have a magnetic sense." He didn't realize that they have two.

As Lohmann suspected, and as he showed in 1991, turtles have a compass. But their other magnetic sense proved to be even more impressive. It hinges on two properties of the geomagnetic field. The first is *inclination*—the angle at which the geomagnetic field lines meet Earth's surface. At the equator, those lines run parallel to the ground; at the magnetic poles, they are perpendicular. The second property is *intensity*—differences in the field's strength. Both inclination and intensity vary around the globe, and most spots in the ocean have a unique combination of the two. Together, they act like coordinates, much like latitude and longitude. They allow the geomagnetic field to act as an oceanic map. And turtles, as Lohmann found, can read that map.

In the mid-1990s, he and his wife, Catherine, took captive loggerhead hatchlings on a magnetic tour of the Atlantic. They exposed the babies to the same inclinations and intensities that they would experience at various places along their long circuit. Amazingly, the turtles knew what to do at each point, and would swim in directions that would keep them within the gyre. This would only be possible if the turtles had both a compass to tell them which way to go and a map to tell them where they were. Only with both senses can they change direction at the appropriate places.[*]

The turtles' abilities are especially impressive because they are in-

---

[*] Over the last 83 million years, the geomagnetic field has reversed 183 times. Magnetic north becomes magnetic south and vice versa. These flips probably occur over thousands of years, so they're unlikely to throw any individual turtle off course. But each turtle species must have

nate. The Lohmanns collected individuals who had only just hatched, kept them in captivity for a single night, and tested them just once. These hatchlings couldn't have learned how to interpret magnetic signals from other turtles. They hadn't even been in the ocean before. Their magnetic maps must be genetically encoded. Lohmann thinks it's unlikely that they're born with a full mental atlas of the entire Atlantic, against which they cross-reference the magnetic readings they feel. Instead, they probably rely on a few instincts that kick in at specific combinations of inclination and intensity that act as magnetic signposts. *When the magnetic field feels like A, head east. When it feels like B, go south.* "The turtle doesn't need to have a conception of where it actually is. It can swim along a pretty elaborate migratory route without needing a lot of information," Lohmann says. "But of course, there's no way of knowing what goes on inside a turtle's head."

Loggerheads that survive their North Atlantic migration end up back in Florida, where they settle down. As they age, they learn, and their magnetic maps get richer. If the Lohmanns captured these older turtles and exposed them to magnetic fields from different parts of the Florida coastline, the animals always swam in directions that would lead them home. They weren't just relying on the sparse signposts that they used as hatchlings. They seemed to know the magnetic topography of their home waters in richer detail.

Magnetic maps have an important limitation: From a given position, a turtle can sense the properties of the magnetic field immediately around it, but it can't tell what the field is like *over there*. To do that, it has to move. And it likely has to move over long distances, because magnetic information isn't especially accurate over short ones. You could use a magnetic sense to travel from Europe to Africa but not to find your bathroom from your bedroom. For this reason, most of the species that convincingly have a map sense use it to travel over long distances.*

experienced many magnetic reversals over their evolutionary history—and their magnetic maps must have adapted accordingly.

* Even apparently simple animals make use of magnetic maps. Caribbean spiny lobsters live in dens within coral reefs but will wander afar in search of food. As long as they don't wind up

Even while blindfolded, Sprouts the harbor seal can track fish by using his whiskers to follow the invisible trails they leave in the water.

Courting peacocks create airflow patterns that they can sense with their crest feathers.

With their sensitive hairs, tiger wandering spiders can detect the air currents created by passing flies.

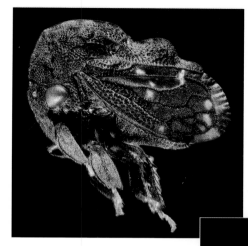

Treehoppers communicate by sending vibrations through the plants on which they stand. When converted into sound, these normally inaudible songs can resemble those of birds, monkeys, or musical instruments.

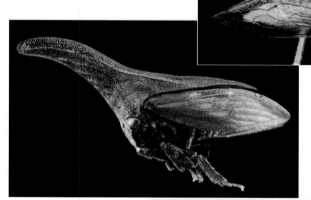

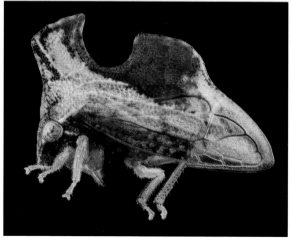

Sand scorpions sense
the footfalls of their prey.
Golden moles detect the
thrums of wind blowing
over termite-rich sand
dunes. Treefrog tadpoles
hatch when they feel
the vibrations of
chewing snakes.

The Nephila *spider's orb web is an extension of its own sensory system and mind—but the small* Argyrodes *spider can hack into it.*

These masters of hearing excel at pinpointing the locations of sound. The barn owl listens for scuttling rodents, while the parasitic Ormia fly listens out for courting crickets.

*The call of the male túngara frog was shaped
by the sensory bias of the female frog's ear.*

*Zebra finches listen for fast details that
humans cannot perceive in their songs.*

Blue whales and Asian elephants can
communicate over long distances with low-pitched
infrasonic calls. In quieter eras, the whales'
calls could carry across entire oceans.

The Philippine tarsier communicates in ultrasonic frequencies that are inaudible to us.

The greater wax moth hears higher frequencies than any other known animal.

Strangely, the blue-throated hummingbird sings ultrasonic notes that it cannot hear.

A big brown bat attacks a luna moth. The colored spectrogram represents echolocation: As the bat closes in, its calls become faster and shorter, providing it with crisper detail.

Dolphins can use their sonar to find buried objects, coordinate formations, and distinguish fish by the shape of their air-filled gas bladders.

The black ghost knife-
fish, the electric eel,
the glass knifefish,
and the Ubangi
elephantfish all produce
their own electric fields,
which they use to
sense the world
around them.

Tiny pores called ampullae of Lorenzini allow sharks
and rays to detect the minute electric fields produced by
their prey. These ampullae are especially common on
the heads of sawfish and hammerhead sharks.

The platypus's bill can sense both pressure and electric fields, which it might combine into a single sense of electrotouch.

Bumblebees can sense the electric fields of flowers.

Bogong moths,
European robins, and
loggerhead turtles can
all navigate over long
distances by sensing
Earth's magnetic
field.

An octopus's arms are partly independent;
they can sense and explore the world without
direction from the central brain.

Some songbirds recognize magnetic signposts on their migration routes, just as turtle hatchlings do. Every winter, thrush nightingales must cross the immense Sahara Desert on their way from Europe to southern Africa. Once they sense the magnetic field of northern Egypt, they react by packing on more fat, in anticipation of the arduous desert crossing ahead. Other migrating songbirds can use these magnetic maps to adjust their bearing if they're blown off course by strong winds—or flown off course by curious scientists. Eurasian reed warblers, for example, normally migrate northeast in the spring, but after Nikita Chernetsov flew some of these birds hundreds of miles to the east, they headed northwest instead.

Many animals, including salmon, turtles, and Manx shearwaters (a kind of seabird), can also imprint on the magnetic signature of their birthplaces, etching it deep within their memory so they can find the same sites as adults. Turtles use these imprints to lay eggs on the same beaches from which they hatched. Their accuracy is uncanny. Green turtles that nest on Ascension Island can find that same tiny nub of land in the middle of the Atlantic after a 1,200-mile journey to and from Brazil. This "natal homing" instinct is so strong that turtles will sometimes swim for hundreds of miles to their beach of birth, even though there's a perfectly good alternative right next to them.* Perhaps that's because good nest sites are hard to find. They must be accessible from the water. The sand grains must be large enough to let oxygen through. The temperature must be exactly right, since turtles develop as males or females depending on how hot or cold their eggs are. "A turtle might say: Well, the one place in the world I know works is the beach

on restaurant plates, they usually end up back in their own dens. Lohmann demonstrated this by capturing lobsters in the Florida Keys, driving them to a marine lab 23 miles away, and doing everything possible to confuse them along the way. He covered their eyes and sealed them in dark plastic containers. He hung swinging magnets above them. He even drove erratically. And yet, once the lobsters were released, they walked off in the exact direction that would take them home.

* The geomagnetic field changes very slightly from year to year, which affects the magnetic signatures of turtle nesting beaches. Lohmann found that in years when the signatures of adjacent beaches converge on each other, nesting turtles crowd together. In years when the signatures diverge, the turtles spread out. These slight variations aren't enough to throw the animals significantly off course, though.

where I developed myself," Lohmann says. And its magnetic map allows it to relocate that sure-bet nursery after years away at sea.

Lohmann is still studying turtles decades after his supposed two-year project.* He has learned so much about their navigational skills, but there is so much left to learn. How quickly can they learn a set of magnetic coordinates? How do their brains represent inclination and intensity? And how do they (or any other animals) even sense magnetic fields at all? I asked Lohmann if he has any thoughts on that last vexing question. He laughs heartily. "Many thoughts and little evidence," he tells me. "I'm optimistic that it'll eventually get solved, but whether it'll be in my lifetime or not is an open question."

IT'S NOT USUALLY DIFFICULT to find sense organs. Their job is to gather stimuli from an animal's surroundings, and, since most stimuli are distorted by the tissues of an animal's body, sense organs are almost always exposed directly to the environment or connected to it by an opening like a pupil or nostril. Such openings can be big clues. Scientists recognized that a rattlesnake's pits, a shark's ampullae of Lorenzini, and a fish's lateral line were sense organs long before working out what they sensed. But researchers who study magnetoreception have no such hints. Magnetic fields can pass unimpeded through biological matter, so the cells that detect them—magnetoreceptors—could be anywhere. They don't need openings like pupils and pits, or focusing structures like lenses and ear flaps. They could be in heads, in toes, or in anything from head to toe. They could be buried deep within flesh. They could even be scattered throughout different body parts and not concentrated into sense organs at all. They could be indistinguishable from the tissues around them. Trying to find them, in the words of Sonke Johnsen, might be like searching for a "needle in a needle stack."

At the time of writing, magnetoreception remains the only sense

---

* When I visit his lab in Raleigh, North Carolina, he is caring for 16 baby loggerheads that were collected in September and will be released the following June. Each year's cohort of turtles is named with a different theme, and this year's is pasta. Lasagne, Ziti, Bowtie, and—my favorite—Turtellini are all swimming around their tanks.

without a known sensor. Magnetoreceptors are "the holy grail of sensory biology," Eric Warrant tells me. "There may even be a Nobel Prize in finding them." Researchers have amassed many important clues about their identity and whereabouts but also several false leads. And without knowing for sure what these receptors are, or even where they are, it is fiendishly difficult to know how they might work. There are, however, three plausible ideas.

The first involves a magnetic iron mineral known as magnetite. In the 1970s, scientists discovered that some bacteria turn themselves into living compass needles by growing chains of magnetite crystals inside their cells. When these microbes are shaken, they tend to swim either north or south. Animals could theoretically build their own magnetite compasses, too. Imagine a magnetite needle that's tethered to a sensory cell. As the animal turns, the needle tugs upon its tether. The cell registers that tension and triggers a nervous signal. In this way, cells could turn an abstract magnetic stimulus into something more tangible— a physical yank. "I think that's an utterly plausible idea," Warrant tells me, "but where those cells are is anybody's guess." Despite several frustrating false leads, nobody's ever found them.*

The second hypothesis for how magnetoreceptors could work involves a phenomenon called electromagnetic induction, which mostly applies to sharks and rays. As a shark swims, it induces weak electric currents in the surrounding water, and the strength of those currents changes depending on the shark's angle relative to the geomagnetic field. By sensing these tiny variations with the electroreceptors we met in the last chapter, the shark could potentially determine its heading.

---

* For decades, many scientists were sure that they had found magnetite-loaded neurons in the beaks of pigeons and other birds. When David Keays started working on magnetoreception, his plan was to study those neurons. But despite using "every method we could think of," he tells me, he couldn't find any. In 2012, Keays published a bombshell study showing that the alleged magnetite neurons that others had found aren't neurons at all. They're macrophages, a type of white blood cell. And while they do contain iron, it isn't in the form of magnetite. That same year, another team developed what looked like a surefire way of identifying magnetite-based receptors. Under a microscope, they saw that some cells in the nose of a trout would spin when placed in a rotating magnetic field. These spinning cells must be magnetic, and they seemed to contain deposits of magnetite. But Keays debunked this finding, too. He showed that the spinning cells just have flecks of iron stuck to their surfaces. They weren't magnetoreceptors. They were just dirty.

Again, no one knows if this actually happens, but it's plausible. A shark's electric sense could double as a magnetic sense.

The induction explanation is often ignored because it's hard to imagine how it would work in animals like birds, which aren't immersed in a conductive fluid like water. But there is a way in which induction might apply to them. The French zoologist Camille Viguier predicted it in 1882, well before magnetoreception had even been confirmed. He noted that a bird's inner ear contains three canals full of conductive fluid. As a bird flies, the geomagnetic field could theoretically induce a detectable voltage in that fluid. Almost 130 years later, David Keays confirmed that he was right. Moreover, he found that these birds have the same protein in their inner ears that sharks use to sense electric fields. "I think induction is a realistic mechanism by which birds can detect magnetic fields, and we're testing it further at the moment," Keays tells me.*

The third explanation for magnetoreception is the most complicated, but also the one that has gained the most momentum. It involves two molecules known as a *radical pair,* whose chemical reaction can be influenced by magnetic fields. To understand this deeply, you must delve into the strange realm of quantum physics. But to understand it well enough, you need only to imagine that the two molecules are dancing. Light triggers the dance, cuing the partners to take hold of each other. Once in this excited state, they can be affected by magnetic fields, which alter the tempo of their dance, and thus its final steps. The partners' final positions offer a record of the magnetic fields that shaped their previous movements. Through their dance, the radical pair transforms a magnetic stimulus that is hard to detect into a chemical stimulus that is simple to assess.†

---

* It's also notable that, in 2011, Le-Qing Wu and David Dickman identified neurons in a pigeon's brain that respond to magnetic fields, and that are connected to the inner ear.

† Here's the longer version. When light hits the two partner molecules, one donates an electron to the other, leaving both of them with an unpaired electron. Molecules with unpaired electrons are called radicals: hence, the radical pair. Electrons have a property called spin, whose exact nature we can leave to the quantum physicists. What matters to biologists is that spin can either be up or down; the radical pair can either have the same spins or opposite ones; they flip between these two states several million times a second; and the magnetic field can

In the 1970s, chemists were mostly studying radical pair reactions in test tubes. But in 1978 the German chemist Klaus Schulten suggested that these obscure reactions might also exist in the cells of birds, and explain their compass-like responses to magnetic fields. He submitted a paper describing this idea to the prestigious journal *Science,* and received a memorable rejection: *A less bold scientist may have designated this idea to the wastepaper basket.* Undeterred, he published the paper anyway. Unfortunately, he placed it in an obscure German journal, and wrote it in a way that was incomprehensible to any biologist who wasn't already well versed in quantum physics—which is to say, almost all of them. In retrospect, however, Schulten was well ahead of his time, and his insight about radical pairs was just the first of several major epiphanies.*

The next occurred when Schulten presented his ideas in a lecture, and an attending Nobel laureate asked: *If radical pair reactions are triggered by light, where is the light in the bird?* Schulten realized that if magnetoreceptors depend on radical pairs, then they can't be found anywhere in an animal's body. Instead, they're probably in the organs best suited to collecting light. A songbird's compass, he suggested, lies in its eyes. This idea lay fallow until 1998, when Schulten read about a new discovery. A group of molecules called *cryptochromes,* which were thought to only exist in animal brains, had also been found in their eyes. "I just fell off my chair," Schulten told me, because he remembered that cryptochromes can form radical pairs with partner molecules called flavins. Here was the missing piece of his theory—a molecule that could take part in the dance he envisioned, and that happened to exist in just the right place.

In 2000, Schulten and his student Thorsten Ritz published a paper arguing that the songbird compass depends on cryptochromes in the eye. It was game-changing. Thanks to Ritz, it was finally comprehensible to biologists. It also gave those biologists something concrete to

---

change the frequency of these flips. So depending on the magnetic field, the two molecules end up in one state or another, which in turn affects how chemically reactive they are.

* I interviewed Klaus Schulten in 2010, well before I had the idea for this book. Schulten died in 2016.

work with—an actual molecule that they could study. Experiment after experiment, researchers confirmed many of Schulten's predictions. The Wiltschkos, for example, discovered that the songbird compass does indeed depend on light—and on blue or green light in particular.*

Henrik Mouritsen, a Danish birdwatcher-turned-biologist who is now one of the leading figures in magnetoreception, also confirmed that light matters.† He placed robins and garden warblers in a moonlit room, and filmed them with infrared cameras. When the birds started showing *Zugunruhe,* Mouritsen looked in their brains to see if any regions were especially active. He found one. Known as cluster N, it sits at the very front of the brain. It's active when and only when migratory songbirds (and not non-migratory ones) are orienting with their compasses at night when they travel (and not during the day when they don't). Cluster N seems to be the magnetic processing center of the bird's brain. And, tellingly, it's also part of the brain's *visual centers*. Cluster N gets information from the retinas, and only buzzes with activity if a bird's eyes are uncovered and if there's some light around.‡ "I think this is one of the strongest pieces of evidence that exists" for the light-dependent radical pair idea, Mouritsen tells me.

These lines of evidence hint at a startling conclusion: Songbirds might be able to *see* Earth's magnetic field, perhaps as a subtle visual cue that overlays their normal field of view. "That's the most likely scenario, but we don't know because we can't ask the birds," Mouritsen says. Perhaps a flying robin always sees a bright spot in the direction of

---

* These wavelengths have exactly the right amount of energy to turn cryptochrome and flavin into a radical pair. Under red light alone, a bird's compass doesn't work.

† Mouritsen has been a birdwatcher since he was 10 years old, and he has seen more than 4,000 species in his lifetime. He originally wanted to be a high school teacher because the vacations were long and would allow him to go on extended birding trips. And even though he ended up as a biology professor, "when I have time to go out, I'm still a birdwatcher," he says. "That's what I miss the most in this coronavirus time: I can't travel anywhere." It is an ironic turn of events for someone who studies animals that migrate over continents.

‡ Robins are nocturnal migrants, so it's odd that they should rely on a light-activated compass. But even at night, there's always a bit of light around. Theoretical calculations suggest that even a moonless and slightly cloudy night has enough light to activate the compass.

north. Perhaps it sees a gradient of shade painted over the landscape. "We have these drawings, and even though they're probably all wrong, they're good for imagining what the birds could be seeing."

While the radical pair idea looks most likely,* all three hypotheses—magnetite, induction, and radical pairs—might be correct. "I think it's very clear that there is more than one mechanism," Keays tells me. And yet, many scientists have formed camps around the different hypotheses, as if one and only one can be correct. As if studying magnetoreception wasn't hard enough, toxic feuds have emerged. One conference infamously descended into farce, as grown adults stood up and screamed at each other. "Everybody wants to be the first to find the magnetoreceptor, which instantly makes people much more competitive and less likely to be nice to one another," Warrant tells me.

It also makes them sloppier.

THROUGHOUT THIS BOOK, WE'VE heard stories of scientists who were mocked or dismissed for ideas about animal senses that ultimately proved to be right. But the opposite phenomenon is just as frequent, if not more so: Discoveries that were thought to be right are later refuted. Such cases are rife in magnetoreception.

A 1997 study claimed that honeybees can detect magnetic fields.†

---

* Even if the radical pair idea is the only correct one, it leaves many unanswered questions. Birds have several cryptochromes, so which is involved in the compass? (One called Cry4 has emerged as a frontrunner; robins mass-produce it during the migratory season, and specifically within the cone cells of their retinas.) How do the final steps of the radical pair dance get converted into a nervous signal? How do the birds separate magnetic information from what they normally see? And why, as Mouritsen showed, can a bird's compass be disrupted by the extremely weak radiofrequency fields of the kind produced by certain electrical equipment or used in AM radio? Such fields carry no useful information and have only become commonplace in the last century of human activity. Birds can't have evolved the ability to sense them—so why are they affected? "We must be missing something major that makes the sensor much more sensitive than we think it should be," says physicist Peter Hore. "This means that our theories aren't fully developed. We haven't come up with the definitive experiment." He and Mouritsen are trying, though. They have started an ambitious project, whose details Hore only tells me about on the strict understanding that I don't write about them.

† This flawed experiment aside, there is good evidence that honeybees can sense magnetic fields.

Two decades later, another group showed that the original team had made such a big statistical error that they might as well have been studying random number generators instead of bees. In 1999, an American team claimed that monarch butterflies have a compass sense; they later retracted their paper when they realized that the insects had actually been orienting to light reflected off their clothes. In 2002, the Wiltschkos published a classic paper claiming that a robin only has a compass in its right eye and cannot orient with its left alone. A decade later, Henrik Mouritsen and his colleagues showed through careful experiments that both eyes contain compasses. In 2015, an American team allegedly found the magnetoreceptor in a nematode worm, while a Chinese group said they had found it in fruit flies. Neither study could be replicated by other researchers, and the fruit fly study was said to conflict with "the basic laws of physics."

To a degree, this is how science is meant to work. Scientists check each other's findings by repeating each other's experiments, building upon what can be replicated and debunking what cannot. But magnetoreception has been plagued by an unusual number of splashy studies that later prove incorrect. Some animals that supposedly have this sense likely don't.* "We have spent a long time chasing other people's assertions, and been very patient," David Keays tells me wearily. "But so many are just fallacious." Science self-corrects, but the science of magnetoreception seems to require more correction than most. Many claims about this sense are wrong. Throughout this book, we have seen

---

* There's even controversy about whether humans have a magnetic sense. In the 1980s, British zoologist Robin Baker drove blindfolded undergraduates on winding routes before asking them to point the way home. They did so more often than expected, but not if they wore magnets on their heads. Baker published his results in *Science,* one of the world's premier journals. But while he repeatedly found the same results, others could not. "We are forced to wonder about the ecological importance of a magnetic sense, the existence of which is so difficult to demonstrate," one duo wrote. More recently, geophysicist Joseph Kirschvink, who was a vocal critic of Baker's experiments, showed that certain brain waves in human volunteers change when an artificial magnetic field is rotated around them. Kirschvink has taken this to mean that humans have magnetoreception. Others aren't convinced. "I guess I can only speak for myself, but I absolutely cannot detect magnetic fields," Keays tells me. "I use an iPhone with a nice compass app, and that's my magnetoreceptor." Kirschvink has argued that humans are unconsciously aware of magnetic stimuli, but he still needs to show that said awareness is useful in some way. Otherwise, so what? Why would it matter for us to have a sense that we are unaware of and that we don't use for anything?

that animal Umwelten are hard to appreciate because they are inherently subjective and because our senses hold us back from making the requisite imaginative leaps. But there's a simpler barrier that stops us from properly understanding other Umwelten: It is easy to study animal senses in misleading ways.

The study of animal behavior is also plagued by human behavior. People tend to see the patterns they want to see. Is that set of scratchy bird footprints really denser in the southwest corner, or are you just interpreting them that way because you expected the bird to head southwest?* Scientists are no less prone to such biases than the average person, but they do have ways of preventing those biases from interfering with their work. For example, they can "blind" a study, withholding key pieces of information until the very last moment, even from themselves. This should be standard practice for all experiments. It isn't.

Making matters worse, the quest to find the elusive magnetoreceptor has become a race. The promise of glory and prizes for the winner has created incentives for fast research and big claims, rather than careful and methodical work. Researchers might run experiments with only a few animals, producing results that might just be flukes. They might tweak their experimental plans on the fly in a bid to find something exciting—a practice known as p-hacking. They might cherry-pick the best data while leaving out findings that don't fit their ideas.

Even if scientists do everything right, they might still flounder because magnetic fields are imperceptible. A researcher who studies vision or hearing would quickly realize if her equipment was accidentally producing bright flashes or loud screeches. But with magnetoreception, "you simply don't notice if you do something stupid," Mouritsen tells me. You might be exposing animals to erratic or unnatural fields, and you'd have no idea unless you were constantly doing checks with the highest-quality equipment. You can dip into the Umwelt of an electric fish or a treehopper using equipment you can buy at a local

---

* To be clear, the early songbird experiments from the 1950s and 1960s, which confirmed that these animals have a magnetic compass, are solid. Those same results have been replicated by many labs, working with many species.

store. But with magnetoreception, "you can't work with cheap equipment," Mouritsen says. "It's very expensive to measure properly."

Magnetic fields are also deeply counterintuitive. As the Insane Clown Posse famously noted, "Fuckin' magnets, how do they work?" Or as Warrant said to me, "I have enough trouble even understanding the stimulus, never mind trying to understand what an animal might perceive from it." Other unusual senses like echolocation and electroreception can at least be compared to more familiar ones like hearing or touch. But I have no idea how to begin thinking about the Umwelt of a loggerhead turtle.

I wonder if this is partly why the radical pair explanation has gained so much traction. Complicated though it is, it brings magnetoreception into the realm of vision, a sense that we can readily appreciate. Similarly, we talk of compasses because they offer a familiar gateway into the abstract world of magnetism. But the compass metaphor can be misleading. Compasses are precise and dependable. They have to point north, and they cannot waver. But Sonke Johnsen, Ken Lohmann, and Eric Warrant suspect that biological compasses are inherently noisy. That is, it might be impossible for them to instantly get a precise, accurate read on Earth's magnetic field because that field is so weak. Animals might have to keep a running average of the signals from their magnetoreceptors over long periods of time. This limitation makes magnetoreception slow, cumbersome, and deeply paradoxical. It detects one of the most pervasive and reliable stimuli on the planet—the geomagnetic field—but does so in an inherently unreliable way. This might explain why so many magnetoreception studies have been hard to replicate. "It could be genuinely difficult to get a consistent result even if you do the same excellent experiment more than once," Warrant tells me.*

Let's say an animal needs five minutes to gather enough information from its erratically swinging compass to determine the right bearing. If experimenters expose it to a magnetic field and record its reactions

---

* Both echolocation and electroreception were discovered at roughly the same time, but neither is plagued by anywhere near the level of irreproducible or controversial results as magnetoreception.

after a minute, the results will be all over the place. I chose these windows of time arbitrarily, but the point is we don't know the right ones. We are used to senses like vision or hearing that offer nigh-instantaneous information. Magnetoreception probably doesn't work like that, but we don't know the timescales over which it does work. Without knowing that, or without even realizing that you need to find out, it's hard to design good experiments. As I wrote in the introduction, a scientist's data are influenced by the questions she asks, which are steered by her imagination, which is delimited by her senses. The boundaries of our own Umwelt corral our ability to understand the Umwelten of others.

The noisy and erratic nature of magnetoreception might also explain why no animal relies on it alone. Instead, they seem to use it as a backup sense in case more reliable ones like vision fail. "If you're a migrating animal, magnetoreception is probably the least important sense, unless you're completely lost," Keays says. In the absence of magnetic cues, bogong moths can still navigate by looking at the pattern of stars in the night sky. Turtle hatchlings ignore magnetic fields when they first enter the water and use the direction of the waves to guide them out to sea.

Animals never use a single sense exclusively. "They use every damn piece of information they can get their hands on," Warrant tells me. "They are multisensory in every possible way."

12.

Every Window at Once

Uniting the Senses

I'M TRYING TO CONVINCE MYSELF THAT I'M NOT REALLY ITCHY. It's just that I'm surrounded by tens of thousands of mosquitoes. They all belong to the same species—*Aedes aegypti,* which is responsible for spreading Zika, dengue, and yellow fever. Mercifully, in the small, sealed room in which I'm standing, the insects are all restrained in white mesh cages. Neuroscientist Krithika Venkataraman pulls one of these cages off a shelf and sets it on the table next to us while she tells me how mosquitoes track their hosts. After talking to her for a few minutes, I look down at the cage and notice, to my horror, that almost all the mosquitoes inside are now perched on the side that's closest to us. They're probing through the mesh with their bloodsucking snouts, which look like a field of black hairs, erupting and subsiding. My itch intensifies. Venkataraman tells me that the mosquitoes are drawn to the carbon dioxide in our breath and the odors emanating from our skin. They can smell us. To demonstrate this, she picks up a different cage, and I exhale along one side of it. Within minutes, almost all the mosquitoes have swarmed onto that side and are probing away.

Leslie Vosshall, who runs the lab where Venkataraman works, spent years trying to protect people from *Aedes aegypti* by befuddling its olfactory abilities. First, she tried to disable a gene called *orco,* which seems to underlie a mosquito's entire sense of smell. This approach worked when Daniel Kronauer, who works down the hall from Vosshall, tried it in clonal raider ants, as we saw earlier. But it failed when Vosshall tried it on mosquitoes: Without *orco,* they ignored human body odor but they were still drawn to carbon dioxide. Switching tactics, Vosshall's team tried to create mutant mosquitoes that could no longer smell carbon dioxide. That didn't work either: The insects could still easily home in on humans. "The results kinda sucked," Vosshall tells me.

Mosquitoes can't be thrown off with any one strategy because they aren't beholden to any one sense. Instead, they use a multitude of cues that interact in complicated ways. They're attracted to the heat of warm-blooded hosts, but only if they first smell carbon dioxide. When Vosshall's student Molly Liu placed the insects in a chamber and slowly heated one of the walls, most of them had buzzed off by the time the surface hit human body temperature. But if Liu sprayed a puff of carbon dioxide into the chamber, the mosquitoes swarmed the hot wall and stayed there. In carbon dioxide's absence, heat is repulsive and a sign of danger. In its presence, heat is attractive and a sign of a meal.* Vosshall still believes she can find a way of cloaking humans from mosquitoes, but she'll need to consider many senses at once—smell, vision, heat, taste, and more. *Aedes aegypti* has "a plan B at every point," she tells me.†

The mosquito's senses have been honed over millennia of evolution. *Aedes aegypti* originally hailed from forests in sub-Saharan Africa, where it drank from a wide variety of animals. But thousands of years ago, one particular lineage got a taste for humans, who had recently started living in densely populated settlements. Drawn to these sites, *Aedes aegypti* transformed into an urban animal that prefers towns over forests,

---

* Our senses undergo similar flips. If you show someone a picture of a dirty sock and let them sniff isovaleric acid, they'll find it disgusting, but pair the same chemical with a picture of fine Époisses cheese and it'll smell delectable.

† After all, that's likely what DEET does. Developed by the U.S. Department of Agriculture in 1944, DEET has a long history of initially protecting troops in tropical countries and then civilians around the world. It works—but no one really knows why. Vosshall originally suspected that it blocks *orco* but now thinks that it bamboozles mosquitoes' sense of smell (and taste) in more complicated ways. If she can duplicate this effect, she hopes to find substances that are more effective than DEET, longer-lasting, and safer for infants.

and a parasite whose Umwelt is tuned to the distinctive cues of our bodies above all else. This mosquito is now among the planet's most effective hunters of humans, and it is extremely picky about anything else. That's why, to feed captive mosquitoes, scientists like Venkataraman often just stick their arms inside their insect cages. "It takes about 10 minutes," she says. "I don't do it regularly, so I still react to the bites, but if you don't scratch, it's fine." It's hard to imagine not scratching.

Imagine, instead, what it might be like to be a mosquito. Flying through a thick soup of tropical air, your antennae slice through plumes of odorants until they catch a whiff of carbon dioxide. Enticed, you turn into the plume, zigzagging when you lose track of it, and surging ahead whenever you pick it up. You spot a dark silhouette and fly over to investigate. You enter into a cloud of lactic acid, ammonia, and sulcatone—molecules released by human skin. Finally, the clincher: an alluring burst of heat. You land, and your feet pick up an explosion of salt, lipids, and other tastes. Your senses, working together, have once again found a human. You find a blood vessel and drink your fill.

In the introduction, we saw that Jakob von Uexküll, pioneer of the Umwelt concept, once compared an animal's body to a house, with many sensory windows overlooking an outlying garden. Over the subsequent 11 chapters, we peered through each of those windows one by one, to better understand what makes each sense unique. Many sensory biologists do the same, looking through a single window over their entire careers. Animals don't. Like the *Aedes* mosquitoes, they combine and cross-reference the information from all of their senses at once. We must follow their lead. To truly appreciate their Umwelten, and to bring our voyage through the senses to a close, we have to consider Uexküll's metaphorical house in its entirety. We must study the architecture of the house itself to see how the form of an animal's entire body defines the nature of its Umwelt. We have to look within the house to see how animals combine the sensory information from the outside world with that from inside their own bodies. And we have to gaze through every window at once, to see how animals use their senses together.

EACH SENSE HAS PROS and cons, and each stimulus is useful in some circumstances and useless in others. That's why animals tap into as many streams of information as their nervous systems can handle, using the strengths of one sense to compensate for the shortcomings of another. No species uses a single sense to the exclusion of every other. Even animals that are paragons of one sensory domain have several at their disposal.

Dogs are masters of smell, but note their large ears. Owls are masters of hearing, but note their large eyes. Jumping spiders depend on their large eyes, but they're also sensitive to surface vibrations traveling through their feet and to airborne sounds that deflect their sensitive, body-wide hairs. Seals use their whiskers to track the hydrodynamic wakes of fish, but their eyes and ears also help them to hunt. The star-nosed mole hunts along its tunnels using touch, but it can also forage underwater, blowing bubbles out of its star and re-inhaling them to detect the odors of prey. Smell dominates the lives of ants, but sounds matter enough that some parasites can inveigle their way into ant nests by mimicking the noises of queens. Smells also guide sharks to their food over mile distances, but vision and the lateral line take over as the distance diminishes, and the electric sense chips in during the final moments of a strike. The Ubangi elephantnose fish creates electric fields to detect small objects close to its body, but its eyes are tuned to spotting large, fast-moving objects like predators that lie beyond the range of its electric sense. Songbirds and bogong moths use Earth's magnetic field to tell them where to go, but they also depend on celestial sights to guide their migrations. Daniel Kish echolocates when he walks around his neighborhood, but he also uses a long cane.

Beyond complementing each other, the senses can also combine. Some people experience synesthesia, where different senses seem to bleed into one another. To some synesthetes, sounds might have textures or colors. To others, words might have tastes. This perceptual blurring is special among humans, but standard to other creatures. The platypus's duck-like bill, for example, contains some receptors that detect electric fields, and others that are sensitive to touch. But in its brain, the neurons that receive signals from the former also receive signals from the latter. The platypus might just have a single sense of electrotouch. As it dives in search of food, it might detect the electric field that a crayfish generates before sensing the flowing water that it stirs up. Some researchers have suggested that the platypus uses the time lag between these signals to judge how far away the crayfish is, just as we can gauge the distance to a storm by the gap between lightning and thunder.

Mosquitoes, meanwhile, have neurons that seem to respond to both temperatures and chemicals. I ask Leslie Vosshall if this means the insects can taste body heat. She shrugs. "The simplest way to sense the world would be to have the senses be separate—to have neurons that taste, or smell, or see," she tells me. "Everything would be very tidy. But the more we look, the more we see that a single cell can do multiple things at the same time." For example, the antennae of ants and other insects are organs of both smell and touch. In an ant's brain, "these probably fuse to produce a single sensation," wrote entomologist William Morton Wheeler in 1910. Imagine if we had delicate noses on our fingertips, he suggested. "If we moved about, touching objects to the right and left along our path, our environment would appear to us to be made up of shaped odors, and we should speak of smells that are spherical, triangular, pointed, etc. Our mental processes would be largely determined by a world of chemical configurations, as they are now by a world of visual (i.e., color) shapes."

Even when the senses don't fuse, they can converge. As we saw in Chapter 9, a dolphin can visually recognize a hidden object that it had previously scanned using echolocation, using one of its senses to build mental representations that are accessible to the others. This feat is called cross-modal object recognition, and it's not limited to big-brained species like dolphins and humans. Electric fish that learn to visually distinguish between crosses and spheres can also tell them apart with their electric sense (and vice versa). Even bumblebees can tell objects apart using touch after learning the visual differences between them.

Some senses also look inward, informing animals about the state of their bodies. There's proprioception, the awareness of the body's position and movement. There's equilibrioception, the sense of balance.* These internal senses are seldom discussed. Aristotle left them out of his five-sense classification, and I have largely ignored them on this journey through nature's Umwelten. But that's not because they are unimportant. It's because they're so important that we take them for granted. We can get by without vision or hearing, but internal senses are non-negotiable. In telling animals about themselves, they help them to make sense of everything else. And they're especially important because animal bodies do something that Uexküll's metaphorical houses do not.

They move.

WHEN ANIMALS MOVE, their sense organs provide two kinds of information. There's *exafference,* signals produced by stuff happening in the world. There's also *reafference,* signals produced by an animal's own actions. I still struggle to remember the difference between these, and if you share that problem, you can think of them as other-produced and self-produced. From my desk, I can see the branches of a tree rustling in the wind. That's exafference—other-produced. But to see those branches, I had to look to my left—a sudden, jarring movement that sent patterns of light sweeping across my retinas. That's reafference—self-produced. Every animal, for each of its senses, has to distinguish between these two kinds of signals. But here's the catch: These signals *are the same* from the point of view of the sense organs.

Consider a simple earthworm. When it burrows through the soil, the touch receptors in its head register pressure. But if you prod the

---

* Millions of people live perfectly well without vision, smell, or hearing. But to lose proprioception is far more debilitating. In 1971, a 19-year-old butcher named Ian Waterman came down with an infection, which triggered an autoimmune attack that robbed him of proprioception. Without feedback from his limbs, he could no longer coordinate his movements. He wasn't paralyzed, but he couldn't stand or walk. If he couldn't see his body, he didn't know where it was. Only after 17 months of intense training did Waterman relearn how to move his body using visual control.

worm in the head, the same touch receptors register the same kind of pressure. So how does the worm know if a given sensation comes from its own movement (reafference) or someone else's (exafference)? How does it know if it is touching something, or if it has been touched? Similarly, if a fish's lateral line detects flowing water, is that because something is swimming toward it, or because it is itself swimming? If you see movement, is that because something around you moved or because your eyes did? If an animal can't tell other-produced signals from self-produced ones, its Umwelt would be an unintelligible mess.

This problem is so fundamental that very different creatures have solved it in the same way.* When an animal decides to move, its nervous system issues a motor command—a set of neural signals that tell its muscles what to do. But on its way to the muscles, this command is duplicated. The copy heads to the sensory systems, which use it to simulate the consequences of the intended movement. When the movement actually occurs, the senses have already predicted the self-produced signals that they are about to experience. And by comparing that prediction against reality, they can work out which signals are actually coming from the outside world and react to them appropriately.† All of this happens unconsciously, and while it isn't intuitive, it is central to our experience of the world. The information detected by the senses is always a mix of self-produced (reafference) and other-produced (exafference), and animals can tell the two apart because their nervous systems are constantly simulating the former.

Philosophers and scholars have speculated about this process for centuries. In 1613, the Belgian physicist François d'Aguilon wrote that

---

* Technically, it's a problem shared only by animals that move. If you are completely immobile, you can be pretty sure that any information from your sense organs is produced by changes in the outside world rather than your own actions. But no animals are completely immobile; even sponges, which have no nervous systems and sit anchored onto rocks, can expel waste from their bodies by "sneezing."

† It's frankly astonishing that this works. Look to your left. Your brain just sent a simple signal that told some of the muscles around your eyeball to contract. How did your nervous system then use that signal to predict how the scene around you would change? We know that it did, but the actual computations that occurred are still a mystery. "How do you go from a motor command to a signal that a sensory structure can work with?" Nate Sawtell, who works with electric fish, asks me. "That's the core problem."

"an internal faculty of the soul perceives the movement of the eye." In 1811, German physician Johann Georg Steinbuch wrote about *Bewegideen,* or "motion ideas"—brain signals that control movements and that interact with sensory information. In 1854, another German physician, Hermann von Helmholtz, referred to the *Bewegidee* as *Willensanstrengung,* or "effort of will." As of 1950, the duplicated motor commands have been called *efference copies* or—my favorite of these terms—*corollary discharges.* * There are subtle differences between these terms, but the underlying idea is the same. Whenever an animal moves, it unconsciously creates a mirror version of its own will, which it uses to predict the sensory consequences of its actions. With every action, the senses are forewarned about what to expect and can prepare themselves accordingly.

Scientists have learned a lot about corollary discharges from studying elephantfish, which use them to coordinate their electric senses. As we saw in Chapter 10, they have three different kinds of electroreceptors. One set detects the elephantfish's own electric pulses. A second detects the communicative signals of *other* elephantfish. And a third detects the weaker electric fields produced by potential prey.† The second and third groups can only work if they ignore the fish's own electric pulses, and they do so through corollary discharges. These are created whenever the electric organ fires, and they prep parts of the brain that receive signals from the second and third groups of receptors to ignore the fish's own pulses. In this way, an elephantfish can tell which signals are being passively produced by potential prey, which are being actively produced by other electric fish, and which are being actively produced by itself.

Electric fish are exceptional creatures, but "almost all animals have some mechanism that's more or less like this," Bruce Carlson tells me. Corollary discharges explain why you can't tickle yourself: You auto-

---

* For a full history of these terms and the idea behind them, there's an excellent paper by Otto-Joachim Grüsser.

† For ampullary receptors and knollenorgans, as for most other sense organs, reafference is noise and exafference is the signal. But for tuberous receptors, which detect the fish's own signals, the opposite is true: Reafference is the signal and exafference is the noise.

matically predict the sensations that your writhing fingers would produce, which cancels out the actual sensations that you feel. They're why your view is stable even though your eyes are constantly darting around.* They're how chirping crickets can block out the sounds of their own calls. They're why fish can sense the flows created by other fish without being confused by their own swimming, and why earthworms can crawl ahead without reflexively recoiling.†

These feats are so profound that they don't feel like feats at all. It feels self-evident that we own our bodies, that we exist within the world, and that we can tell the former from the latter. But these are not axiomatic properties. Distinguishing self from other isn't a given; it's a difficult problem that nervous systems have to solve. "This is largely what sentience is," neuroscientist Michael Hendricks tells me. "And perhaps it's *why* sentience is: It's the process of sorting perceptual experiences into self-generated and other-generated."

That sorting process doesn't require consciousness, or any advanced mental abilities. "It isn't some fancy, late-added thing in evolution," Hendricks says. It exists in nervous systems with a few hundred neurons and those with tens of billions. It's a foundational condition of animal existence, which flows from the simplest acts of sensing and moving. Animals cannot make sense of what's around them without first making sense of themselves. And this means that an animal's Umwelt is the product not just of its sense organs but of its entire nervous system acting in concert. If the sense organs acted alone, nothing would make sense. Throughout this book, we have explored the senses

---

* Corollary discharges apply to other senses, too. A brain area that controls the movement of your diaphragm sends signals to the olfactory bulb—the smell center of the brain. That bulb processes signals differently depending on whether you are inhaling or exhaling.

† Some scientists have suggested that schizophrenia is fundamentally a disorder of corollary discharges. People with the condition might experience hallucinations and delusions because they can't distinguish their own inner speech from the voices around them. A failure to sort self from other might also explain some of schizophrenia's stranger symptoms, like the ability to tickle yourself. Might there be schizophrenic elephantfish that can't tell their own discharges from those of other fish? "It's certainly possible," Carlson tells me. "I would expect dramatically disrupted behavior."

as separate parts. But to truly understand them, we need to think about them as part of a unified whole.

IN JUNE 2019, DURING a panel discussion on animal intelligence at the World Science Festival, psychologist Frank Grasso brought a two-spot octopus named Qualia onto the stage. He then offered the animal a black-lidded jar containing a tasty crab. He hoped that she would unscrew the lid and extract the crab—a party trick that many octopuses are capable of, and that's often offered as evidence for their intelligence. Qualia had unscrewed many jars in her time, but Grasso warned the audience that she may instead "decide to have a little pout and hang out in the corner." Sure enough, that's what she did. She's still doing that a month later, when I visit Grasso at his New York lab.

Qualia used to swim to the front of her tank when strangers entered, but in her old age, she hunches in a corner. Ra, another two-spot octopus, has taken her place as the lab's attention hog. She's actively sidling across her tank, suckers pressing against the glass. Two of Grasso's undergraduate students drop a jar with a crab into her tank, and she quickly descends upon it. The web of her arms envelops the lid, her skin darkens in color . . . and then nothing happens. She seems to lose interest and jets off. Later, she extends a single arm and touches the jar but then retracts it. The lid remains unscrewed; the crab, uneaten. "There was a time when both these animals were avidly opening bottles," Grasso tells me. But now they don't bother. They'll readily pounce upon a loose crab, and they can certainly get at the bottled ones. They just don't. Grasso now wonders if the octopuses can even see the bottled crabs at all. "It might be that all the jar opening that we've been seeing is a result of them just being curious about this novel object," he tells me, and "they can't see through the rounded glass to know if there's a crab in there."

To work out why an octopus would unscrew a jar and why they would stop, we need to understand their Umwelt. We can start by exploring their eyes, their suckers, and their other sense organs in turn.

But we must then understand how the octopus's entire nervous system works, how it controls a body of almost unfettered flexibility, and how its brain and body combine to create not just one Umwelt but arguably two.

An octopus's central nervous system contains around 500 million neurons—a total that dwarfs that of all other invertebrates and that's comparable to the number found in small mammals.* But only a third of these neurons are located in the animal's head, within the central brain and the adjacent optic lobes that receive information from the eyes. The remaining 320 million are in the arms. Each arm "has a large and relatively complete nervous system, which seems barely to communicate with the other arms," Robyn Crook once wrote. "An octopus effectively has nine brains that have their own agendas."

Even the 300 suckers on each arm are somewhat independent. Once a sucker makes contact with something, it reshapes itself to create a seal and then sticks by creating suction. Meanwhile, it simultaneously touches and tastes using 10,000 mechanoreceptors and chemoreceptors on its rim. Our tongues perceive flavor and mouthfeel as separate qualities, but given the wiring of the sucker, an octopus likely doesn't. Its sensations of taste and touch "are probably inextricably fused" in a way that resembles synesthesia, Grasso tells me. Depending on the flavors it feels, or the textures it tastes, the sucker might continue sucking or let go. And it can make that decision on its own, since each sucker is served by its own mini-brain—a dedicated cluster of neurons called the sucker ganglion. The suckers' independence is obvious when watching disembodied arms, which are often found stuck to the sides of fish, but will never stick to other arms from the same octopus.

Each sucker ganglion connects to another cluster of neurons in the center of the arm called the brachial ganglion. All the brachial ganglia are then connected in a long row running down the arm: Think of them as a string of fairy lights, and the sucker ganglia as their bulbs.

---

* In humans, the central nervous system includes everything in the brain and spine, while the peripheral nervous system includes the nerves in our limbs, organs, and other body parts. But in the octopus, this distinction breaks down. The nerves in the brachial and sucker ganglia are very much part of the central nervous system, even though they exist in the arms.

The sucker ganglia don't communicate with each other, but the brachial ganglia do.* They coordinate the individual suckers and allow the entire arm to act in an organized way. And they can also accomplish a lot on their own, without involving the central brain. The arm contains all the circuitry it needs to reach out, grab objects, and pull them back in. For example, neurobiologist Binyamin Hochner found that when the arm touches an object, two waves of neural signals travel down its length, one from the contact point and one from the base. Where these waves meet, the arm forms a temporary elbow, bending to draw the object toward the octopus's mouth. "There's so much information and behavior that's stored in the arms," Grasso tells me.†

The central brain *can* control the arms, but it's a relaxed boss. It doesn't like to micromanage but coordinates its team of eight when needed. A single arm can snake its way through an opaque maze, using taste-touch to find the right route with no input from the rest of the animal. But Hochner's colleague Tamar Gutnick has shown that octopuses can also solve problems that stump individual arms. She set up a transparent maze in which the correct path forced the arm out of the water, depriving it of chemical cues. The octopuses could still find that path by guiding their arms with their eyes, but it didn't come naturally to them. It took a while for them to learn how to do it, and one individual out of seven never did.

Letizia Zullo, another member of Hochner's team, found more evidence of the arms' autonomy in the way the central brain is organized. The human brain contains rough maps of the body. Tactile sensations from different body parts, like each finger, are processed by separate clusters of neurons. Similarly, distinct parts of the brain drive specific movements: Stimulate the right spot, and your arm might rise or your hand might reach out. But Zullo found that the octopus has no

---

* Between them, each sucker ganglion and its corresponding brachial ganglion contain around 10,000 neurons. That's roughly as many as in an entire leech or sea slug. A single octopus arm contains roughly as many neurons as a lobster.

† In the 1950s and 1960s, Martin Wells removed large parts of the brain from some octopuses and showed that these "decerebrate" animals could still use their suckers to manipulate objects, open clamshells, and feed.

such maps. Whenever she stimulated a part of the brain that made one arm extend, other arms would stretch out, too. Would an octopus be aware if the twentieth sucker on its first arm touches a crab, just as I know when my right index finger has just pressed the Y key? Maybe not! It's possible that the animal simply knows that arm number one has found food, while delegating the specifics to the arm itself. Does an octopus even know where its limbs are, just as I can visualize my body without looking at it? Again, maybe not! The arms certainly contain proprioceptors, which help them to coordinate their movements, but that coordination might be entirely local. Martin Wells, a late pioneer of octopus research, was convinced that these animals don't really have a sense of where their limbs are, or an internal image of their shape.

Perhaps that's just as well. Controlling a human body is relatively simple for a human brain because our bones and joints constrain our movements. There are only so many ways in which, for example, you can pick up a mug. But as philosopher Peter Godfrey-Smith wrote in *Other Minds,* an octopus has "a body of pure possibility." Aside from its hard beak, it is soft, malleable, and free to contort. Its skin can change color and texture at a whim. Its arms can extend, contract, bend, and rotate anywhere along their lengths, and have practically infinite ways of performing even simple movements. How could a brain, even a large one, keep track of such boundless options? The question turns out to be irrelevant. The brain doesn't have to. It can mostly let the arms sort themselves out, while imposing the occasional guiding nudge.*

The octopus, then, arguably has *two* distinct Umwelten. The arms live in a world of taste and touch. The head is dominated by vision. There's undoubtedly some cross-talk between these sides, but Grasso suspects that the information exchanged between the head and the arms is simplified. To extend Uexküll's metaphor of animal bodies as houses with sensory windows, the octopus's body consists of two semidetached houses with utterly different architectural styles and a

---

* Godfrey-Smith marvelously compares the central brain to a conductor and the arms to "jazz players, inclined to improvisation, who will accept only so much direction."

small connecting door between them. Never mind what it's like to be a bat, as Nagel pondered. How can we possibly know what it is like to be an octopus? Its unusual senses challenge our imagination, but so does the way it brings those senses together. Its component threads are unfamiliar, the weave is exotic, and the tapestry that results is utterly alien.

THE ACT OF SENSING creates an illusion that, ironically, makes it harder to appreciate how the senses work. When I looked at Qualia and Ra, I didn't have any conscious awareness of the photoreceptors firing in my eyes. I simply saw. When I touched their tanks, I didn't feel the mechanoreceptors in my fingers reacting to pressure. I simply felt. Our experiences of the world feel disconnected from the very sense organs that produce them, which makes it easy to believe that they are purely mental constructs divorced from physical reality. That's why our stories and myths are so full of characters who can transfer their consciousness into the bodies of animals—the Norse god Odin, for example, or Bran from the once-popular series *Game of Thrones*. Such feats, in which humans literally step into the sensory worlds of other animals, feel like the ultimate form of Umwelt-appreciation. But they also fundamentally misunderstand the concept. An animal's sensory world is the result of solid tissues that detect real stimuli and produce cascades of electrical signals. It is not separate from the body, but of it. You can't simply imagine how a human mind would work in a bat's body or an octopus's, because it *wouldn't work*.

When Qualia and Ra began opening crab-filled jars, they looked like they were deliberately solving a problem in pursuit of a goal. But were their central brains even involved, or were their arms simply exploring new objects on their own? If the latter is true, is their behavior any less intelligent than it seemed, or does the octopus's intelligence manifest through the autonomous curiosity of its limbs? (Can an octopus's arms be curious?) When Qualia and Ra stopped opening those jars, were *they* getting bored or were *their arms*? (Can an octopus's arms get bored?) Was there some conflict between their dual Umwelten—

between what their eyes were seeing and what their arms were feel-tasting?

These questions are extraordinarily hard to answer, but they become impossibly hard if we look at each part of the octopus separately. The workings of its suckers or its eyes can't tell us what the whole animal perceives. The movements of its body can be easily misinterpreted without knowing the structure of its nervous system. This is why Nagel's challenge about imagining another creature's conscious experience is so vexing: To stand any chance of knowing what it is like to be another animal, we need to know almost everything about that animal. We need to know about all of its senses, its nervous system and the rest of its body, its needs and its environment, its evolutionary past and its ecological present. We should approach this work humbly, recognizing how easily our intuitions can lead us astray. We should move forward hopefully, knowing that even a partially successful attempt will reveal wonders that were previously hidden to us. And we should act quickly, knowing that our time is running out.

13.

Save the Quiet, Preserve the Dark

Threatened Sensescapes

WITHIN THE 310,000 ACRES OF WYOMING'S GRAND TETON National Park, one of the largest parking lots is in the village of Colter Bay. Beyond its far edge, nestled among some trees, is a foul-smelling sewage-processing building that Jesse Barber calls the Shiterator. Beneath its metal awning, sitting quietly within a crevice and illuminated by Barber's flashlight, there is a little brown bat. And on the bat's back, there is a white device the size of a rice grain. "That's the radio tag," Barber tells me. He'd previously affixed it to the bat so that he could track its movements. He has returned tonight to tag a few more.

From inside the Shiterator, I can hear the chirps of other roosting bats. As the sun sets, they start to emerge. Navigating more through memory than echolocation, they fail to notice the large mist net that Barber has strung between two trees. A few become entangled. Barber frees them, and his students Hunter Cole and Abby Krahling carefully examine each one to check that they're healthy and heavy enough to carry a tag. One individual opens its mouth, filling the air with a stream of sonar pulses that I can't hear. Cole daubs a spot of surgical cement between its shoulder blades. He attaches the tiny tag, and waits for the cement to dry. "It's a little bit of an art project, the tagging of a bat," Barber tells me. After a few minutes, Cole places the bat on the trunk of the nearest tree. It crawls upward and takes off, carrying $175 worth of radio equipment into the woods.

As the hours wear on, the darkness intensifies. The echolocating bats don't mind. Neither does the sharp-eared owl that flies overhead, nor the carbon-dioxide-tracking mosquitoes that bite me through my shirt. But Barber and his students can only continue their work using their headlamps, whose beams have attracted clouds of insects. Ironically, that's why Barber is here. He's one of a growing number of sensory biologists who fear that humans are polluting the world with too much light, to the detriment of other species. Even here, in the middle of a national park, light intrudes upon the darkness. It spews forth from the headlamps of passing vehicles, from the fluorescent bulbs of the visitor center, and from the lampposts encircling the parked cars. "The parking lot is lit up like a Walmart because no one thought about the implications for wildlife," Barber tells me.

Through centuries of effort, people have learned much about the sensory worlds of other species. But in a fraction of the time, we have upended those worlds. We now live in the Anthropocene—a geological epoch defined and dominated by the deeds of our species. We have changed the climate and acidified the oceans by releasing titanic amounts of greenhouse gases. We have shuffled wildlife across continents, replacing indigenous species with invasive ones. We have instigated what some scientists have called an era of "biological annihilation," comparable to the five great mass extinction events of prehistory. And amid this already dispiriting ledger of ecological sins, there is one that should be especially easy to appreciate and yet is often ignored—sensory pollution. Instead of stepping into the Umwelten of other animals, we have forced them to live in ours by barraging them with stimuli of our own making. We have filled the night with light, the silence with noise, and the soil and water with unfamiliar molecules. We have distracted animals from what they actually need to sense, drowned out the cues they depend upon, and lured them, like moths to a flame, into sensory traps.

Many flying insects are fatally attracted to streetlights, mistaking them for celestial lights and hovering below them until they succumb to exhaustion. Some bats exploit their confusion, feasting on the disoriented swarms. Other, slow-moving species, like the little brown bats that Barber tagged, stay clear of the light, perhaps because it makes them easier prey for owls. Lights reshape the animal communities around them, drawing some in and pushing others away, with consequences that are hard to predict. Could the light-averse bats do badly because their habitable zones have shrunk and their insect prey have been pulled away? Might the light-attracted bats temporarily benefit but eventually suffer as the local insect populations crash? To find out, Barber convinced the National Park Service to let him try an unusual experiment.

In 2019, he refitted all 32 streetlights in the Colter Bay parking lot with special bulbs that can change color. They can either produce white light, which strongly affects the behavior of insects and bats, or red light, which doesn't seem to.* Every three days, Barber's team flips their color. Funnel-shaped traps hanging below the lamps collect the gathering insects, while radio transponders pick up the signals from the tagged bats. These data should reveal how normal white lights affect the local animals, and whether red lights can help to rewild the night sky.

Cole gives me a little demonstration by flipping the lights to red. At first, the parking lot looks disquietingly infernal, as if we have stepped into a horror movie. But as my eyes adjust, the red hues feel less dramatic and become almost pleasant. It is amazing how much we can still see. The cars and the surrounding foliage are all visible. I look up, and notice that fewer insects seem to be gathered beneath the lamps. I look up even further, and see the stripe of the Milky Way, cutting across the sky. It's an achingly beautiful sight, which I have never before seen in the Northern Hemisphere.

In 2001, when astronomer Pierantonio Cinzano and his colleagues created the first global atlas of light pollution, they calculated that two-thirds of the world's population lived in light-polluted areas,

---

* A team of Dutch scientists led by Kamiel Spoelstra discovered this pattern in 2017. In response, a neighborhood within the town of Nieuwkoop, which sits next to a nature reserve, switched its streetlights to bat-friendly red LEDs.

where the nights were at least 10 percent brighter than natural darkness. Around 40 percent of humankind is permanently bathed in the equivalent of perpetual moonlight, and around 25 percent constantly experiences an artificial twilight that exceeds the full moon. " 'Night' never really comes for them," the researchers wrote. In 2016, when the team updated their atlas, they found that the problem was even worse. By then, around 83 percent of people—and more than 99 percent of Americans and Europeans—were living under light-polluted skies. Every year, the proportion of the planet covered by artificial light gets 2 percent bigger and 2 percent brighter. A luminous fog now smothers a quarter of Earth's surface and is thick enough in many places to blot out the stars. Over a third of humanity, and almost 80 percent of North Americans, can no longer see the Milky Way. "The thought of light traveling billions of years from distant galaxies only to be washed out in the last billionth of a second by the glow from the nearest strip mall depresses me no end," vision scientist Sonke Johnsen once wrote.

At Colter Bay, Cole flips the lights back to white, and I wince. The extra illumination feels harsh and unpleasant. The Milky Way seems fainter now, and consequently, the world feels smaller. Sensory pollution is the pollution of disconnection. It detaches us from the cosmos. It drowns out the stimuli that link animals to their surroundings and to each other. In making the planet brighter and louder, we have also fragmented it. While razing rainforests and bleaching coral reefs, we have also endangered sensory environments. That must now change. We have to save the quiet, and preserve the dark.

EVERY YEAR, ON SEPTEMBER 11, the sky above New York City is pierced by two columns of intense blue light. This annual art installation, known as Tribute in Light, commemorates the terrorist attacks of 2001, with the ascending beams standing in for the fallen Twin Towers. Each is produced by 44 xenon bulbs with 7,000-watt intensities. Their light can be seen from 60 miles away. From closer up, onlookers often notice small flecks, dancing amid the beams like gentle flurries of snow. Those flecks are birds. Thousands of them.

This annual ritual unfortunately occurs within the autumn migratory season, when billions of small songbirds undergo long flights through North American skies. Navigating under cover of darkness, they fly in such large numbers that they show up on radar. And by analyzing radar images, Benjamin van Doren showed that the Tribute in Light, across seven nights of operation, waylaid around 1.1 million birds. The beams reach so high that even at altitudes of several miles, passing birds are drawn into them. Warblers and other small species congregate within the light at densities up to 150 times their normal levels. They circle slowly, as if trapped within an incorporeal cage. They call frequently and intensely. They occasionally crash into nearby buildings.

Migrations are grueling affairs that push small birds to their physiological limit. Even a nightlong detour could prematurely sap their energy reserves to fatal effect. So whenever a thousand birds or more are caught within the Tribute in Light, the bulbs are turned off for 20 minutes to let them regain their bearing. But that's just one source of light among many, and though intense and vertical, it only shines once a year. At other times, light pours out of sports stadia and tourist attractions, oil rigs and office buildings. It pushes back the dark and pulls in migrating birds. In 1886, shortly after Edison commercialized the electric lightbulb, nearly 1,000 birds died after colliding with an electrically illuminated tower in Decatur, Illinois. Over a century later, environmental scientist Travis Longcore and his colleagues calculated that almost 7 million birds a year die in the United States and Canada after flying into communication towers.* The red lights of those towers are meant to warn aircraft pilots, but they also disrupt the orientation of nocturnal avian fliers, which then veer into wires or each other. Many of these deaths could be avoided simply by replacing steady lights with blinking ones.

"We too quickly forget that we don't perceive the world in the

---

* As we've seen, migrating birds use a variety of senses to guide their way. Collisions with communication towers seem to happen when all of their senses are befuddled at once—when bad weather prevents them from seeing visual landmarks, and when red lights disable their compasses.

same way as other species, and consequently, we ignore impacts that we shouldn't," Longcore tells me. Our eyes are among the sharpest in the animal kingdom, but their high resolution comes with the inescapable cost of low sensitivity. Unlike most other mammals, our vision fails us at night, and our culture reflects our diurnal Umwelt. Light has come to symbolize safety, progress, knowledge, hope, and good. Darkness epitomizes danger, stagnation, ignorance, despair, and evil. From campfires to computer screens, we have craved more light, not less.* It is jarring for us to think of light as a pollutant, but it becomes one when it creeps into times and places where it doesn't belong.

Many of the other planetary changes we have wrought have natural counterparts: Modern climate change is unquestionably the result of human influence, but the planet's climate does change naturally over much slower timescales. Light at night, however, is a uniquely anthropogenic force. The daily and seasonal rhythms of bright and dark remained inviolate throughout all of evolutionary time—a 4-billion-year streak that began to falter in the nineteenth century. Astronomers and physicists were among the first to talk about light pollution, which dimmed their view of the stars. Biologists only started seriously paying attention in the 2000s, Longcore tells me.† In part, that's because biologists are themselves diurnal. At night, while they sleep, the dramatic changes that occur around them go unstudied. But "the problem is right in front of you once you open your eyes to look for it," Longcore says.

When sea turtle hatchlings emerge from their nests, they crawl away from the dark shapes of dune vegetation toward the brighter oceanic horizon. But lit roads and beach resorts can steer them in the wrong direction, where they are easily picked off by predators or

---

* Scientific studies on light pollution tend to use the acronym *ALAN* to refer to artificial light at night. Unfortunately, this means that a lot of papers read like they are passive-aggressively shouting at some guy called Alan, who is single-handedly screwing things up for wildlife. "ALAN may affect a diverse array of nocturnally active animals," says one. "The biological impact of even low intensities of ALAN may be marked," claims another.

† There had been earlier accounts of birds crashing into lit buildings and turtle hatchlings heading toward lit cities. But Longcore says that an international conference in 2002 marked a moment when a smattering of concerned researchers started becoming a coherent field.

squashed by vehicles. In Florida alone, artificial lights kill baby turtles in the thousands every year. They've wandered into active baseball games and, more horrifyingly, abandoned beach fires. The caretaker of one property found hundreds of dead hatchlings piled beneath a single mercury-vapor lamp.

Artificial lights can also fatally attract insects and might be contributing to their alarming global declines. A single streetlamp can lure moths from 25 yards away, and a well-lit road might as well be a prison. Many of the insects that gather around streetlamps will likely be eaten or dead from exhaustion by sunrise. Those that zoom toward vehicle headlights probably won't last that long. The consequences of these losses can ripple across ecosystems and into the day. In 2014, as part of an experiment, ecologist Eva Knop installed streetlamps in seven Swiss meadows. After sunset, she then prowled these fields with night-vision goggles, peering into flowers to search for moths and other pollinators. By comparing these sites to others that had been kept dark, Knop showed that the illuminated flowers received 62 percent fewer visits from pollinating insects. One plant produced 13 percent less fruit even though it was also visited by a day shift of bees and butterflies.

It's not just the presence of light that matters but also its nature. Insects with aquatic larvae like mayflies and dragonflies will fruitlessly lay their eggs on wet roads, windows, and car roofs, because these reflect horizontally polarized light in the same way as bodies of water. Flickering lightbulbs can cause headaches and other neurological problems in humans, even though our eyes are usually too slow to detect these changes; what, then, would they do to animals with faster vision, like insects and small birds?

Colors matter, too. Red can disrupt migrating birds but is better for bats and insects.* Yellow doesn't bother insects and turtles but can disrupt salamanders. No wavelength is perfect, Longcore says, but blue and white are worst of all. Blue light disrupts body clocks and strongly attracts insects. It is also easily scattered, increasing the spread of light

---

* The red lights that Barber used in the Grand Tetons shouldn't be a problem because they're not high enough to waylay migrating birds.

pollution. It is, however, cheap and efficient to produce. The new generation of energy-efficient white LEDs contain a lot of blue light, and, if the world switches to them from traditional yellow-orange sodium lights, the amount of global light pollution would increase by two or three times. "We can make better choices by tuning lights with intention," Longcore says. "And we shouldn't use full-spectrum at night. We shouldn't want to give everything the signal that it's constantly daytime."

After talking to Longcore at his office in Los Angeles, I return home on a red-eye flight. As the plane takes off, I peer out the window at the illuminated city. The twinkling grid of lights still stirs the same primordial awe that comes from watching a starry sky or a moonlit sea. Humans equate light with knowledge. We draw lightbulbs to symbolize ideas, we describe intelligent people as bright sparks and luminaries, and we illuminated a path out of the Dark Ages. But as Los Angeles recedes beneath my window, that familiar awe is now tinged with unease. Light pollution is no longer just an urban problem, either. Light travels, metastasizing even into protected places that are otherwise untouched by human influence. The light from Los Angeles reaches Death Valley, one of the largest national parks in the continental United States, 200 miles away. True darkness is increasingly hard to find.

So is true silence.

IT'S A SUNNY APRIL morning in Boulder, Colorado, and I've hiked up to a rocky hillside, about 6,000 feet above sea level. The world feels wider here, not just because of the panoramic view over conifer forests but also because it is blissfully quiet. Away from urban ruckus, quieter sounds are unmasked and become audible over greater distances. On the hillside, a chipmunk is rustling. Grasshoppers snap their wings together as they fly. A woodpecker pounds its beak against a nearby trunk. Wind rushes past. The longer I sit, the more I seem to hear.

Two men puncture the tranquility. I can't see them, but they're somewhere on the trail below, intent on broadcasting their opinions to all of Colorado. Further away, I can hear vehicles zooming along a highway beyond the trees. Denver hums in the distance, an ambient backdrop that I had all but blocked out. I notice a plane flying overhead, engines roaring. "I've been backpacking since the mid-sixties, and in that time, the number of aircraft has increased by a factor of six or seven," says Kurt Fristrup, whom I meet after my hike. "One of my favorite parlor tricks when friends visit is to ask, at the end of the hike, if they heard any aircraft. People will say they remember one or two. And I'll say there were twenty-three jets and two helicopters."

Fristrup works at the Natural Sounds and Night Skies Division of the National Park Service, a group that endeavors to safeguard (among other things) the United States' natural soundscapes. To protect them, the team first had to map them, and, unlike light, sound can't be detected by satellites. Fristrup and his colleagues spent years lugging recording equipment to almost 500 sites around the country, capturing nearly 1.5 million samples of audio. They found that human activity has doubled the background noise levels in 63 percent of protected spaces, and increased them tenfold in 21 percent. In the latter places, "if you could have heard something 100 feet away, now you can only hear it 10 feet away," Rachel Buxton of the NPS tells me. Aircraft and roads are the main culprits, but so are industries like oil and gas extraction, mining, and forestry. Even the most heavily protected areas are under acoustic siege.

In towns and cities, the problem is worse, and not just in the United States. Two-thirds of Europeans are immersed in ambient noise equivalent to perpetual rainfall. Such conditions are difficult for the many animals that communicate through calls and songs. In 2003, Hans Slabbekoorn and Margriet Peet found that noisy neighborhoods in Leiden, Netherlands, compel great tits to sing at higher frequencies, so their notes don't get masked by the city's low-pitched hubbub. A year later, Henrik Brumm found that the nightingales of Berlin, Germany, are forced to belt out their tunes more loudly to be heard over the urban din. These influential studies spurred a wave of research into noise pollution, which showed that urban and industrial noise can also change the timing of a bird chorus, suppress the complexity of their songs, and prevent them from finding mates. Even for city birds, noise hurts.

Noise pollution masks not only the sounds that animals deliberately make but also the "web of unintended sounds that ties communities together," Fristrup tells me. He means the gentle rustles that tell owls where their prey are, or the faint flaps that warn mice about impending doom. "They are the most vulnerable parts of the soundscape to intrusion, and we're cutting them off," Fristrup says. Sound levels are measured in decibels, where a soft whisper is usually 30 decibels, normal conversation is around 60, and a rock concert is about 110. Every extra 3 decibels can halve the range over which natural sounds can be heard. Noise shrinks an animal's perceptual world. And while some species like great tits and nightingales stay and make the best of it, others just leave.

In 2012, Jesse Barber, Heidi Ware, and Christopher McClure built a phantom road. On a ridge in Idaho that acts as a stopover for migrating birds, the team set up a half-mile corridor of speakers and played looped recordings of passing cars. At the sound of these disembodied noises, a third of the usual birds stayed away. Many of those that stayed paid a price for persisting. With tires and horns drowning out the sounds of predators, the birds spent more time looking for danger and less time looking for food. They put on less weight, and were weaker as they continued their arduous migrations. The phantom road experiment was pivotal in showing that wildlife could be deterred by noise and noise alone, detached from the sight of vehicles or the stench of exhaust. Hundreds of studies have come to similar conclusions.* In noisy conditions, prairie dogs spend more time underground. Owls flub their attacks. Parasitic *Ormia* flies struggle to find their cricket hosts. Sage grouse abandon their breeding sites (and those that stay are more stressed).

Sounds can travel over long distances, at all times of day, and through solid obstacles. These qualities make them excellent stimuli for animals but also pollutants par excellence. The concept of pollution calls forth images of chemicals billowing from smokestacks, scum-

---

* In one experiment, ladybird beetles ate fewer aphids when exposed to either urban sounds or the music of AC/DC, disproving the band's hypothesis that "rock and roll ain't noise pollution."

covered rivers, and other visible signs of degradation. But noise can degrade habitats that look otherwise idyllic, and make otherwise livable places unlivable. It can act as an invisible bulldozer that pushes animals out of their normal ranges.* And where will they go? More than 83 percent of the continental United States lies within a kilometer of a road.

Even the seas can't offer silence. Although Jacques Cousteau once described the ocean as a silent world, it is anything but. It naturally teems with the sounds of breaking waves and blowing winds, bubbling hydrothermal vents and calving icebergs, all of which carry further and travel faster underwater than in air. Marine animals are noisy, too. Whales sing, toadfish hum, cod grunt, and bearded seals trill. Thousands of snapping shrimps, which stun passing fish with the shockwaves produced by their large claws, fill coral reefs with what sounds like sizzling bacon, or Rice Krispies popping in milk. Some of this soundscape has been muted as humans have netted, hooked, and harpooned the oceans' residents. Other natural noises have been drowned out by those that we added: the scrapes of nets that trawl the seafloor; the staccato beats of seismic charges used to scout for oil and gas; the pings of military sonar; and, as a ubiquitous backing track for all this hubbub, the sound of ships.†

"Think about where your shoes come from," marine mammal expert John Hildebrand says as we talk in his office. I look; unsurprisingly, it's China. Some tanker carried those shoes across the Pacific, belching out a wake of sound that radiated for miles. Between World

---

* In the summer of 2017, ecologist Justin Suraci did a version of Barber's experiment by playing the sound of human speech through speakers set up in the Santa Cruz Mountains. Whether it was Suraci reading poetry or Rush Limbaugh spewing bile, mountain lions, bobcats, and other predators moved away when they heard the voices. This isn't an issue of noise pollution in the classic sense, though. It's more that humans are terrifying superpredators, whose very voices are enough to unnerve other hunters.

† Beaked whales have repeatedly stranded en masse after exposure to naval sonar, prompting waves of research and litigation. *War of the Whales* by Joshua Horwitz offers a masterly account of the events that connected naval sonar to whale strandings, and the legal battles that ensued. "Indisputably, you can use sonar and get a beaked whale to strand," John Hildebrand tells me. "Why they do that is still a mystery." It's unclear if the sound physically injures them or causes them to swim erratically and get the bends. Either way, sonar clearly disturbs them.

War II and 2008, the global shipping fleet more than tripled, and began moving 10 times more cargo at higher speeds. Together, they raised the levels of low-frequency noise in the oceans by 32 times—a 15-decibel increase over levels that Hildebrand suspects were already around 15 decibels louder than in primordial pre-propeller seas. Since giant whales can live for a century or more, there are likely individuals alive today who have personally witnessed this growing underwater racket and who now only hear over a tenth of their former range. As ships pass in the night, humpback whales stop singing, orcas stop foraging, and right whales become stressed. Crabs stop feeding, cuttlefish change colors, damselfish are more easily caught. "If I said that I'm going to increase the noise level in your office by 30 decibels, OSHA would come in and say you'd need to wear earplugs," Hildebrand tells me. "We're conducting an experiment on marine animals by exposing them to these high levels of noise, and it's not an experiment we'd allow to be conducted on ourselves."

THE PREVIOUS 12 CHAPTERS of this book represent centuries of hard-won knowledge about the sensory worlds of other species. But in the time it took to accumulate that knowledge, we have radically re-molded those worlds. We are closer than ever to understanding what it is like to be another animal, but we have made it harder than ever for other animals to be.

Senses that have served their owners well for millions of years are now liabilities. Smooth vertical surfaces, which don't exist in nature, return echoes that sound like open air; perhaps that's why bats so often crash into windows. DMS, the seaweed-y chemical that once reliably guided seabirds to food, now also guides them to the millions of tons of plastic waste that humans have dumped into the oceans; perhaps that's why an estimated 90 percent of seabirds eventually swallow plastic. The currents produced by objects moving in the water can be detected by the body-wide hairs of manatees, but not with enough notice to avoid a fast-moving speedboat; boat collisions are responsible for at least a quarter of deaths among Florida's manatees. Odorants in river water can guide salmon back to their streams of birth, but not if pesticides in that same water weaken their sense of smell. Weak electric fields at the bottom of the sea can guide sharks to buried prey, but also to high-voltage cables.

Some animals have come to tolerate the sights and sounds of modernity. Others even flourish among them. Some urban moths have evolved to become less attracted to light. Some urban spiders have gone in the opposite direction, spinning webs beneath streetlights to feast on the attracted insects. In the towns of Panama, nighttime lights drive frog-eating bats away, allowing male túngara frogs to add more sexy chucks to their songs without the risk of attracting predators. Animals can adapt, either by changing their behavior over an individual lifetime or by evolving new behaviors over many generations.

But adaptation is not always possible. Species with slow lives and long generations can't evolve quickly enough to keep pace with levels of light and noise pollution that double every few decades. Creatures that have already been confined to narrow corners of shrinking habitats can't just up and leave. Those that rely on specialized senses can't just retune their entire Umwelt. Coping with sensory pollution isn't a simple matter of habituation. "I don't think people quite understand that if you can't hear something, you don't suddenly become able to hear it," Clinton Francis tells me. "When your sensory organ *cannot perceive a signal,* you don't just get used to that."

Our influence is not inherently destructive, but it is often homogenizing. In pushing out sensitive species that cannot abide our sensory onslaughts, we leave behind smaller and less diverse communities. We flatten the undulating sensescapes that have generated the wondrous variety of animal Umwelten. Consider Lake Victoria in East Africa. Once, it was home to over 500 species of cichlid fish, almost all of which were found nowhere else. That extraordinary diversity arose partly because of light. In deeper parts of the lake, light tends to be yellow or orange, while blue is more plentiful in shallower waters. These differences affected the eyes of the local cichlids and, in turn, their mating choices. Evolutionary biologist Ole Seehausen found that female cichlids from deeper waters prefer redder males, while those in the shallows have their eyes set on bluer ones. These diverging penchants acted like physical barriers, splitting the cichlids into a spectrum of differently colored forms. Diversity in light led to diversity in vision, in colors, and in species. But over the last century, runoff from farms, mines, and sewage filled the lake with nutrients that spurred the growth of clouding, choking algae. The old light gradients flattened in some places, the cichlids' colors and visual proclivities no longer mattered, and the number of species collapsed. By turning off the light in the lake, humans also switched off the sensory engine of diversity, leading to what Seehausen has called "the fastest large-scale extinction event ever observed."*

A cynic might ask why it matters that a lake has fewer species of similar fish. Why get worked up about a woodland that has 21 species of birds instead of 32? In 2020, science writer Maya Kapoor pondered these questions in a story about the Yaqui catfish, an endangered species from the western United States that's similar to the extremely common channel catfish. "I wondered whether the loss of a species that looked just like one of the most common fish species on the planet really mattered," Kapoor wrote. "Only later did it occur to me that . . . their seeming interchangeability said more about my limited understanding than it did about their limited distinctions." Her epiphany also applies to the cichlids, and to the many groups of animals where closely related members can have starkly different senses. As those species go extinct, so too do their Umwelten. With every creature that vanishes, we lose a way of making sense of the world. Our sensory bubbles shield us from the knowledge of those losses. But they don't protect us from the consequences.

In the woodlands of New Mexico, Clinton Francis and Catherine Ortega found that the Woodhouse's scrub-jay would flee from the noise of compressors used in extracting natural gas. The scrub-jay spreads the seeds of the pinyon pine tree, and a single bird can bury

---

* Lake Victoria's cichlids also suffered because of overfishing and exploding numbers of the invasive Nile perch. But even when the perch declined and cichlid numbers bounced back, the diversity of cichlid species remained much lower in cloudy waters. Note that light conditions are just one of several factors that explained the incredible diversity of Lake Victoria's cichlids.

between 3,000 and 4,000 pine seeds a year. They are so important to the forests that in quiet areas where they still thrive, pine seedlings are four times more common than in noisy areas that they have abandoned. Pinyon pines are the foundation of the ecosystem around them—a single species that provides food and shelter for hundreds of others, including Indigenous Americans. To lose three-quarters of them would be disastrous. And since they grow slowly, "noise might have hundred-plus-year consequences for the entire ecosystem," Francis tells me.

A BETTER UNDERSTANDING OF the senses can show us how we're defiling the natural world. It can also point to ways of saving it. In 2016, marine biologist Tim Gordon traveled to Australia's Great Barrier Reef to begin his PhD work. He should have spent months swimming among the corals' vivid splendor. Instead, "I watched in horror as my study site got completely obliterated," he tells me. A heat wave had forced the corals to expel the symbiotic algae that give them nutrients and colors. Without these partners, the corals starved and whitened in the worst bleaching event on record, and the first of several to come. Snorkeling through the rubble, Gordon found that the reefs had been not only bleached but also silenced. Snapping shrimps no longer snapped. Parrotfish no longer crunched. Those sounds normally help to guide baby fish back to the reef after their first vulnerable months out at sea. Soundless reefs were much less attractive. Gordon feared that if fish avoided the degraded reefs, the seaweed they normally eat would run amok, overgrowing the bleached corals and preventing them from rebounding. But in 2017, "we went back and thought: Can we flip that on its head?" he says.

He and his colleagues set up loudspeakers that continuously played recordings of healthy reefs over patches of coral rubble. The team would dive every few days to survey the local animals. "And on day 30," Gordon says, "I remember moseying around with my dive buddies and saying, 'There's a big pattern here, isn't there?'" After 40 days, he ran the numbers and saw that the acoustically enriched reefs had twice as many young fish as silent ones and 50 percent more species. They had not only been attracted by the sounds but stayed and formed a community. "It was a lovely experiment to do," Gordon says. It showed what conservationists can accomplish by "seeing the world through the perceptions of the animals you're trying to protect."*

Realistically, this is a small-scale solution: Loudspeakers are expensive, and coral reefs are big. Without reducing carbon emissions and forestalling climate change, reefs are in for a grim future, no matter how attractive they sound. Still, with half of the Great Barrier Reef already dead, corals need all the help they can get. Restoring their natural sounds might give them a fighting chance and make the task of saving them a little less Herculean.

Gordon's experiment was only possible because the team could still find healthy, unbleached reefs whose sounds they could record. Natural sensescapes still exist. There is still time to preserve and restore them before the last echo of the last reef fades into memory. In most cases, instead of adding stimuli that we have removed, we can simply remove those that we added—a luxury that doesn't apply to most pollutants. Radioactive waste can take millennia to degrade. Persistent chemicals like the pesticide DDT can thread their way through the bodies of animals long after they are banned. Plastics will continue to despoil the oceans for centuries even if all plastic production halts tomorrow. But light pollution ceases as soon as lights are turned off. Noise pollution abates once engines and propellers wind down. Sensory pollution is an ecological gimme—a rare example of a planetary problem that can be immediately and effectively addressed. And in the spring of 2020, the world did unknowingly address it.

As the COVID-19 pandemic spread, public spaces closed. Flights were grounded. Cars stayed parked. Cruise ships stayed docked. Around 4.5 billion people—almost three-fifths of the world's population—were told or encouraged to stay at home. As a result,

---

* Conversely, conservation attempts can backfire if they fail to account for different Umwelten. Wire cages that are sometimes put up to protect turtle nests from raccoons and foxes could distort the magnetic fields around those nests and disrupt the hatchlings' ability to learn the magnetic signatures of their home beaches.

many places became substantially darker and quieter. With fewer planes and cars on the move, the night skies around Berlin, Germany, were half as bright as normal. Seismic vibrations around the world were half as intense for months—the longest such reduction on record. Alaska's Glacier Bay, a sanctuary for humpback whales, was half as loud as the previous year, as were cities in California, New York, Florida, and Texas.* Sounds that would normally be muffled became clearer. City-dwellers around the world suddenly noticed singing birds. "People realized that there are all these animals around them that they hadn't sensed before," Francis tells me. "The sensory worlds of people in their backyards are huge compared to pre-COVID."†

In a multitude of ways, the pandemic revealed the problems that societies had come to tolerate and the changes they were actually prepared to make. It showed that sensory pollution *can* be reduced if people are sufficiently motivated. Such reductions are possible without the debilitating consequences of a global lockdown. In the summer of 2007, Kurt Fristrup and his colleagues did a simple experiment at Muir Woods National Monument in California. On a random schedule, they stuck up signs that declared one of the most popular parts of the park a quiet zone and encouraged visitors to silence their phones and lower their voices. These simple steps, with no accompanying enforcement, reduced the noise levels in the park by 3 decibels, equivalent to 1,200 fewer visitors.

But personal responsibility cannot compensate for societal irresponsibility. To truly make a dent in sensory pollution, bigger steps are needed. Lights can be dimmed or switched off when buildings and streets are not in use. They can be shielded so that they stop shining above the horizon. LEDs can be changed from blue or white to red. Quiet pavements with porous surfaces can absorb the noise from pass-

---

* Behavioral ecologist Elizabeth Derryberry found that the songs of white-crowned sparrows in the Bay Area were a third quieter during the lockdown of spring 2020, when they had less urban noise to contend with.

† Similar reductions in noise pollution followed other recent disasters. Oceanic noise fell in the waters off California after the financial collapse of 2008, and in the Bay of Fundy, Canada, after the September 11 terrorist attacks of 2001. The latter change seemed to reduce stress among right whales.

ing vehicles. Sound-absorbing barriers, including berms on land and bubble nets in the water, can soften the din of traffic and industry. Vehicles can be diverted from important areas of wilderness, or they can be forced to slow down: In 2007, when commercial ships in the Mediterranean began slowing down by just 12 percent, they produced half as much noise. Such vessels can also be fitted with quieter hulls and propellers, which are already used to muffle military ships (and would make commercial ones more fuel-efficient). Many helpful technologies already exist, but the economic incentives to make them cheaper or to deploy them en masse are lacking. We could regulate industries causing sensory pollution, but there's not enough societal will. "Plastic pollution in the sea looks hideous and everyone is worried, but noise pollution in the sea is something we don't experience, so no one's up in arms about it," Gordon tells me.

We normalize the abnormal, and accept the unacceptable. Remember that more than 80 percent of people live under light-polluted skies, and that two-thirds of Europeans are immersed in noise equivalent to constant rainfall. Many people have no idea what true darkness or quiet feels like. Within that inexperience, vicious cycles begin to spin. As we desecrate sensory environments, we become accustomed to the results. As we push animals away, we get used to their absence. As the problem of sensory pollution grows, our willingness to address it subsides. How do we solve a problem that we don't realize exists?

IN 1995, ENVIRONMENTAL HISTORIAN William Cronon wrote that "the time has come to rethink wilderness." In a searing essay, he argued that the concept of wilderness, especially as perceived in the United States, had become unjustly synonymous with grandeur. Eighteenth-century thinkers believed that vast and magnificent landscapes reminded people of their own mortality and brought them closer to glimpsing the divine. "God was on the mountaintop, in the chasm, in the waterfall, in the thundercloud, in the rainbow, in the sunset," Cronon wrote. "One has only to think of the sites that

Americans chose for their first national parks—Yellowstone, Yosemite, Grand Canyon, Rainier, Zion—to realize that virtually all of them fit one or more of these categories. Less sublime landscapes simply did not appear worthy of such protection; not until the 1940s, for instance, would the first swamp be honored, in Everglades National Park, and to this day there is no national park in the grasslands."

Equating wilderness with otherworldly magnificence treats it as something remote, accessible only to those with the privilege to travel and explore. It imagines that nature is something separate from humanity rather than something we exist within. "Idealizing a distant wilderness too often means not idealizing the environment in which we actually live, the landscape that for better or worse we call home," Cronon wrote.

I couldn't agree more. The majesty of nature is not restricted to canyons and mountains. It can be found in the wilds of perception—the sensory spaces that lie outside our Umwelt and within those of other animals. To perceive the world through other senses is to find splendor in familiarity, and the sacred in the mundane. Wonders exist in a backyard garden, where bees take the measure of a flower's electric fields, leafhoppers send vibrational melodies through the stems of plants, and birds behold the hidden palettes of rurples and grurples. In writing this book, I have found the sublime while confined to my home by a pandemic, watching tetrachromatic starlings gathering in the trees outside and playing sniffing games with my dog, Typo. Wilderness is not distant. We are continually immersed in it. It is there for us to imagine, to savor, and to protect.

IN 1934, AFTER CONSIDERING the senses of ticks, dogs, jackdaws, and wasps, Jakob von Uexküll wrote about the Umwelt of the astronomer. "Through gigantic optical aids," he wrote, this unique creature has eyes that "are capable of penetrating outer space as far as the most distant stars. In its [Umwelt], suns and planets circle at a solemn pace." The tools of astronomy can capture stimuli that no animal can natu-

rally sense—X-rays, radio waves, and gravitational waves from colliding black holes. They extend the human Umwelt across the extent of
the universe and back to its very beginning.

The tools of biologists are more modest in scale, but they too offer
a glimpse into the infinite. Eric Warrant used night-vision goggles to
watch nocturnal bees finding their nests in the dark. Almut Kelber used
night-vision goggles to watch elephant hawkmoths drinking from
flowers in the dark. Paloma Gonzalez-Bellido used high-speed cameras
to determine how fast killer flies see, and Ken Catania used them to work
out how star-nosed moles hunt by touch. With lasers, Kurt Schwenk
visualized the vortices that snakes make when they flick their tongues.
With an ultrasound detector, Donald Griffin discovered the sonar of
bats. Laser vibrometers and clip-on microphones allow Rex Cocroft to
eavesdrop on leafhoppers. The Navy's SOSUS hydrophones allowed
Chris Clark to confirm how far blue whale calls can travel. With simple electrodes, Eric Fortune and other electric fish researchers can listen in on the pulses of knifefish and elephantfish. With microscopes,
cameras, speakers, satellites, recorders, and even paper-lined cages with
inkpads at their bases, people have explored other sensory worlds. We
have used technology to make the invisible visible and the inaudible
audible.

This ability to dip into other Umwelten is our greatest sensory
skill. Think back to the hypothetical room that we envisioned at the
start of this book, with the elephant, the rattlesnake, and all the rest.
Among that imaginary menagerie, the human—Rebecca—lacked ultraviolet vision, magnetoreception, echolocation, and an infrared
sense. But she was the only creature capable of knowing what the others were sensing and, perhaps, the only one who might care.

A bogong moth will never know what a zebra finch hears in its
song, a zebra finch will never feel the electric buzz of a black ghost
knifefish, a knifefish will never see through the eyes of a mantis shrimp,
a mantis shrimp will never smell the way a dog can, and a dog will
never understand what it is like to be a bat. We will never fully do any
of these things either, but we are the only animal that can even come
close. We may not ever know what it is to be an octopus, but at least

we know that octopuses exist, and that their experiences differ from ours. Through patient observation, through the technologies at our disposal, through the scientific method, and, above all else, through our curiosity and imagination, we can try to step into their worlds. We must choose to do so, and to have that choice is a gift. It is not a blessing we have earned, but it is one we must cherish.

Acknowledgments

At the end of 2018, I was sitting in a London café with my wife, Liz Neeley, telling her that while I very much wanted to write a second book, my well of ideas had run dry. Liz listened patiently, and then gently suggested that I could write about how animals perceive the world. This kind of thing happens a lot.

The idea drew from our shared interest in nature. It flowed naturally from our entire careers: Liz had started her marine biology PhD on the visual systems of coral reef fish, and I had written about sensory biology for over a decade. It reflected our desire to tell the stories of those whose lives often go overlooked or unheard. I'm profoundly grateful to Liz not just for seeding the idea of this book and supporting me through its creation, but for embodying its values and instilling them in me. She is relentlessly joyful, curious, and empathetic, and she brings those same qualities out in people who have the privilege to know her. To spend time with Liz is to see the world and its inhabitants in new ways—exactly the feeling that I hope you, dear reader, got from the preceding pages.

For shepherding this book from concept to finished product, my deepest thanks go to: Will Francis, my British agent and friend, who saw the promise in this idea from the start and helped to nurture it into life; PJ Mark, my American agent; Hilary Redmon, my American publisher and intellectual co-conspirator, who edited the early drafts into shape; and Stuart Williams, my British publisher, who also provided incisive notes on the manuscript. All four of them were also involved in my first book, *I Contain Multitudes,* and getting to work with them again was like coming home.

Sarah Laskow and Ross Andersen, my editors at *The Atlantic,* deserve huge credit for everything they've taught me about writing over the last years; they didn't directly work on this book, but their influence in these pages runs deep. They, together with Robert Brenner, Meehan Crist, Tom Cunliffe, Rose Eveleth, Natalie Omundsen, Sarah Ramey, Rebecca Skloot, Beck Smith, Maddie Sofia, and Maryam Zaringhalam, also kept me afloat in a very difficult year, when I turned my attention from the delightful realms of animal senses to the more grueling and tragic world of the COVID-19 pandemic.

I spoke to more scientists in the course of writing this book than I could reasonably list, many of whom were incredibly generous with their time. My deepest thanks to Jesse Barber, Bruce Carlson, Rex Cocroft, Robyn Crook, Heather Eisthen, Ken Lohmann, Colleen Reichmuth, Cassie Stoddard, and Eric Warrant for crucial feedback on various chapters and deep discussions. Thanks also to most of the above, and to Whitlow Au, Gordon Bauer, Adriana Briscoe, Astra Bryant, Rulon Clark, Tom Cronin, Molly Cummings, Elena Gracheva, Frank Grasso, Alexandra Horowitz, Martin How, Elizabeth Jakob, Sonke Johnsen, Suzanne Amador Kane, Daniel Kish, Daniel Kronauer, Travis Longcore, Malcolm MacIver, Justin Marshall, Beth Mortimer, Cindy Moss, Paul Nachtigall, Dan-Eric Nilsson, Thomas Park, Daniel Robert, Nicholas Roberts, Mike Ryan, Nate Sawtell, Kurt Schwenk, Jim Simmons, Daphne Soares, Amy Streets, Leslie Vosshall, Karen Warkentin, and George Wittemyer for variously showing me their labs, their animals, or their lives. Special thanks to Matthew Cobb for early encouragement and a very useful set of slides, Catherine Williams for helping me think through the chapter on pain in the early stages, Michael Hendricks for his help in shaping the chapter on unifying the senses, Eleanor Caves for creating a bespoke figure based on her acuity work; and Brian Branstetter, Ken Catania, Kurt Fristrup, Amanda Melin, Nate Morehouse, and Aude Pacini for especially helpful discussions.

I'm also profoundly grateful to Ashley Shew, a brilliant thinker on the intersection between technology and disability, for giving the manuscript a thorough sensitivity read, and helping me to avoid the insidiously ableist language and ideas that characterize so much writing about the senses. (Any lingering errors in the text are mine and mine alone.)

It was a pleasure to meet Finn the dog, Margaret the rattlesnake, Sprouts the harbor seal, Hugh and Buffett the manatees, Zipper the big brown bat, Blubby the electric catfish, Qualia and Ra the octopuses, and the unnamed mantis shrimp who punched me in the finger. And finally, thanks to Moro, Ellers, Athena, Ruby, Midge, Ezra, Bingo, Nellie, Bennet, Margaux, Canela, Dolly, Tim, Janet, Clarence, Zako, Whiskey, Caleb, Posey, Tesla, Crosby, Bing, Bear, Buddy, Mickey, and especially to my dearest Typo for teaching me to have animals in my heart and home as well as in my head. To all the other very good dogs (and cats) whom I'm sure I have forgotten, I am so sorry. It's a good thing you can't read.

Notes

22 **In front of each animal** (Duranton and Horowitz, 2019)

24 **Some dislike that dogs get treated** (Pihlström et al., 2005)

24 **In some cases, humans do *better*** (Laska, 2017)

24 **McGann traced the origin** (McGann, 2017)

24 **In 2019, Tali Weiss identified** (Weiss et al., 2020)

25 **"of extremely slight service"** (Darwin, 1871, volume 1, p. 24)

25 **"smell does not allow itself"** (Kant, 2007, p. 270)

25 **The English language confirms his view** (Majid, 2015)

25 **"is the one without words"** (Ackerman, 1991, p. 6)

25 **The Jahai people of Malaysia** (Majid et al., 2017; Majid and Kruspe, 2018)

25 **In 2006, neuroscientist Jess Porter** (Porter et al., 2007)

26 **Their signals can then be detected** (Silpe and Bassler, 2019)

26 **Chemicals, then, are the most ancient** (Dusenbery, 1992)

27 **The variation among possible odorants** An excellent review on the basics of olfaction is Keller and Vosshall (2004b).

27 **When mixed, some pairs of odors** (Keller and Vosshall, 2004b)

27 **Noam Sobel, a neurobiologist** (Ravia et al., 2020)

27 **Their noses are kings of infinite space** Reviews on smell: Eisthen (2002); Ache and Young (2005); Bargmann (2006).

27 **In work that would earn them a Nobel** (Firestein, 2005)

28 **One widely popularized theory** (Keller and Vosshall, 2004a)

28 **For example, the OR7D4 gene** (Keller et al., 2007)

29 **Male moths, for example** (Vogt and Riddiford, 1981)

29 **Smell is so important to them** (Kalberer, Reisenman, and Hildebrand, 2010)

29 **Moths have been described as** (Atema, 2018)

29 **By mimicking female moth odors** (Haynes et al., 2002)

30 **The chemicals they use are pheromones** A review on animal pheromones is Wyatt (2015a).

30 **Indeed, despite the existence of pheromone parties** (Wyatt, 2015b)

30 **Human pheromones likely exist** (Wyatt, 2015b)

30 **Ant pheromones are another story** (Leonhardt et al., 2016)

31 **Leafcutter ants are so sensitive** (Tumlinson et al., 1971)

31 **Known as cuticular hydrocarbons** (Sharma et al., 2015)

31 **Queens also use these substances** (Monnin et al., 2002)

31 **Red ants will look after** (Lenoir et al., 2001)

31 **Army ants are so committed** (Schneirla, 1944)

31 **In September 2020, I noted** (Yong, 2020)

31 **Many ants use pheromones to discern dead** (Wilson, Durlach, and Roth, 1958)

31 **"The ant world is a tumult"** (Treisman, 2010)

32 **Ant civilizations are among the most impressive** (D'Ettorre, 2016)

32 **Ants are essentially a group of** (Moreau et al., 2006)

32 **Along the way, their repertoire of odorant receptor genes** (McKenzie and Kronauer, 2018)

32 **Why? Here are three clues** (McKenzie and Kronauer, 2018)

32 **When Kronauer deprived his clonal raiders** (Trible et al., 2017)

33 **Back in 1874, the Swiss scientist** (Forel, 1874)

33 **Female lobsters urinate into the faces** (Atema, 2018)

33 **Male mice produce a pheromone** (Roberts et al., 2010)

33 **The early spider-orchid deceives male bees** (Schiestl et al., 2000)

33 **"We live, all the time"** (Wilson, 2015)

34 **You don't need to know about an elephant's** (Niimura, Matsui, and Touhara, 2014)

34 **African elephants can use their trunks** (McArthur et al., 2019)

34 **They can learn unfamiliar smells** (Miller, Hensman, et al., 2015)

34   **Two of those same elephants** (von Dürckheim et al., 2018)
34   **Asian elephants are no slouches, either** (Plotnik et al., 2019)
35   **When the animals approached washed garments** (Bates et al., 2007)
35   **When African elephants reunite** (Moss, 2000)
35   **Few people have done more to study elephant odors** (Hurst et al., 2008)
35   **In 1996, after 15 years of work** (Rasmussen et al., 1996)
36   **Rasmussen eventually discovered that elephants** (Rasmussen and Schulte, 1998)
36   **As they walk the time-worn trails** (Hurst et al., 2008)
36   **In 2007, Lucy Bates found** (Bates et al., 2008)
37   **Elephants that have returned to postwar Angola** (Miller, Hensman, et al., 2015)
37   **They've been known to dig wells** (Ramey et al., 2013)
37   **Rasmussen once speculated** (Rasmussen and Krishnamurthy, 2000)
37   **Salmon can return** (Wisby and Hasler, 1954)
38   **Whip spiders use the smell sensors** (Bingman et al., 2017)
38   **Polar bears might** (Owen et al., 2015)
38   **These examples are so common** (Jacobs, 2012)
38   **John James Audubon, the avid naturalist** (Stager, 1964; Birkhead, 2013; Eaton, 2014)
38   **These birds, he claimed in 1826** (Audubon, 1826)
39   **Ornithologist Kenneth Stager** (Stager, 1964)
39   **Betsy Bang revitalized it** A historical look at Bang and Wenzel's influence is Nevitt and Hagelin (2009).
39   **Concerned that the textbooks were spouting misinformation** (Bang, 1960; Bang and Cobb, 1968)
39   **For them, "olfaction is of primary importance"** (Nevitt and Hagelin, 2009)
39   **By scanning the skulls** (Zelenitsky, Therrien, and Kobayashi, 2009)
39   **Elsewhere in California, Bernice Wenzel** (Sieck and Wenzel, 1969)
40   **She repeated that test** (Wenzel and Sieck, 1972)
40   **Both Bang and Wenzel** (Nevitt and Hagelin, 2009)
41   **In the varying levels of the chemical** (Nevitt, 2000)
41   **Once back on her feet, Nevitt** (Nevitt, Veit, and Kareiva, 1995)
41   **She calculated that they can detect** (Nevitt and Bonadonna, 2005)
41   **She showed that some tubenoses** (Bonadonna et al., 2006; Van Buskirk and Nevitt, 2008)
41   **Henri Weimerskirch fitted wandering albatrosses** (Nevitt, Losekoot, and Weimerskirch, 2008)
42   **The smellscapes that seabirds track** (Nevitt, 2008; Nevitt, Losekoot, and Weimerskirch, 2008)
42   **She transported a few Cory's shearwaters** (Gagliardo et al., 2013)
42   **"What may be featureless to us"** (Nicolson, 2018, p. 230)
42   **By comparing the odorants** (Sobel et al., 1999)
43   **Whatever the case, serious scholars** (Schwenk, 1994)
44   **Using its tongue, a male garter snake** (Shine et al., 2003)
44   **By comparing what she deposited** (Ford and Low, 1984)
44   **Schwenk reasoned that the fork** (Schwenk, 1994)
44   **Rulon Clark, whom we'll meet** (Clark, 2004; Clark and Ramirez, 2011)
44   **Aside from lethal toxins** (Durso, 2013)
44   **The snakes can use these aromas** (Chiszar et al., 1983, 1999; Chiszar, Walters, and Smith, 2008)
45   **Chuck Smith, one of Schwenk's former students** (Smith et al., 2009)
45   **Bill Ryerson, another of Schwenk's students** (Ryerson, 2014)
46   **For some reason, humans lost our vomeronasal** (Baxi, Dorries, and Eisthen, 2006)
46   **Without it, garter snakes stop following** (Kardong and Berkhoudt, 1999)
46   **In other animals, the organ is a mystery** (Baxi, Dorries, and Eisthen, 2006)

48 **Adults vary so much** (Pain, 2001)

48 **And while smell can be put** (Yarmolinsky, Zuker, and Ryba, 2009)

49 **When a python swallows a pig** (Secor, 2008)

49 **Bees can detect the sweetness** (de Brito Sanchez et al., 2014)

49 **Flies can taste the apple** (Thoma et al., 2016)

49 **Parasitic wasps can use taste sensors** (Van Lenteren et al., 2007)

49 **But if that arm is covered with bitter-tasting DEET** (Dennis, Goldman, and Voss-hall, 2019)

50 **Some have taste receptors on their wings** (Raad et al., 2016)

50 **Flies will start grooming themselves** (Yanagawa, Guigue, and Marion-Poll, 2014)

50 **The most extensive sense of taste** (Atema, 1971; Caprio et al., 1993)

50 **They have taste buds** (Kasumyan, 2019)

50 **They're exquisitely sensitive to amino acids** (Caprio, 1975)

50 **So in the mid-1990s** (Caprio et al., 1993)

50 **Cats, spotted hyenas** (Jiang et al., 2012)

50 **Vampire bats, which drink only blood** (Shan et al., 2018)

51 **Other leaf-eating specialists, like koalas** (Johnson et al., 2018)

51 **In 2014, evolutionary biologist Maude Baldwin** (Toda et al., 2021)

51 **Baldwin also showed that hummingbirds** (Baldwin et al., 2014)

52 **This is how all animals see** (Nilsson, 2009)

CHAPTER 2

53 **The *Portia* species are famed** (Cross et al., 2020)

54 **And unlike other spiders** (Morehouse, 2020)

54 **The late British neurobiologist Mike Land** Land wrote great accounts of his own work in Land (2018).

54 **In 1968, he developed an ophthalmoscope** (Land, 1969a, 1969b)

55 **"an exhilarating but very weird"** (Land, 2018, p. 107)

56 **And here's the truly bizarre part** (Jakob et al., 2018)

56 **The eyes of the giant squid** (Nilsson et al., 2012; Polilov, 2012)

56 **Squid, jumping spiders, and humans** A review of animal eyes is Nilsson (2009).

57 **Animal eyes can be bifocal** (Stowasser et al., 2010; Thomas, Robison, and Johnsen, 2017)

57 **They can have lenses** (Li et al., 2015)

57 **Jakob's colleague Nate Morehouse** (Goté et al., 2019)

57 **vision "is about light"** (Johnsen, 2012, p. 2)

57 **Every animal that sees does** (Porter et al., 2012)

58 **In 2012, evolutionary biologist Megan Porter** (Porter et al., 2012)

58 **Vision is diverse** The textbook *Visual Ecology* is a fantastic and very readable primer on vision and its many uses (Cronin et al., 2014).

58 **The biologist Dan-Eric Nilsson** (Nilsson, 2009)

58 **The hydra, a relative of jellyfish** (Plachetzki, Fong, and Oakley, 2012)

58 **Olive sea snakes have photoreceptors** (Crowe-Riddell, Simões, et al., 2019)

58 **Octopuses, cuttlefish, and other cephalopods** (Kingston et al., 2015)

59 **The Japanese yellow swallowtail butterfly** (Arikawa, 2001)

59 **This flurry of evolutionary innovation** (Parker, 2004)

60 **"To suppose that the eye"** (Darwin, 1958, p. 171)

60 **The jellyfish alone have evolved** (Picciani et al., 2018)

60 **In 1994, Nilsson and Susanne Pelger** (Nilsson and Pelger, 1994)

60 **As we saw in the introduction** (Garm and Nilsson, 2014)

61 **Consider the freshwater bacterium *Synechocystis*** (Schuergers et al., 2016)

61 **The warnowiids, a group of single-celled algae** (Gavelis et al., 2015)

61   **Caro had become the latest** (Caro, 2016)

62   **She and Caro calculated that** (Melin et al., 2016)

62   **Caro has a definitive answer: to ward off bloodsucking flies** (Caro et al., 2019)

62   **An animal's visual acuity** An excellent review of visual acuity in animals is Caves, Brandley, and Johnsen (2018).

62   **The current record, at 138 cycles per degree** (Reymond, 1985; Mitkus et al., 2018)

62   **One oft-quoted study from the 1970s** (Fox, Lehmkuhle, and Westendorf, 1976)

62   **Sensory biologist Eleanor Caves** (Caves, Brandley, and Johnsen, 2018)

63   **Octopuses (46 cpd)** (Veilleux and Kirk, 2014; Caves, Brandley, and Johnsen, 2018)

64   **Robber flies manage** (Feller et al., 2021)

64   **For a fly's eye** (Kirschfeld, 1976)

65   **Each half of a scallop's** (Mitkus et al., 2018)

65   **It's even stranger that those eyes** (Land, 1966)

65   **And both are found in the scallop** (Speiser and Johnsen, 2008a)

66   **He strapped their shells** (Speiser and Johnsen, 2008b)

67   **In 1964, Mike Land** (Land, 2018)

67   **Guanine crystals don't naturally form squares** (Palmer et al., 2017)

67   **Chitons are mollusks** (Li et al., 2015)

67   **Fan worms look like** (Bok, Capa, and Nilsson, 2016)

67   **Giant clams look like** (Land, 2003)

68   **In 2018, Lauren Sumner-Rooney** (Sumner-Rooney et al., 2018)

68   **Like brittle stars, sea urchins** (Ullrich-Luter et al., 2011)

68   **Weirder still, it's only an eye** (Sumner-Rooney et al., 2020)

69   **In one Spanish province alone** (Carrete et al., 2012)

69   **In 2012, Martin and his colleagues** (Martin, Portugal, and Murn, 2012)

70   **A soaring vulture** See Martin (2012), which also reviews and cites Martin's many papers on bird visual fields.

70   **"The human visual world"** (Martin, 2012)

71   **Many animals have an area** (Moore et al., 2017; Baden, Euler, and Berens, 2020)

71   **When a chicken investigates** (Stamp Dawkins, 2002)

71   **Many birds of prey** (Mitkus et al., 2018)

71   **When a peregrine falcon** (Potier et al., 2017)

71   **The left half of a chick's brain** A wide range of experiments is reviewed in Rogers (2012).

72   **A seal's visual field** (Hanke, Römer, and Dehnhardt, 2006)

72   **Cows and other livestock** (Hughes, 1977)

72   **The same is true** An excellent review of regionalization in animal retinas is Baden, Euler, and Berens (2020).

72   **Elephants, hippos, rhinos, whales** (Mass and Supin, 1995; Baden, Euler, and Berens, 2020)

72   **A whale's pupil doesn't constrict** (Mass and Supin, 2007)

72   **Chameleons don't have to turn** (Katz et al., 2015)

73   **Many male flies focus upward** (Perry and Desplan, 2016)

73   **The fish *Anableps anableps*** (Owens et al., 2012)

73   **The brownsnout spookfish** (Partridge et al., 2014)

73   **So can the cock-eyed squid** (Thomas, Robison, and Johnsen, 2017)

73   **Meanwhile, the deep-sea crustacean *Streetsia*** (Meyer-Rochow, 1978)

74   **If you can coax a killer fly** (Simons, 2020)

74   **By filming these pursuits** (Wardill et al., 2013)

75   **Their ultrafast hunts are guided** (Gonzalez-Bellido, Wardill, and Juusola, 2011)

75   **Compared to the photoreceptors of a fruit fly** (Gonzalez-Bellido, Wardill, and Juusola, 2011)

75   **By contrast, it takes between 30** (Masland, 2017)

76 **In general, animals tend to have higher CFFs** (Laughlin and Weckström, 1993)
76 **Compared to human vision** Several values of animal CFFs can be found in Healy et al. (2013); Inger et al. (2014).
76 **Those of swordfish** (Fritsches, Brill, and Warrant, 2005)
76 **Many birds have naturally fast vision** (Boström et al., 2016)
76 **Traditional fluorescent lights flicker at 100 Hz** (Evans et al., 2012)
76 **And those insects have eyes** (Ruck, 1958)
77 **By filming the insect** (Warrant et al., 2004)
77 **The first is obvious** (O'Carroll and Warrant, 2017)
78 **The second challenge is less intuitive** (O'Carroll and Warrant, 2017)
78 **It takes a lot of energy** (Niven and Laughlin, 2008; Moran, Softley, and Warrant, 2015)
78 **Others unsubscribe from vision entirely** (Porter and Sumner-Rooney, 2018)
78 **There are many ways to break an eye** (Porter and Sumner-Rooney, 2018)
78 **Some use neural tricks** (Warrant, 2017)
79 **The structure of a reindeer's tapetum** (Stokkan et al., 2013)
79 **The tarsiers—small primates** (Collins, Hendrickson, and Kaas, 2005)
79 **To dive into the ocean** (Warrant and Locket, 2004)
79 **At 10 meters down** Two great reviews about vision in the ocean are Warrant and Locket (2004); Johnsen (2014).
80 **To be more respectful of deep-sea** (Widder, 2019)
81 **The footage was unmistakable** (Johnsen and Widder, 2019)
81 **But no other creature** (Nilsson et al., 2012)
81 **Sonke Johnsen, Eric Warrant, and Dan-Eric Nilsson** (Nilsson et al., 2012)
82 **The first natural footage was captured in 2012** (Schrope, 2013)
83 **Then, in 2002, Eric Warrant** (Kelber, Balkenius, and Warrant, 2002)

CHAPTER 3

84 **One textbook claimed that** (Tansley, 1965)
84 **And yet, very few species** (Neitz, Geist, and Jacobs, 1989)
85 **Dogs do see color** (Neitz, Geist, and Jacobs, 1989)
85 **Light comes in a range** For excellent primers on color vision, check out Osorio and Vorobyev (2008); Cuthill et al. (2017); and Chapter 7 of Cronin et al. (2014).
86 *Daphnia* **water fleas** A review of unusual color vision is Marshall and Arikawa (2014).
86 **Consider the story of the artist** (Sacks and Wasserman, 1987)
87 **Some, like sloths and armadillos** (Emerling and Springer, 2015)
87 **Others, like raccoons and sharks** (Peichl, 2005; Hart et al., 2011)
87 **Whales have just one cone, too** (Peichl, Behrmann, and Kröger, 2001)
87 **Surprisingly, the cephalopods** (Hanke and Kelber, 2020)
87 **The firefly squid** (Seidou et al., 1990)
87 **Physiologist Vadim Maximov suggested** (Maximov, 2000)
88 **Dogs have two cones** (Neitz, Geist, and Jacobs, 1989)
88 **This means that horses struggle** (Paul and Stevens, 2020)
88 **Color-blind people might be confused** (Colour Blind Awareness, n.d.)
89 **The first primates** (Carvalho et al., 2017)
89 **That's exactly what happened** (Carvalho et al., 2017)
90 **Each extra opsin increases** (Pointer and Attridge, 1998; Neitz, Carroll, and Neitz, 2001)
90 **Since the nineteenth century** (Mollon, 1989; Osorio and Vorobyev, 1996; Smith et al., 2003)
90 **More recently, some researchers** (Dominy and Lucas, 2001; Dominy, Svenning, and Li, 2003)
90 **In 1984, Gerald Jacobs** (Jacobs, 1984)

90   **These monkeys never developed** (Jacobs and Neitz, 1987)
91   **Howler monkeys** (Saito et al., 2004)
91   **Females might inherit two** (Jacobs and Neitz, 1987)
91   **Neither group, she found** (Fedigan et al., 2014)
91   **The trichromats are indeed better** (Melin et al., 2007, 2017)
91   **In 2007, the Neitzes** (Mancuso et al., 2009)
92   **In the 1880s, John Lubbock** (Lubbock, 1881)
92   **There's only a narrow Goldilocks** (Dusenbery, 1992)
93   **At the time, some scientists** For an excellent overview on UV vision and its history, see Cronin and Bok (2016).
93   **But after another half century** (Goldsmith, 1980)
93   **Still wrong: In 1991** (Jacobs, Neitz, and Deegan, 1991)
93   **Not so: In the 2010s** (Douglas and Jeffery, 2014)
93   **This happened to the painter Claude Monet** (Zimmer, 2012)
93   **Most animals that can see color** (Tedore and Nilsson, 2019)
94   **Some scientists think** (Marshall, Carleton, and Cronin, 2015)
94   **Reindeer can quickly** (Tyler et al., 2014)
94   **Flowers use dramatic UV patterns** (Primack, 1982)
94   **Crab spiders lurk** (Herberstein, Heiling, and Cheng, 2009)
94   **In 1998, two independent teams** (Andersson, Ornborg, and Andersson, 1998; Hunt et al., 1998)
94   **The same is true** (Eaton, 2005)
94   **The swordtail fish** (Cummings, Rosenthal, and Ryan, 2003)
95   **But Ulrike Siebeck found** (Siebeck et al., 2010)
95   **Scientists have often attributed** (Stevens and Cuthill, 2007)
95   **In 1995, a Finnish team** (Viitala et al., 1995)
95   **In 2013, she and her colleagues** (Lind et al., 2013)
96   **Exploiting the hummingbirds' natural instinct** (Stoddard et al., 2020)
97   **Picture trichromatic human vision** A classic paper on visualizing color vision is Kelber, Vorobyev, and Osorio (2003).
97   **Stoddard found that these non-spectral** (Stoddard et al., 2020)
97   **Many supposedly "white" bird feathers** (Stoddard et al., 2019)
98   **Reptiles, insects, and freshwater fish** (Neumeyer, 1992)
98   **By looking at tetrachromats** (Collin et al., 2009)
100 **In any one place, these two species** (Hines et al., 2011)
100 **But in 2010, Briscoe discovered** (Briscoe et al., 2010)
100 **Even birds, with their single UV opsin** (Finkbeiner et al., 2017)
100 **In 2016, Briscoe's student** (McCulloch, Osorio, and Briscoe, 2016)
101 **Somewhere in Newcastle, England** (Jordan et al., 2010)
101 **Around one in eight women** (Greenwood, 2012; Jordan and Mollon, 2019)
103 **At least three kinds** (Zimmermann et al., 2018)
103 **Kentaro Arikawa has found** (Koshitaka et al., 2008; Chen et al., 2016; Arikawa, 2017)
105 **The clubs of a large smasher** (Patek, Korff, and Caldwell, 2004)
106 **When Marshall looked at the midband** (Marshall, 1988)
106 **And to their shock** (Cronin and Marshall, 1989a, 1989b)
106 **The midband consists** An excellent review of mantis shrimp vision is Cronin, Marshall, and Caldwell (2017).
107 **Mantis shrimps have more classes** (Marshall and Oberwinkler, 1999; Bok et al., 2014)
107 *The Oatmeal* (Inman, 2013)
107 **In 2014, Marshall's student** (Thoen et al., 2014)
109 **Nigel's eyes are constantly moving** (Daly et al., 2018)
110 **Mantis shrimps do something similar** (Marshall, Land, and Cronin, 2014).
110 **When it spots something** (Land et al., 1990)

112 **Humans are largely oblivious** (Marshall et al., 2019b)
112 **Cephalopods are more sensitive** (Temple et al., 2012)
112 **And as Marshall's postdoc** (Chiou et al., 2008)
112 **They can also rotate their eyes** (Daly et al., 2016)
113 **One species reflects it** (Gagnon et al., 2015)
114 **Tom Cronin thinks** (Cronin, 2018)
114 **The red faces** (Hiramatsu et al., 2017; Moreira et al., 2019)
114 **But the fish themselves** (Marshall et al., 2019a)
115 **But as Molly Cummings** (Maan and Cummings, 2012)
115 **In 1992, Lars Chittka** (Chittka and Menzel, 1992)
115 **Their style of trichromacy** (Chittka, 1997)

CHAPTER 4

117 **I highly recommend the paper** (Braude et al., 2021)
117 **Naked mole-rats are so weird** (Park, Lewin, and Buffenstein, 2010; Braude et al., 2021)
117 **Their lower incisors** (Catania and Remple, 2002)
117 **Their sperm are misshapen** (Van der Horst et al., 2011)
117 **They can survive for up** (Park et al., 2017)
118 **They've also been forced** (Zions et al., 2020)
118 **Park demonstrated this with an arena** (Park et al., 2017)
118 **They'll sniff strong vinegary fumes** (LaVinka and Park, 2012)
118 **They don't register drops of acid** (Park et al., 2008)
118 **They dislike pinches and burns** (Poulson et al., 2020)
118 **Our experience of pain depends** The basics of nociception are reviewed in Kavaliers (1988); Lewin, Lu, and Park (2004); Tracey (2017).
119 **But theirs are fewer in number** (Smith, Park, and Lewin, 2020)
119 **Those that would normally be activated** (Smith et al., 2011)
119 **Several hibernating mammals** (Liu et al., 2014)
119 **Birds that carry the seeds** (Jordt and Julius, 2002)
119 **Humans are insensitive to nepetalactone** (Melo et al., 2021)
119 **The grasshopper mouse** (Rowe et al., 2013)
120 **In the early 1900s** (Sherrington, 1903)
120 **Over a century later** Excellent reviews of nociception and pain are Sneddon (2018); Williams et al. (2019).
121 **Other people are congenitally indifferent** (Cox et al., 2006; Goldberg et al., 2012)
121 **One Pakistani boy** (Cox et al., 2006)
121 **I highly recommend Leigh Cowart's** (Cowart, 2021)
121 **People (and especially women)** *The Lady's Handbook for Her Mysterious Illness* by Sarah Ramey (2020) and *Doing Harm* by Maya Dusenbery (2018) are excellent books on this topic.
122 **It is so widespread and consistent** A review of pain in animals is Sneddon (2018).
122 **The signs of pain** (Bateson, 1991)
122 **For many historical thinkers** (Sullivan, 2013)
123 **But fierce debates are still raging** (Sneddon et al., 2014)
123 **Until the 1980s** (Anand, Sippell, and Aynsley-Green, 1987)
123 **That distinction "is a relic"** (Broom, 2001)
123 **Humans have taste receptors** (Li, 2013; Lu et al., 2017)
124 **In the early 2000s, Lynne Sneddon** (Sneddon, Braithwaite, and Gentle, 2003a, 2003b)
125 **When fish nociceptors fire** (Dunlop and Laming, 2005; Reilly et al., 2008)
125 **Sure enough, when the animals** (Bjørge et al., 2011; Mettam et al., 2011)
125 **In one experiment, Sneddon showed** (Sneddon, 2013)
125 **In another study, Sarah Millsopp** (Millsopp and Laming, 2008)

125   **"There is as much evidence"** (Braithwaite, 2010)
126   **But a group of vocal critics** (Rose et al., 2014; Key, 2016)
126   **For a sense of the debate** (Rose et al., 2014; Key, 2016; Sneddon, 2019)
126   **"Fishes are neurologically equipped"** (Rose et al., 2014)
126   **Ironically, this argument** (Braithwaite and Droege, 2016)
126   **And by the same faulty logic** (Dinets, 2016)
126   **For perspective, crabs and lobsters** (Marder and Bucher, 2007)
127   **What matters is not just the total tally** (Garcia-Larrea and Bastuji, 2018)
127   **But such links are much sparser** (Adamo, 2016, 2019)
128   **But Elwood and his colleague** (Appel and Elwood, 2009; Elwood and Appel, 2009)
128   **These data, Elwood says** (Elwood, 2019)
128   **But notably, Adamo, Sneddon, and Elwood** (Sneddon et al., 2014)
129   **Evolution has pushed** (Chittka and Niven, 2009)
129   **Some scientists suggest** (Bateson, 1991; Elwood, 2011)
129   **Engineers have designed robots** (Stiehl, Lalla, and Breazeal, 2004; Lee-Johnson and Carnegie, 2010; Ikinamo, 2011)
130   **But they also have** (Hochner, 2012)
130   **And, as the EU noted** (European Parliament, Council of the European Union, 2010)
130   **She began to bridge that gap** (Crook et al., 2011)
131   **Even more surprisingly, Crook found** (Crook, Hanlon, and Walters, 2013)
131   **By setting their entire bodies** (Crook et al., 2014)
131   **Crook confirmed this** (Alupay, Hadjisolomou, and Crook, 2014)
132   **Octopuses will sometimes break off** (Alupay, Hadjisolomou, and Crook, 2014)
132   **In her latest study** (Crook, 2021)
133   **"We could simply accept"** (Chatigny, 2019)
133   **Insects, for example** (Eisemann et al., 1984)
133   **These behaviors "strongly suggest"** (Eisemann et al., 1984)

CHAPTER 5

135   **Hibernation isn't sleep** (Geiser, 2013)
135   **The two processes are so different** (Daan, Barnes, and Strijkstra, 1991)
136   **Its heart, which beats** (Andrews, 2019)
136   **A thirteen-lined ground squirrel** (Matos-Cruz et al., 2017)
136   **Vanessa Matos-Cruz, who worked with Gracheva** (Matos-Cruz et al., 2017)
137   **The limits of that zone vary** The temperature ranges that animals tolerate are reviewed in McKemy (2007); Sengupta and Garrity (2013).
137   **Animals use a variety** (Matos-Cruz et al., 2017; Hoffstaetter, Bagriantsev, and Gracheva, 2018)
138   **In a rat, that set point** (Hoffstaetter, Bagriantsev, and Gracheva, 2018)
138   **Fish seem to lack TRPM8 altogether** (Gracheva and Bagriantsev, 2015)
138   **Matos-Cruz found that** (Matos-Cruz et al., 2017)
138   **There's a version of the human TRPM8** (Key et al., 2018)
139   **The TRPV1 sensor** (Hoffstaetter, Bagriantsev, and Gracheva, 2018)
139   **In Gracheva's hotplate tests** (Laursen et al., 2016)
139   **The Saharan silver ant** (Gehring and Wehner, 1995; Ravaux et al., 2013)
139   **Snow flies are active** (Hartzell et al., 2011)
140   **The air 5 millimeters above** (Corfas and Vosshall, 2015)
140   **If it landed on my head** (Heinrich, 1993)
140   **Neuroscientist Marco Gallio demonstrated** (Simões et al., 2021)
141   **Fish, from tiny larvae** (Wurtsbaugh and Neverman, 1988; Thums et al., 2013)
141   **Sulfide worms that live** (Bates et al., 2010)
141   **Butterflies that are warming** (Tsai et al., 2020)

141   **Turtle embryos can even** (Du et al., 2011)

141   **At 11:20 A.M. on August 10, 1925** (Schmitz and Bousack, 2012)

142   **These black, half-inch-long insects** (Linsley, 1943)

142   **One summer, Linsley saw them** (Linsley and Hurd, 1957)

142   **Arriving at a fire** (Schmitz, Schmitz, and Schneider, 2016)

143   **And though their antennae** (Schütz et al., 1999)

143   **The atoms and molecules** (Dusenbery, 1992; Schmitz, Schmitz, and Schneider, 2016)

143   **When zoologist Helmut Schmitz** (Schmitz and Bleckmann, 1998)

144   **Based on this distance** (Schmitz and Bousack, 2012)

144   **During flight, their beating wings** (Schneider, Schmitz, and Schmitz, 2015)

145   **The 11 species of *Melanophila*** (Schmitz, Schmitz, and Schneider, 2016)

145   **The exceptions include the species** (Bisoffi et al., 2013)

146   **So is heat** (Bryant and Hallem, 2018; Bryant et al., 2018)

146   **It's likely that the majority** (Windsor, 1998; Forbes et al., 2018)

147   **It's no surprise that at least** (Lazzari, 2009; Chappuis et al., 2013; Corfas and Vosshall, 2015)

147   **The vampire's heat sensors** (Kürten and Schmidt, 1982)

147   **Elena Gracheva studied those neurons** (Gracheva et al., 2011)

148   **Ann Carr and Vincent Salgado** (Carr and Salgado, 2019)

150   **Heat-sensitive pits have evolved** (Goris, 2011)

150   **And yet, Elena Gracheva found** (Gracheva et al., 2010)

150   **No one hit on the right answer** (Ros, 1935)

150   **Rattlesnakes will strike at warm objects** (Noble and Schmidt, 1937)

150   **Even a congenitally blind rattlesnake** (Kardong and Mackessy, 1991)

151   **They'll respond if the membrane** (Bullock and Diecke, 1956)

151   **This astonishing sensitivity means** (Ebert and Westhoff, 2006)

151   **There, the two streams are combined** (Hartline, Kass, and Loop, 1978; Newman and Hartline, 1982)

151   **"It is a fallacy"** (Goris, 2011)

152   **When confronted, they raise their tails** (Rundus et al., 2007)

152   **He got grainy images** (Bakken and Krochmal, 2007)

152   **Sidewinders tend to point** (Schraft, Bakken, and Clark, 2019)

152   **And on China's Shedao Island** (Shine et al., 2002)

152   **Chinese herpetologist Yezhong Tang** (Chen et al., 2012)

153   **The nerves in their membranes** (Goris, 2011)

153   **Ecologist Burt Kotler** (Bleicher et al., 2018; Embar et al., 2018)

153   **But Schraft found that blindfolded sidewinders** (Schraft and Clark, 2019)

153   **Schraft presented them with lizard carcasses** (Schraft, Goodman, and Clark, 2018)

154   **In 2013, Viviana Cadena** (Cadena et al., 2013)

154   **In a refreshing act of** (Bakken et al., 2018)

154   **Kröger found that** (Gläser and Kröger, 2017; Kröger and Goiricelaya, 2017)

155   **His team successfully trained three dogs** (Bálint et al., 2020)

CHAPTER 6

157   **Orphaned and stranded** (Monterey Bay Aquarium, 2016)

157   **They do have the densest fur** (Kuhn et al., 2010)

157   **To stay warm** (Costa and Kooyman, 2011)

157   **They're always diving** (Yeates, Williams, and Fink, 2007)

157   **The sensitivity of their paws** (Radinsky, 1968)

157   **Different sections of the somatosensory cortex** (Wilson and Moore, 2015)

158   **To measure what these mittens** (Strobel et al., 2018)

158   **Likewise, Strobel found that humans** (Strobel et al., 2018)

158 **It will only stay submerged** (Thometz et al., 2016)
159 **Touch is one of the mechanical senses** A review of touch is Prescott and Dürr (2015).
159 **These cells come in several varieties** The various kinds of touch sensors are reviewed in Zimmerman, Bai, and Ginty (2014); Moayedi, Nakatani, and Lumpkin (2015).
160 **But exactly how this happens** (Walsh, Bautista, and Lumpkin, 2015)
160 **In one experiment** (Carpenter et al., 2018)
160 **In another test** (Skedung et al., 2013)
161 **These incredible feats are possible** (Prescott, Diamond, and Wing, 2011)
161 **Mark Rutland, who led** (Skedung et al., 2013)
161 **The star-nosed mole** Catania's account of his work with the star-nosed mole is Catania (2011).
162 **Scientists have long speculated** (Catania, 1995b)
162 **A star-nosed mole embryo** (Catania, Northcutt, and Kaas, 1999)
162 **The mole's somatosensory cortex** (Catania et al., 1993)
163 **Around 5 percent of star-nosed moles** (Catania and Kaas, 1997b)
163 **The 11th pair of rays** (Catania, 1995a)
163 **By filming the mole** (Catania and Kaas, 1997a)
164 **By analyzing such footage** (Catania and Remple, 2004, 2005)
165 **In chickens, which rely heavily** (Gentle and Breward, 1986)
165 **But in some ducks** (Schneider et al., 2014, 2017)
165 **"Imagine being given a bowl"** (Birkhead, 2013, p. 78)
166 **Compared to other ducks** (Schneider et al., 2019)
166 **But in 1995, Theunis Piersma** (Piersma et al., 1995)
166 **This simple experiment revealed** (Piersma et al., 1998)
167 **Ibises use the technique** (Cunningham, Castro, and Alley, 2007; Cunningham et al., 2010)
167 **Ram Gal and Frederic Libersat** (Gal et al., 2014)
168 **The remoras, or suckerfishes** (Cohen et al., 2020)
168 **The round goby** (Hardy and Hale, 2020)
168 **The whiskered auklet** (Seneviratne and Jones, 2008)
168 **When Sampath Seneviratne placed** (Seneviratne and Jones, 2008)
168 **It's more likely that they're touch sensors** (Cunningham, Alley, and Castro, 2011)
168 **It's clear that birds evolved** (Persons and Currie, 2015)
169 **Mammalian hair might have** (Prescott and Dürr, 2015)
169 **They're called vibrissae** A review of mammalian vibrissae is Prescott, Mitchinson, and Grant (2011).
169 **This action, delightfully known as whisking** (Bush, Solla, and Hartmann, 2016)
169 **The rodent constantly scans** (Grant, Breakell, and Prescott, 2018)
169 **If it senses something** (Grant, Sperber, and Prescott, 2012)
169 **If we turn our head** (Arkley et al., 2014)
170 **Mammals have been using whiskers** (Mitchinson et al., 2011)
170 **Grant showed that the opossum** (Mitchinson et al., 2011)
171 **The disk is muscular** (Marshall, Clark, and Reep, 1998)
171 **There are around 2,000** The vibrissae of manatees are reviewed in Reep and Sarko (2009); Bauer, Reep, and Marshall (2018).
172 **But when it's time to eat** (Marshall et al., 1998)
172 **In 2012, Bauer tested Hugh** (Bauer et al., 2012)
172 **A few other mammals** (Crish, Crish, and Comer, 2015; Sarko, Rice, and Reep, 2015)
172 **Manatees use these body-wide whiskers** (Reep, Marshall, and Stoll, 2002)
173 **Bauer and his colleagues** (Gaspard et al., 2017)
174 **Sprouts has around a hundred facial whiskers** (Hanke and Dehnhardt, 2015)
174 **Sprouts can use them** (Murphy, Reichmuth, and Mann, 2015)
174 **The seals actively keep the whiskers** (Dehnhardt, Mauck, and Hyvärinen, 1998)

174 **This ability was only discovered in 2001** (Dehnhardt et al., 2001)

176 **The Rostock team showed** (Hanke et al., 2010)

176 **From those impressions alone** (Wieskotten et al., 2010)

176 **It can discriminate between the wakes** (Wieskotten et al., 2011)

176 **In one experiment, Henry** (Niesterok et al., 2017)

177 **The lateral line is found** A review of the lateral line is Montgomery, Bleckmann, and Coombs (2013).

177 **After describing the pores** (Dijkgraaf, 1989)

177 **In the 1930s, the biologist** (Dijkgraaf, 1989)

177 **In 1908, the ichthyologist Bruno Hofer** (Hofer, 1908)

177 **If it swims toward an aquarium** (Dijkgraaf, 1963)

177 **In 1963, Dijkgraaf summarized** (Dijkgraaf, 1963)

178 **With the lateral line** (Webb, 2013; Mogdans, 2019)

178 **Schooling fish use their lateral lines** (Partridge and Pitcher, 1980)

178 **Blind fish can still school** (Pitcher, Partridge, and Wardle, 1976)

178 **Though all fish share** (Webb, 2013)

178 **Surface-feeding fish** (Mogdans, 2019)

178 **Halfbeaks have massive underbites** (Montgomery and Saunders, 1985)

178 **Blind cavefish have lost their sight** (Yoshizawa et al., 2014; Lloyd et al., 2018)

178 **Some blind cavefish have evolved** (Patton, Windsor, and Coombs, 2010)

179 **Examining it under a microscope** (Haspel et al., 2012)

179 **Soares showed that the joysticks** (Haspel et al., 2012)

180 **The bumps, she discovered** (Soares, 2002)

180 **Crocodilians—alligators, crocodiles** (Soares, 2002)

181 **And yet, they are covered** (Leitch and Catania, 2012)

181 **Many snakes have thousands** (Crowe-Riddell, Williams, et al., 2019)

181 *Spinosaurus,* **an enormous sail-backed dinosaur** (Ibrahim et al., 2014)

181 *Daspletosaurus,* **a close relative of** *Tyrannosaurus* (Carr et al., 2017)

182 **To find out if they** (Kane, Van Beveren, and Dakin, 2018)

182 **These results suggest that a peahen** (Kane, Van Beveren, and Dakin, 2018)

183 **But filoplumes are especially important** (Necker, 1985; Clark and de Cruz, 1989)

183 **This rarely happens, in part because filoplumes** (Brown and Fedde, 1993)

183 **They are covered with a smattering** (Sterbing-D'Angelo et al., 2017)

184 **When Sterbing treated bat wings** (Sterbing-D'Angelo and Moss, 2014)

184 **In 1960, a shipment of bananas** Barth's account of his work with the tiger wandering spider is Barth (2002).

184 **Its legs are covered in hundreds** (Barth, 2015)

185 **If it is running** (Seyfarth, 2002)

185 **Even air that's moving** (Barth and Höller, 1999)

185 **It grabs the fly from the air** (Klopsch, Kuhlmann, and Barth, 2012, 2013)

186 **Many have airflow sensors** (Casas and Dangles, 2010)

186 **It is fast, but Casas found** (Dangles, Casas, and Coolen, 2006; Casas and Steinmann, 2014)

186 **"Spiders can detect danger"** (Di Silvestro, 2012)

186 **These hairs are a hundred times** (Shimozawa, Murakami, and Kumagai, 2003)

187 **For example, in 1978, Jürgen Tautz** (Tautz and Markl, 1978)

187 **Thirty years later, Tautz showed** (Tautz and Rostás, 2008)

CHAPTER 7

189 **They collected batches of eggs** (Warkentin, 1995)

189 **Their experiments showed** (Cohen, Seid, and Warkentin, 2016)

189 **By recording different vibrations** (Warkentin, 2005; Caldwell, McDaniel, and Warkentin, 2010)

190   **They can clearly sense the world** A review of environmentally cued hatching in embryos is Warkentin (2011).

190   **But they don't respond to snakes** (Jung et al., 2019)

190   **So Jung built a jury-rigged** (Jung et al., 2019)

190   **By watching them with infrared cameras** (Caldwell, McDaniel, and Warkentin, 2010)

191   **Male fiddler crabs** (Takeshita and Murai, 2016)

191   **Termite soldiers drum their heads** (Hager and Kirchner, 2013)

191   **Water striders—insects that skate** (Han and Jablonski, 2010)

191   **Scientists call these substrate-borne vibrations** (Hill, 2009; Hill and Wessel, 2016; Mortimer, 2017)

191   **Surface waves, by contrast** (Hill, 2014)

192   **"We have encountered it, but"** A seminal text by Peggy Hill about vibrational communication is Hill (2008). The quote appears on page 2.

193   **By rapidly contracting muscles** Insect vibrational communication is reviewed in Cocroft and Rodríguez (2005); Cocroft (2011).

193   **Insects exploit that property** (Cokl and Virant-Doberlet, 2003)

194   **Cocroft now has a library** It can be found at treehoppers.insectmuseum.org.

194   **A treehopper can produce** (Cocroft and Rodríguez, 2005)

195   **Some babies produce synchronized vibrations** (Cocroft, 1999)

195   **Some mothers produce vibrations** (Hamel and Cocroft, 2012)

195   **They court each other** (Legendre, Marting, and Cocroft, 2012)

195   **Many duetting insects will jam** (Eriksson et al., 2012; Polajnar et al., 2015)

195   **Masked birch caterpillars scrape** (Yadav, 2017)

195   **Acacia ants vigorously defend** (Hager and Krausa, 2019)

196   **In 1949, three decades before** (Ossiannilsson, 1949)

197   **"Gentle disturbances of the sand"** Brownell's account of his sand scorpion work is Brownell (1984).

197   **Brownell and Farley tested this idea** (Brownell and Farley, 1979c)

198   **Its sensors lie in its feet** (Brownell and Farley, 1979a)

198   **The first time this happens** (Brownell and Farley, 1979b)

198   **Can animals sense earthquakes** (Woith et al., 2018)

199   **The footfalls of an ant** (Fertin and Casas, 2007; Martinez et al., 2020)

199   **It reacts by tossing sand** (Mencinger-Vračko and Devetak, 2008)

199   **Ken Catania—the same man** (Catania, 2008; Mitra et al., 2009)

200   **"if the ground is beaten"** (Darwin, 1890)

201   **But it is highly sensitive** (Mason, 2003)

201   **The golden mole forages at night** (Lewis et al., 2006)

201   **Peter Narins has suggested** (Narins and Lewis, 1984; Mason and Narins, 2002)

201   **The golden mole's version** (Mason, 2003)

201   **"The lying about"** (Hill, 2008, p. 120)

202   **In the early 1990s, Caitlin O'Connell** O'Connell's account of her own elephant work is O'Connell (2008).

202   **The animals seemed to be listening** (O'Connell-Rodwell, Hart, and Arnason, 2001)

202   **In 2002, O'Connell returned** (O'Connell-Rodwell et al., 2006)

202   **"All these years of planning"** (O'Connell, 2008, p. 180)

202   **A few years later, she repeated** (O'Connell-Rodwell et al., 2007)

203   **The vibrations can travel over** (O'Connell, Arnason, and Hart, 1997; Günther, O'Connell-Rodwell, and Klemperer, 2004)

203   **As we spread around the globe** (Smith et al., 2018)

203   **Between 30 and 60 million bison** (Phippen, 2016)

204   **"The Lakota . . . loved the earth"** (Standing Bear, 2006, p. 192)

205   **Spiders have been around** An excellent book on spider silk and its evolution is Brunetta and Craig (2012).

205 **Though light and elastic** (Agnarsson, Kuntner, and Blackledge, 2010)
205 **The orb web is a trap** (Blackledge, Kuntner, and Agnarsson, 2011)
206 **From this position** (Masters, 1984)
206 **They can probably work out** (Landolfa and Barth, 1996)
206 **They can assess the size** (Robinson and Mirick, 1971; Suter, 1978)
206 **If the prey stops moving** (Klärner and Barth, 1982)
206 **The small dewdrop spider** *Argyrodes* (Vollrath, 1979a, 1979b)
207 **Some assassin bugs** (Wignall and Taylor, 2011)
207 *Portia,* **a jumping spider** (Wilcox, Jackson, and Gentile, 1996)
207 **"a small woven world"** (Barth, 2002, p. 19)
207 **By using gas guns** (Mortimer et al., 2014)
207 **It can do this every time it builds** (Mortimer et al., 2016)
207 **Zoologist Takeshi Watanabe showed** (Watanabe, 1999, 2000)
208 **Orb-weavers will also** (Nakata, 2010, 2013)
208 **The web, then, is not just** A great review of spiderwebs as examples of extended cognition is Japyassú and Laland (2017).
208 **Biophysicist Natasha Mhatre showed that** (Mhatre, Sivalinghem, and Mason, 2018)

CHAPTER 8

210 **To test this idea** Payne's account of his own work on barn owls is Payne (1971).
210 **Over the next four years** (Payne, 1971)
211 **If a mouse rustles** (Dusenbery, 1992)
212 **The barn owl's ear** (Konishi, 1973, 2012)
212 **Their hair cells regenerate** (Krumm et al., 2017)
212 **Masakazu Konishi and Eric Knudsen** (Knudsen, Blasdel, and Konishi, 1979)
213 **An owl's ears, however** (Payne, 1971)
213 **The owl's brain uses** (Carr and Christensen-Dalsgaard, 2015, 2016)
214 **William Stebbins once encapsulated this beautifully** An old but good review of animal hearing is Stebbins (1983). The quote is from page 1.
214 **Fortunately, the owl has soft feathers** (Weger and Wagner, 2016; Clark, LePiane, and Liu, 2020)
214 **The noise it does make** (Konishi, 2012)
215 **These little hopping rodents** (Webster and Webster, 1980)
215 **So they're especially difficult** (Webster, 1962; Stangl et al., 2005)
215 **They can even hear the sounds** (Webster and Webster, 1971)
215 **They have also evolved ears** Insect ears are reviewed in Fullard and Yack (1993); Göpfert and Hennig (2016).
215 **After all, the first insects** (Göpfert and Hennig, 2016)
215 **They had to evolve ears** (Robert, Mhatre, and McDonagh, 2010)
216 **Ears exist on the knees** (Göpfert, Surlykke, and Wasserthal, 2002; Montealegre-Z et al., 2012)
216 **Mosquitoes hear with their antennae** (Menda et al., 2019)
216 **Monarch caterpillars hear** (Taylor and Yack, 2019)
216 **The bladder grasshopper** (Yager and Hoy, 1986; Van Staaden et al., 2003)
216 **"In later years some further ears"** (Pye, 2004)
216 **Insect ears are so diverse** (Fullard and Yack, 1993)
217 **Accordingly, many insects seem** (Strauß and Stumpner, 2015)
217 **Many butterflies, including** (Lane, Lucas, and Yack, 2008)
217 **Instead, Jayne Yack has shown** (Fournier et al., 2013)
217 **Fossilized insects that have** (Gu et al., 2012)
218 **But Daniel Robert is so familiar** (Robert, Amoroso, and Hoy, 1992)

218 **It can turn toward a singing cricket** (Mason, Oshinsky, and Hoy, 2001; Müller and Robert, 2002)

218 **But Robert and his mentor** (Miles, Robert, and Hoy, 1995)

219 **Through several painstaking studies, Barbara Webb** (Webb, 1996)

219 **Webb even built a simple robot** (Webb, 1996)

219 **This happened within 20 generations** (Zuk, Rotenberry, and Tinghitella, 2006; Schneider et al., 2018)

220 **the túngara frog** (Ryan, 1980)

221 **Ryan knows this because he spent** (Ryan, 1980)

221 **Females almost always go for males** (Ryan et al., 1990)

221 **Ryan found that the frog's inner ear** (Ryan and Rand, 1993)

221 **Ryan discovered the actual story** (Ryan and Rand, 1993)

222 **This discovery flipped Ryan's narrative** (Ryan and Rand, 1993)

222 **Ryan calls this phenomenon "sensory exploitation"** His account of his work on túngara frogs is Ryan (2018).

222 **But Alexandra Basolo found** (Basolo, 1990)

223 **Tuttle and Ryan showed** (Tuttle and Ryan, 1981)

223 **One of Ryan's students, Rachel Page** (Page and Ryan, 2008)

223 **Another of Ryan's students, Ximena Bernal** (Bernal, Rand, and Ryan, 2006)

224 **Bird enthusiasts have long suspected** Bird hearing is reviewed in Dooling and Prior (2017).

224 **A mockingbird doesn't need** (Birkhead, 2013)

224 **In the 1960s, before his work** (Konishi, 1969)

225 **From the 1970s onward** (Dooling, Lohr, and Dent, 2000)

225 **Dooling confirmed this through** (Dooling et al., 2002)

226 **When Beth Vernaleo** (Vernaleo and Dooling, 2011)

226 **They completely shuffled the order** (Lawson et al., 2018)

226 **A zebra finch's song** (Dooling and Prior, 2017)

226 **Not all species** (Fishbein et al., 2020)

227 **But Dooling's colleague Nora Prior** (Prior et al., 2018)

228 **He and his colleagues placed electrodes** (Lucas et al., 2002)

228 **Likewise, ears can have exceptional** (Henry et al., 2011)

229 **Lucas found that in the fall** (Lucas et al., 2007)

229 **The hearing of the white-breasted nuthatch** (Lucas et al., 2007)

229 **This might explain why** (Noirot et al., 2009)

229 **Lucas and his colleague Megan Gall** (Gall, Salameh, and Lucas, 2013)

230 **It might get duller with age** (Caras, 2013)

230 **Males of the plainfin midshipman fish** (Sisneros, 2009)

230 **Green tree frogs** (Gall and Wilczynski, 2015)

231 **In the 1960s** (Kwon, 2019)

231 **One, based on recordings that Payne** (Payne and McVay, 1971)

231 **The second showed that fin whales** (Payne and Webb, 1971)

231 **The actual source only became clear** (Schevill, Watkins, and Backus, 1964)

231 **Below those frequencies** (Narins, Stoeger, and O'Connell-Rodwell, 2016)

232 **Knowing that fin whales** (Payne and Webb, 1971)

232 **Amid the spectrograms** (Clark and Gagnon, 2004)

232 **On his first day, Clark** (Costa, 1993)

233 **Geophysicists can certainly use** (Kuna and Nábělek, 2021)

233 **He also suspects that the animals** (Tyack and Clark, 2000)

234 **That might seem preposterous** (Goldbogen et al., 2019)

234 **Those ancestral creatures** (Mourlam and Orliac, 2017)

234 **The mysticetes achieved their huge size** (Shadwick, Potvin, and Goldbogen, 2019)

235 **In May 1984, Katy Payne** Her account of her own elephant research is Payne (1999).

235 **"It had been like the feeling"** (Payne, 1999, p. 20)

236 **But when Payne sped up the recordings** (Payne, Langbauer, and Thomas, 1986)

236 **She accepted, and in 1986** (Poole et al., 1988)

236 **At close range** (Poole et al., 1988)

237 **A few hours after sunset** (Garstang et al., 1995)

238 **Her work strongly suggests** (Ketten, 1997)

238 **These superlatively big animals** (Miles, Robert, and Hoy, 1995)

238 **In the winter of 1877** (Sidebotham, 1877)

238 **Roughly a century later** (Noirot, 1966; Zippelius, 1974; Sales, 2010)

239 **Pups that are separated** (Sewell, 1970)

239 **Rats that are tickled** (Panksepp and Burgdorf, 2000)

239 **Richardson's ground squirrels** (Wilson and Hare, 2004)

239 **Male mice that sniff** (Holy and Guo, 2005)

239 **Females attracted to these serenades** (Neunuebel et al., 2015)

239 **It refers to sound waves** A review of ultrasonic communication is Arch and Narins (2008).

239 **A dog can hear 45 kHz** (Heffner, 1983; Heffner and Heffner, 1985, 2018; Kojima, 1990; Ridgway and Au, 2009; Reynolds et al., 2010)

239 **Rickye and Henry Heffner** (Heffner and Heffner, 2018)

240 **Subterranean animals are a striking exception** (Heffner and Heffner, 2018)

240 **This means that ultrasonic calls** (Arch and Narins, 2008)

240 **This is also why devices** (Aflitto and DeGomez, 2014)

241 **That's what Marissa Ramsier noticed** (Ramsier et al., 2012)

241 **The blue-throated hummingbird** (Pytte, Ficken, and Moiseff, 2004)

241 **Several other hummingbirds** (Duque, Rodríguez-Saltos, and Wilczynski, 2018; Olson et al., 2018)

241 **This orange frog is insensitive** (Goutte et al., 2017)

241 **More than half of the 160,000 species** The battle between insects and bats is reviewed in Conner and Corcoran (2012).

241 **The greater wax moth** (Moir, Jackson, and Windmill, 2013)

242 **Some moths do make ultrasonic** (Nakano et al., 2009, 2010)

242 **The most likely answer** (Kawahara et al., 2019)

242 **Moth ears almost always evolved** (Kawahara et al., 2019)

CHAPTER 9

243 **By listening for the returning echoes** Echolocation is thoroughly reviewed in Surlykke et al. (2014).

244 **During the day, sharp-eyed predators** (Boonman et al., 2013)

244 **It's actually the other way round** (Kalka, Smith, and Kalko, 2008)

245 **In the 1790s, the Italian priest** The history of echolocation research is reviewed in Griffin (1974); Grinnell, Gould, and Fenton (2016).

245 **The meaning of these observations** Donald Griffin's classic work on echolocation and his research is Griffin (1974).

246 **For over a century** (Griffin, 1974)

246 **"We were surprised and delighted"** (Griffin, 1974, p. 67)

246 **A year later, Griffin** (Griffin and Galambos, 1941; Galambos and Griffin, 1942)

246 **But the duo did mean it** (Griffin, 1944a)

247 **Sitting by a pond near Ithaca** (Griffin, 1953)

247 **It's also how bats hunt** (Griffin, Webster, and Michael, 1960)

247 **"Our scientific imaginations"** (Griffin, 2001)

248 **The origins of echolocation** (Jones and Teeling, 2006)

248 **The basic process seems straightforward** (Schnitzler and Kalko, 2001; Fenton et al., 2016; Moss, 2018)

249 **An average bat can only** (Surlykke and Kalko, 2008)
249 **Anything farther away is probably imperceptible** (Holderied and von Helversen, 2003)
249 **That's because bats concentrate** (Jakobsen, Ratcliffe, and Surlykke, 2013)
249 **The big brown bat** (Ghose, Moss, and Horiuchi, 2007)
249 **Annemarie Surlykke showed that** (Hulgard et al., 2016)
249 **Even the so-called whispering bats** (Brinkløv, Kalko, and Surlykke, 2009)
249 **This desensitizes their hearing** (Henson, 1965; Suga and Schlegel, 1972)
249 **This is called acoustic gain control** (Kick and Simmons, 1984)
249 **John Ratcliffe showed that** (Elemans et al., 2011; Ratcliffe et al., 2013)
250 **And as James Simmons** (Simmons, Ferragamo, and Moss, 1998)
251 **All of these frequencies** (Simmons and Stein, 1980; Moss and Schnitzler, 1995)
251 **It knows the insect's position** (Zagaeski and Moss, 1994)
251 **A bat must constantly adjust its sonar** (Moss and Surlykke, 2010; Moss, Chiu, and Surlykke, 2011)
252 **Bats can race through rugged caves** (Grinnell and Griffin, 1958)
252 **These messy spaces pose special problems** (Surlykke, Simmons, and Moss, 2016)
252 **She also found that** (Chiu, Xian, and Moss, 2009)
252 **They also tend to group their calls** (Moss et al., 2006; Kothari et al., 2014)
252 **The chestnut sac-winged bat** (Jung, Kalko, and von Helversen, 2007)
253 **Inga Geipel found that the bat** (Geipel, Jung, and Kalko, 2013; Geipel et al., 2019)
253 **Big browns do this by** (Chiu and Moss, 2008; Chiu, Xian, and Moss, 2008)
253 **Some bats can recognize the sonar** (Yovel et al., 2009)
253 **The greater bulldog bat** (Suthers, 1967)
253 **Researchers have called this the "cocktail party nightmare"** (Ulanovsky and Moss, 2008; Corcoran and Moss, 2017)
253 **This explains the many historical incidents** (Griffin, 1974)
254 **an entire section to "bumbling bats"** (Griffin, 1974, p. 160)
254 **They can distinguish two grades of sandpaper** (Zagaeski and Moss, 1994)
255 **But around 160 species** (Schnitzler and Denzinger, 2011; Fenton, Faure, and Ratcliffe, 2012)
256 **Hans-Ulrich Schnitzler, who** (Kober and Schnitzler, 1990; von der Emde and Schnitzler, 1990; Koselj, Schnitzler, and Siemers, 2011)
257 **The greater horseshoe bat** (Schuller and Pollak, 1979; Schnitzler and Denzinger, 2011)
257 **Other species have their own signature** (Grinnell, 1966; Schuller and Pollak, 1979)
257 **But as Schnitzler discovered in 1967** (Schnitzler, 1967)
257 **And they do this (quite literally)** (Schnitzler, 1973)
258 **A horseshoe bat can throw its attention** (Hiryu et al., 2005)
258 **When not inflaming the airways** (Ntelezos, Guarato, and Windmill, 2016; Neil et al., 2020)
258 **This acoustic armor** (Conner and Corcoran, 2012)
259 **So while bats can hear** (Surlykke and Kalko, 2008)
259 **Others talk back** (Dunning and Roeder, 1965)
259 **Dorothy Dunning and Kenneth Roeder** (Dunning and Roeder, 1965)
259 **Many tiger moths are full** (Barber and Conner, 2007)
259 **In 2009, Aaron Corcoran** (Corcoran, Barber, and Conner, 2009)
259 **The clicks overlapped** (Corcoran et al., 2011)
259 **But unlike the tiger moths** (Barber and Kawahara, 2013)
259 **With their stealthy whispers** (Goerlitz et al., 2010; ter Hofstede and Ratcliffe, 2016)
260 **On average, a luna moth** (Barber et al., 2015)
260 **Moths have evolved long tails** (Rubin et al., 2018)
260 **Donald Griffin once described** (Griffin, 2001)
261 **Both groups did so by evolving echolocation** Echolocation in whales and bats is compared in Au and Simmons (2007); Surlykke et al. (2014).

261 **After watching porpoises** (Schevill and McBride, 1956)
261 **Ken Norris carried out** (Norris et al., 1961)
262 **So researchers who study dolphins** Dolphin echolocation research is reviewed in Au (2011); Nachtigall (2016).
262 **A field station in Hawaii's** Whitlow Au's seminal work on dolphin sonar is Au (1993).
262 **At Kāneʻohe Bay, where bottlenose dolphins** (Au, 1993)
262 **Dolphins could discriminate between different objects** (Au and Turl, 1983)
263 **The animal, Kina, could use** (Brill et al., 1992)
263 **Dolphins can also echolocate on** (Pack and Herman, 1995; Harley, Roitblat, and Nachtigall, 1996)
264 **At the top of the dolphin's head** (Cranford, Amundin, and Norris, 1996)
264 **The sperm whale** (Madsen et al., 2002)
264 **At 236 decibels** (Møhl et al., 2003)
265 **Odontocetes also intercept their own echoes** (Mooney, Yamato, and Branstetter, 2012)
265 **When they need more information** (Finneran, 2013)
265 **They can adjust the sensitivity** (Nachtigall and Supin, 2008)
265 **In an early experiment, Au showed** (Au, 1993)
266 **Later studies showed that echolocating dolphins** (Ivanov, 2004; Finneran, 2013)
266 **Sound also interacts differently** (Madsen and Surlykke, 2014)
266 **If a dolphin echolocates on you** (Au, 1996)
266 **It can pick out the air-filled swim bladders** (Au et al., 2009)
266 **The American shad** (Popper et al., 2004)
266 **Pavel Gol'din has suggested** (Gol'din, 2014)
267 **But despite their rarity** (Tyack, 1997; Tyack and Clark, 2000)
267 **One way to find out** (Johnson, Aguilar de Soto, and Madsen, 2009)
267 **Since 2003, one team of researchers** (Johnson et al., 2004; Arranz et al., 2011; Madsen et al., 2013)
268 **Kelly Benoit-Bird and Whitlow Au showed** (Benoit-Bird and Au, 2009a, 2009b)
268 **When Daniel Kish clicks** (Thaler et al., 2017)
269 **Now in his 50s, Kish** (Kish, 2015)
269 **Small mammals might make ultrasonic clicks** (Gould, 1965; Eisenberg and Gould, 1966; Siemers et al., 2009)
269 **Certain fruit bats** (Boonman, Bumrungsri, and Yovel, 2014)
269 **The oilbird, a large South American** (Brinkløv and Warrant, 2017; Brinkløv, Elemans, and Ratcliffe, 2017)
270 **Swiftlets, small insect-eating birds** (Brinkløv, Fenton, and Ratcliffe, 2013)
270 **And as Kish and** (Thaler and Goodale, 2016)
270 **Why don't seals echolocate?** (Schusterman et al., 2000)
271 **Since at least 1749** (Diderot, 1749; Supa, Cotzin, and Dallenbach, 1944; Kish, 1995)
272 **In the 1940s** (Supa, Cotzin, and Dallenbach, 1944)
272 **Supa referenced the bat studies** (Griffin, 1944a)
274 **Neuroscientist Lore Thaler** (Thaler, Arnott, and Goodale, 2011)
274 **Without vision, the brain** (Norman and Thaler, 2019)
275 **His memory, his cane** (Thaler et al., 2020)

CHAPTER 10

276 **Around 350 species of fish** For primers on electric fish, see Hopkins (2009); Carlson et al. (2019).
276 **Around 5,000 years ago** For a history of electric fish, see Wu (1984); Zupanc and Bullock (2005); Carlson and Sisneros (2019).
276 **For a fuller account, try** (Finger and Piccolino, 2011)

277   **None do this better than electric eels** (Catania, 2019)

277   **In 1800, Chayma fishers** (Catania, 2016)

277   **And a team led by** (de Santana et al., 2019)

277   **Their discharges are so faint** (Hopkins, 2009)

278   **"It is impossible to conceive"** (Darwin, 1958, p. 178)

279   **Hans Lissmann was a Ukrainian-born** Lissmann's eventful life is detailed in Alexander (1996).

279   **During a fateful visit** (Turkel, 2013)

279   **In 1951, Lissmann used electrodes** (Lissmann, 1951)

279   **And by sensing those distortions** (Lissmann, 1958)

279   **Lissmann and Machin published their results** (Lissmann and Machin, 1958)

280   **The fish can detect these differences** Good reviews on active electrolocation include Lewis (2014); Caputi (2017).

280   **The black ghost knifefish** (von der Emde, 1990, 1999; von der Emde et al., 1998; Snyder et al., 2007)

281   **"A most interesting consequence"** (Hopkins, 2009)

281   **It is also omnidirectional** (Snyder et al., 2007)

282   **To double the range** (Salazar, Krahe, and Lewis, 2013)

282   **Elephantfish eyes seem to be tuned** (von der Emde and Ruhl, 2016)

283   **The omnidirectional nature of electrolocation** (Caputi et al., 2013)

283   **They'll wrap their bodies** (Caputi, Aguilera, and Pereira, 2011)

283   **Angel Caputi has argued** (Caputi et al., 2013)

283   **The electric sense** *evolved from* (Baker, 2019)

283   **Electroreceptors grow from the same** (Modrell et al., 2011; Baker, Modrell, and Gillis, 2013)

284   **Electric fields aren't blocked by barriers** (Lewis, 2014)

284   **They are sensitive not only to conductance** (von der Emde, 1990)

285   **For decades, scientists have studied** (Carlson and Sisneros, 2019)

285   **But the animals' actual worlds** For some of the challenges of field research, see Hagedorn (2004).

285   **Such electrodes have improved over time** (Henninger et al., 2018; Madhav et al., 2018)

285   **They court mates, claim territory** For more on electrocommunication, see Zupanc and Bullock (2005); Baker and Carlson (2019).

286   **The shape of these pulses** (Hopkins, 1981; McGregor and Westby, 1992; Carlson, 2002)

286   **One rhythm might be as attractive** (Hopkins and Bass, 1981)

286   **The neuroscientist Ted Bullock** (Bullock, Behrend, and Heiligenberg, 1975)

286   **In** *Sensory Exotica* (Hughes, 2001)

286   **By minutely changing the frequencies** (Bullock, 1969)

286   **By briefly and sharply increasing** (Hagedorn and Heiligenberg, 1985)

287   **If two** *Eigenmannia* **meet** (Bullock, Behrend, and Heiligenberg, 1975)

287   **The mormyrins have altered** (Carlson and Arnegard, 2011; Vélez, Ryoo, and Carlson, 2018)

287   **Carlson suspects that these changes** (Baker, Huck, and Carlson, 2015)

288   **One species, the Ubangi elephantfish** (Nilsson, 1996; Sukhum et al., 2016)

288   **Carlson has shown that one mormyrid** (Arnegard and Carlson, 2005)

289   **They probe the sandy floor** (Amey-Özel et al., 2015)

290   **In 1960, biologist R. W. Murray** (Murray, 1960)

290   **A few years later, Sven Dijkgraaf** (Dijkgraaf and Kalmijn, 1962)

290   **The jelly inside the ampullae** (Josberger et al., 2016)

290   **It turns out that** *all* **living things** (Kalmijn, 1974)

291   **These fields are thousands of times** (Kalmijn, 1974; Bedore and Kajiura, 2013)

291 **Kalmijn proved as much in 1971** (Kalmijn, 1971)

291 **Wild sharks will also bite** (Kalmijn, 1982)

291 **Some do so from birth** (Kajiura, 2003)

291 **The shark's electric sense** For reviews on passive electroreception, see Hopkins (2005, 2009).

291 **Some stingrays use electric fields** (Tricas, Michael, and Sisneros, 1995)

291 **And some embryonic sharks** (Kempster, Hart, and Collin, 2013)

292 **A shark's electric sense only works** (Kajiura and Holland, 2002)

292 **Over mile distances, a shark** (Gardiner et al., 2014)

292 **It's also why electric fields trigger** (Dijkgraaf and Kalmijn, 1962)

292 **When cuttlefish see the looming shapes** (Bedore, Kajiura, and Johnsen, 2015)

292 **Instead of conical snouts, hammerhead sharks** (Kajiura, 2001)

293 **It greatly extends the sawfish's** (Wueringer, Squire, et al., 2012a)

293 **She showed that the saw** (Wueringer, Squire, et al., 2012b)

293 **Wueringer founded an organization** (Wueringer, 2012)

293 **The ability to detect electric fields** Electroreception is reviewed in Collin (2019); Crampton (2019).

293 **Among vertebrates, around one in six** (Albert and Crampton, 2006)

293 **At least one species of dolphin** (Czech-Damal et al., 2012)

293 **Similarly, it's unclear how the echidnas** (Gregory et al., 1989)

294 **Their close relative, the platypus** (Pettigrew, Manger, and Fine, 1998; Proske and Gregory, 2003)

294 **This extensive cabal of electroreceptive critters** (Baker, Modrell, and Gillis, 2013)

294 **The knifefishes and elephantfishes are special cases** (Lavoué et al., 2012)

294 **These events happened at roughly** (Lavoué et al., 2012)

295 **Air, by contrast, is an insulator** (Czech-Damal et al., 2013)

295 **This is the atmospheric potential gradient** (Feynman, 1964)

296 **Attracted by their opposing charges** (Corbet, Beament, and Eisikowitch, 1982; Vaknin et al., 2000)

296 **In 2013, Robert and his colleagues** (Clarke et al., 2013)

296 **The bees also learned to more quickly** (Clarke et al., 2013)

297 **Instead, their electroreceptors are** (Sutton et al., 2016)

297 **The mere possibility of widespread aerial electroreception** Aerial electroreception is reviewed in Clarke, Morley, and Robert (2017).

298 **In 2018, Robert's colleague Erica Morley** (Morley and Robert, 2018)

298 **another scientist had suggested that spiders** (Blackwall, 1830)

299 **The rival won** It was resurrected in Gorham (2013).

CHAPTER 11

301 **Every spring, billions of bogongs** (Warrant et al., 2016)

301 **Warrant realized that** (Dreyer et al., 2018)

302 **Their ability, known as magnetoreception** For reviews of magnetoreception, see Johnsen and Lohmann (2005); Mouritsen (2018).

303 **Merkel and his students** (Merkel and Fromme, 1958; Pollack, 2012)

303 **In 1859, the zoologist** (Middendorff, 1855)

303 **Absent such proof, even Donald Griffin** (Griffin, 1944b)

304 **Merkel and Wiltschko provided that evidence** (Wiltschko and Merkel, 1965; Wiltschko, 1968)

304 **At roughly the same time** (Brown, 1962; Brown, Webb, and Barnwell, 1964)

304 **Earth's magnetic field** (Johnsen and Lohmann, 2005)

305 **Many scientists, including Wiltschko** (Wiltschko and Wiltschko, 2019)

305 **Since Merkel's robins took** (Lohmann et al., 1995; Deutschlander, Borland, and Phillips, 1999; Sumner-Rooney et al., 2014; Scanlan et al., 2018)

305 **After a busy night of insect-catching** (Holland et al., 2006)

305 **After an early life** (Bottesch et al., 2016)

305 **Mole-rats use their compass** (Kimchi, Etienne, and Terkel, 2004)

305 **And bogong moths** (Dreyer et al., 2018)

305 **To see if they do, Granger** (Granger et al., 2020)

306 **Robins can also be sent off course** (Bianco, Ilieva, and Åkesson, 2019)

306 **Few migrations are as treacherous** A review of sea turtle migrations is Lohmann and Lohmann (2019).

307 **By the 1990s, no one** (Carr, 1995)

307 **As Lohmann suspected** (Lohmann, 1991)

307 **In the mid-1990s** (Lohmann and Lohmann, 1994, 1996)

307 **But each turtle species** (Lohmann, Putman, and Lohmann, 2008)

307 **The turtles' abilities are especially impressive** (Lohmann et al., 2001)

308 **Loggerheads that survive** (Lohmann et al., 2004)

309 **Lohmann demonstrated this by capturing lobsters** (Boles and Lohmann, 2003)

309 **Every winter, thrush nightingales** (Fransson et al., 2001)

309 **Eurasian reed warblers** (Chernetsov, Kishkinev, and Mouritsen, 2008)

309 **Many animals, including salmon** (Putman et al., 2013; Wynn et al., 2020)

309 **Turtles use these imprints** (Lohmann, Putman, and Lohmann, 2008)

309 **Green turtles that nest on Ascension Island** (Mortimer and Portier, 1989)

309 **The geomagnetic field changes very slightly** (Brothers and Lohmann, 2018)

310 **Trying to find them** (Johnsen, 2017)

310 **At the time of writing** (Nordmann, Hochstoeger, and Keays, 2017)

311 **The first involves a magnetic iron mineral** (Wiltschko and Wiltschko, 2013; Shaw et al., 2015)

311 **In the 1970s, scientists discovered** (Blakemore, 1975)

311 **For decades, many scientists were sure** (Fleissner et al., 2003, 2007)

311 **In 2012, Keays published a bombshell study** (Treiber et al., 2012)

311 **That same year, another team** (Eder et al., 2012)

311 **But Keays debunked this finding** (Edelman et al., 2015)

311 **As a shark swims** (Paulin, 1995)

312 **The French zoologist Camille Viguier** (Viguier, 1882)

312 **Almost 130 years later, David Keays** (Nimpf et al., 2019)

312 **It's also notable that, in 2011** (Wu and Dickman, 2012)

312 **It involves two molecules** A good review of the radical pair hypothesis is Hore and Mouritsen (2016).

313 **He submitted a paper** (Schulten, personal communication, 2010)

313 **Undeterred, he published the paper** (Schulten, Swenberg, and Weller, 1978)

313 **In 2000, Schulten and his student** (Ritz, Adem, and Schulten, 2000)

314 **Known as cluster N** (Mouritsen et al., 2005)

314 **Cluster N gets information** (Heyers et al., 2007; Zapka et al., 2009)

315 **One called Cry4** (Einwich et al., 2020; Hochstoeger et al., 2020)

315 **And why, as Mouritsen showed** (Engels et al., 2014)

315 **A 1997 study claimed that honeybees** (Kirschvink et al., 1997)

316 **Two decades later, another group** (Baltzley and Nabity, 2018)

316 **In 1999, an American team** (Etheredge et al., 1999)

316 **In 2002, the Wiltschkos** (Wiltschko et al., 2002)

316 **A decade later, Henrik Mouritsen** (Hein et al., 2011; Engels et al., 2012)

316 **In 2015, an American team** (Vidal-Gadea et al., 2015; Qin et al., 2016)

316 **Neither study could be replicated** (Meister, 2016; Winklhofer and Mouritsen, 2016; Friis, Sjulstok, and Solov'yov, 2017; Landler et al., 2018)

316   **Baker published his results** (Baker, 1980)

316   **More recently, geophysicist Joseph Kirschvink** (Wang et al., 2019)

317   **They might tweak their experimental plans** A review of the many issues with irre-
       producible science is Aschwanden (2015).

318   **But Sonke Johnsen, Ken Lohmann** (Johnsen, Lohmann, and Warrant, 2020)

319   **Instead, they seem to use it** Magnetoreception and other means of animal navigation
       are reviewed in Mouritsen (2018).

CHAPTER 12

320   **Venkataraman tells me that the mosquitoes** The sensory cues that mosquitoes use to
       find their hosts are reviewed in Wolff and Riffell (2018).

321   **But it failed when Vosshall** (DeGennaro et al., 2013)

321   **Switching tactics, Vosshall's team tried** (McMeniman et al., 2014)

321   **When Vosshall's student Molly Liu** (Liu and Vosshall, 2019)

321   **After all, that's likely what DEET** (Dennis, Goldman, and Vosshall, 2019)

321   **But thousands of years ago** (McBride et al., 2014; McBride, 2016)

323   **Jumping spiders depend on** (Shamble et al., 2016)

323   **The star-nosed mole hunts** (Catania, 2006)

323   **Smell dominates the lives of ants** (Barbero et al., 2009)

323   **Smells also guide sharks** (Gardiner et al., 2014)

323   **The Ubangi elephantnose fish** (von der Emde and Ruhl, 2016)

323   **Songbirds and bogong moths** (Dreyer et al., 2018; Mouritsen, 2018)

323   **Some people experience synesthesia** (Ward, 2013)

323   **The platypus's duck-like bill** (Pettigrew, Manger, and Fine, 1998)

324   **"these probably fuse"** (Wheeler, 1910, p. 510)

324   **Electric fish that learn** (Schumacher et al., 2016)

324   **Even bumblebees can tell** (Solvi, Gutierrez Al-Khudhairy, and Chittka, 2020)

325   **There's proprioception, the awareness** Proprioception is reviewed in Tuthill and
       Azim (2018).

325   **In 1971, a 19-year-old butcher** (Cole, 2016)

325   **When animals move, their sense organs** The concepts of exafference, reafference,
       and corollary discharges are reviewed in Cullen (2004); Crapse and Sommer (2008).

325   **Consider a simple earthworm** (Merker, 2005)

326   **But no animals are completely immobile** (Ludeman et al., 2014)

326   **Philosophers and scholars have speculated** For a full history of this idea, see Grüsser
       (1994).

327   **As of 1950, the duplicated motor commands** (von Holst and Mittelstaedt, 1950;
       Sperry, 1950)

327   **For a full history** (Grüsser, 1994)

327   **Scientists have learned a lot** Corollary discharges in electric fish are reviewed in Saw-
       tell (2017); Fukutomi and Carlson (2020).

328   **They're how chirping crickets** (Poulet and Hedwig, 2003)

328   **Some scientists have suggested that schizophrenia** (Pynn and DeSouza, 2013)

330   **An octopus's central nervous system** The neurobiology of the octopus is reviewed in
       Grasso (2014); Levy and Hochner (2017).

330   **"An octopus effectively has nine brains"** (Crook and Walters, 2014)

330   **Meanwhile, it simultaneously touches and tastes** (Graziadei and Gagne, 1976)

330   **The suckers' independence is obvious** (Nesher et al., 2014)

331   **Between them, each sucker ganglion** (Grasso, 2014)

331   **For example, neurobiologist Binyamin Hochner** (Sumbre et al., 2006)

331   **But Hochner's colleague Tamar Gutnick** (Gutnick et al., 2011)

331   **Letizia Zullo, another member of Hochner's team** (Zullo et al., 2009; Hochner, 2013)

332 **"a body of pure possibility"** (Godfrey-Smith, 2016, p. 48)
332 **Godfrey-Smith marvelously compares** (Godfrey-Smith, 2016, p. 105)
332 **The octopus, then, arguably has** *two* (Grasso, 2014)

CHAPTER 13

336 **We have instigated** The sixth extinction of wildlife is documented in Kolbert (2014); Ceballos, Ehrlich, and Dirzo (2017).
336 **Instead of stepping into the Umwelten** Sensory pollution is reviewed in Swaddle et al. (2015); Dominoni et al. (2020).
337 **Other, slow-moving species** (Spoelstra et al., 2017)
337 **A team of Dutch scientists** (D'Estries, 2019)
337 **In 2001, when astronomer Pierantonio Cinzano** (Cinzano, Falchi, and Elvidge, 2001)
338 **In 2016, when the team updated** (Falchi et al., 2016)
338 **Every year, the proportion** (Kyba et al., 2017)
338 **"The thought of light"** (Johnsen, 2012, p. 57)
339 **And by analyzing radar images** (Van Doren et al., 2017)
339 **In 1886, shortly after Edison** (Longcore and Rich, 2016)
339 **Over a century later, environmental scientist** (Longcore et al., 2012)
339 **Many of these deaths** (Gehring, Kerlinger, and Manville, 2009)
340 **Light at night** Light pollution and its effects on wildlife are reviewed in Sanders et al. (2021).
340 **In part, that's because biologists** (Gaston, 2019)
340 **When sea turtle hatchlings emerge** (Witherington and Martin, 2003)
341 **Artificial lights can also fatally attract** (Owens et al., 2020)
341 **A single streetlamp** (Degen et al., 2016)
341 **In 2014, as part of an experiment** (Knop et al., 2017)
341 **Insects with aquatic larvae** (Horváth et al., 2009)
341 **Flickering lightbulbs can cause headaches** (Inger et al., 2014)
342 **The new generation of energy-efficient white LEDs** (Falchi et al., 2016; Longcore, 2018)
343 **To protect them, the team first** (Buxton et al., 2017)
343 **Even the most heavily protected areas** Noise pollution and its effects are reviewed in Barber, Crooks, and Fristrup (2010); Shannon et al. (2016).
343 **Two-thirds of Europeans** (Swaddle et al., 2015)
343 **In 2003, Hans Slabbekoorn** (Slabbekoorn and Peet, 2003)
343 **A year later, Henrik Brumm** (Brumm, 2004)
343 **These influential studies spurred** (Leonard and Horn, 2008; Gross, Pasinelli, and Kunc, 2010; Montague, Danek-Gontard, and Kunc, 2013; Gil et al., 2015)
344 **Every extra 3 decibels** (Francis et al., 2017)
344 **In 2012, Jesse Barber, Heidi Ware** (Ware et al., 2015)
344 **In one experiment, ladybird beetles** (Barton et al., 2018)
344 **In noisy conditions, prairie dogs** (Shannon et al., 2014)
344 **Owls flub their attacks** (Senzaki et al., 2016)
344 **Parasitic** *Ormia* **flies struggle** (Phillips et al., 2019)
344 **Sage grouse abandon** (Blickley et al., 2012)
345 **In the summer of 2017** (Suraci et al., 2019)
345 **More than 83 percent** (Riitters and Wickham, 2003)
345 **Even the seas can't offer silence** Natural and anthropogenic noises in the ocean are reviewed in Duarte et al. (2021).
345 *War of the Whales* (Horwitz, 2015)
345 **Either way, sonar clearly disturbs them** (DeRuiter et al., 2013; Miller, Kvadsheim, et al., 2015)

345   **Between World War II and 2008** (Frisk, 2012)
346   **Since giant whales can live** (Payne and Webb, 1971)
346   **As ships pass in the night** (Rolland et al., 2012; Erbe, Dunlop, and Dolman, 2018; Tsujii et al., 2018; Erbe et al., 2019)
346   **Crabs stop feeding** (Kunc et al., 2014; Simpson et al., 2016; Murchy et al., 2019)
346   **"We're conducting an experiment"** For more on shipping noise, see Hildebrand (2005); Malakoff (2010).
346   **Smooth vertical surfaces** (Greif et al., 2017)
346   **DMS, the seaweed-y chemical** (Wilcox, Van Sebille, and Hardesty, 2015; Savoca et al., 2016)
346   **The currents produced by objects** (Rycyk et al., 2018)
346   **Odorants in river water** (Tierney et al., 2008)
347   **Weak electric fields** (Gill et al., 2014)
347   **Some urban moths** (Altermatt and Ebert, 2016)
347   **Some urban spiders** (Czaczkes et al., 2018)
347   **In the towns of Panama** (Halfwerk et al., 2019)
347   **That extraordinary diversity arose** (Seehausen et al., 2008)
348   **By turning off the light** (Seehausen, van Alphen, and Witte, 1997)
348   **Lake Victoria's cichlids also suffered** (Witte et al., 2013)
348   **In 2020, science writer Maya Kapoor** (Kapoor, 2020)
348   **In the woodlands of New Mexico** (Francis et al., 2012)
349   **In 2016, marine biologist Tim Gordon** (Gordon et al., 2018, 2019)
350   **Wire cages that are** (Irwin, Horner, and Lohmann, 2004)
351   **With fewer planes and cars** (Jechow and Hölker, 2020)
351   **Seismic vibrations around the world** (Lecocq et al., 2020)
351   **Alaska's Glacier Bay** (Calma, 2020; Smith et al., 2020)
351   **Behavioral ecologist Elizabeth Derryberry** (Derryberry et al., 2020)
351   **In the summer of 2007** (Stack et al., 2011)
351   **To truly make a dent** Ways of reducing sensory pollution are reviewed in Longcore and Rich (2016); Duarte et al. (2021).
352   **In 1995, environmental historian William Cronon** (Cronon, 1996)
353   **In 1934, after considering** (Uexküll, 2010, p. 133)

Bibliography

Ache, B. W., and Young, J. M. (2005) Olfaction: Diverse species, conserved principles, *Neuron*, 48(3), 417–430.

Ackerman, D. (1991) *A natural history of the senses*. New York: Vintage Books.

Adamo, S. A. (2016) Do insects feel pain? A question at the intersection of animal behaviour, philosophy and robotics, *Animal Behaviour*, 118, 75–79.

Adamo, S. A. (2019) Is it pain if it does not hurt? On the unlikelihood of insect pain, *The Canadian Entomologist*, 151(6), 685–695.

Aflitto, N., and DeGomez, T. (2014) Sonic pest repellents, College of Agriculture, University of Arizona (Tucson, AZ). Available at: repository.arizona.edu/handle/10150/333139.

Agnarsson, I., Kuntner, M., and Blackledge, T. A. (2010) Bioprospecting finds the toughest biological material: Extraordinary silk from a giant riverine orb spider, *PLOS One*, 5(9), e11234.

Albert, J. S., and Crampton, W. G. R. (2006) Electroreception and electrogenesis, in Evans, D. H., and Claiborne, J. B. (eds), *The physiology of fishes*, 3rd ed., 431–472. Boca Raton, FL: CRC Press.

Alexander, R. M. (1996) Hans Werner Lissmann, 30 April 1909–21 April 1995, *Biographical Memoirs of Fellows of the Royal Society*, 42, 235–245.

Altermatt, F., and Ebert, D. (2016) Reduced flight-to-light behaviour of moth populations exposed to long-term urban light pollution, *Biology Letters*, 12(4), 20160111.

Alupay, J. S., Hadjisolomou, S. P., and Crook, R. J. (2014) Arm injury produces long-term behavioral and neural hypersensitivity in octopus, *Neuroscience Letters*, 558, 137–142.

Amey-Özel, M., et al. (2015) More a finger than a nose: The trigeminal motor and sensory innervation of the Schnauzenorgan in the elephant-nose fish Gnathonemus petersii, *Journal of Comparative Neurology*, 523(5), 769–789.

Anand, K. J. S., Sippell, W. G., and Aynsley-Green, A. (1987) Randomised trial of fentanyl anaesthesia in preterm babies undergoing surgery: Effects on the stress response, *The Lancet*, 329(8527), 243–248.

Andersson, S., Ornborg, J., and Andersson, M. (1998) Ultraviolet sexual dimorphism and assortative mating in blue tits, *Proceedings of the Royal Society B: Biological Sciences*, 265(1395), 445–450.

Andrews, M. T. (2019) Molecular interactions underpinning the phenotype of hibernation in mammals, *Journal of Experimental Biology*, 222(Pt 2), jeb160606.

Appel, M., and Elwood, R. W. (2009) Motivational trade-offs and potential pain experience in hermit crabs, *Applied Animal Behaviour Science*, 119(1), 120–124.

Arch, V. S., and Narins, P. M. (2008) "Silent" signals: Selective forces acting on ultrasonic communication systems in terrestrial vertebrates, *Animal Behaviour*, 76(4), 1423–1428.

Arikawa, K. (2001) Hindsight of butterflies: The *Papilio* butterfly has light sensitivity in the genitalia, which appears to be crucial for reproductive behavior, *BioScience*, 51(3), 219–225.

Arikawa, K. (2017) The eyes and vision of butterflies, *Journal of Physiology*, 595(16), 5457–5464.

Arkley, K., et al. (2014) Strategy change in vibrissal active sensing during rat locomotion, *Current Biology*, 24(13), 1507–1512.

Arnegard, M. E., and Carlson, B. A. (2005) Electric organ discharge patterns during group hunting by a mormyrid fish, *Proceedings of the Royal Society B: Biological Sciences*, 272(1570), 1305–1314.

Arranz, P., et al. (2011) Following a foraging fish-finder: Diel habitat use of Blainville's beaked whales revealed by echolocation, *PLOS One*, 6(12), e28353.

Aschwanden, C. (2015) Science isn't broken, *FiveThirtyEight*. Available at: fivethirtyeight.com/features/science-isnt-broken/.

Atema, J. (1971) Structures and functions of the sense of taste in the catfish (*Ictalurus natalis*), *Brain, Behavior and Evolution*, 4(4), 273–294.

Atema, J. (2018) Opening the chemosensory world of the lobster, *Homarus americanus*, *Bulletin of Marine Science*, 94(3), 479–516.

Au, W. W. L. (1993) *The sonar of dolphins*. New York: Springer-Verlag.

Au, W. W. L. (1996) Acoustic reflectivity of a dolphin, *Journal of the Acoustical Society of America*, 99(6), 3844–3848.

Au, W. W. L. (2011) History of dolphin biosonar research, *Acoustics Today*, 11(4), 10–17.

Au, W. W. L., et al. (2009) Acoustic basis for fish prey discrimination by echolocating dolphins and porpoises, *Journal of the Acoustical Society of America*, 126(1), 460–467.

Au, W. W. L., and Simmons, J. A. (2007) Echolocation in dolphins and bats, *Physics Today*, 60(9), 40–45.

Au, W. W., and Turl, C. W. (1983) Target detection in reverberation by an echolocating Atlantic bottlenose dolphin (*Tursiops truncatus*), *Journal of the Acoustical Society of America*, 73(5), 1676–1681.

Audubon, J. J. (1826) Account of the habits of the turkey buzzard (*Vultur aura*), particularly with the view of exploding the opinion generally entertained of its extraordinary power of smelling, *Edinburgh New Philosophical Journal*, 2, 172–184.

Baden, T., Euler, T., and Berens, P. (2020) Understanding the retinal basis of vision across species, *Nature Reviews Neuroscience*, 21(1), 5–20.

Baker, C. A., and Carlson, B. A. (2019) Electric signals, in Choe, J. C. (ed), *Encyclopedia of animal behavior*, 2nd ed., 474–486. Amsterdam: Elsevier.

Baker, C. A., Huck, K. R., and Carlson, B. A. (2015) Peripheral sensory coding through oscillatory synchrony in weakly electric fish, *eLife*, 4, e08163.

Baker, C. V. H. (2019) The development and evolution of lateral line electroreceptors: Insights from comparative molecular approaches, in Carlson, B. A., et al. (eds), *Electroreception: Fundamental insights from comparative approaches*, 25–62. Cham: Springer.

Baker, C. V. H., Modrell, M. S., and Gillis, J. A. (2013) The evolution and development of vertebrate lateral line electroreceptors, *Journal of Experimental Biology*, 216(13), 2515–2522.

Baker, R. R. (1980) Goal orientation by blindfolded humans after long-distance displacement: Possible involvement of a magnetic sense, *Science*, 210(4469), 555–557.

Bakken, G. S., et al. (2018) Cooler snakes respond more strongly to infrared stimuli, but we have no idea why, *Journal of Experimental Biology*, 221(17), jeb182121.

Bakken, G. S., and Krochmal, A. R. (2007) The imaging properties and sensitivity of the facial pits of pitvipers as determined by optical and heat-transfer analysis, *Journal of Experimental Biology*, 210(16), 2801–2810.

Baldwin, M. W., et al. (2014) Evolution of sweet taste perception in hummingbirds by transformation of the ancestral umami receptor, *Science*, 345(6199), 929–933.

Bálint, A., et al. (2020) Dogs can sense weak thermal radiation, *Scientific Reports*, 10(1), 3736.

Baltzley, M. J., and Nabity, M. W. (2018) Reanalysis of an oft-cited paper on honeybee magnetoreception reveals random behavior, *Journal of Experimental Biology*, 221(Pt 22), jeb185454.

Bang, B. G. (1960) Anatomical evidence for olfactory function in some species of birds, *Nature*, 188(4750), 547–549.

Bang, B. G., and Cobb, S. (1968) The size of the olfactory bulb in 108 species of birds, *The Auk*, 85(1), 55–61.

Barber, J. R., et al. (2015) Moth tails divert bat attack: Evolution of acoustic deflection, *Proceedings of the National Academy of Sciences*, 112(9), 2812–2816.

Barber, J. R., and Conner, W. E. (2007) Acoustic mimicry in a predator-prey interaction, *Proceedings of the National Academy of Sciences*, 104(22), 9331–9334.

Barber, J. R., Crooks, K. R., and Fristrup, K. M. (2010) The costs of chronic noise exposure for terrestrial organisms, *Trends in Ecology & Evolution*, 25(3), 180–189.

Barber, J. R., and Kawahara, A. Y. (2013) Hawkmoths produce anti-bat ultrasound, *Biology Letters*, 9(4), 20130161.

Barbero, F., et al. (2009) Queen ants make distinctive sounds that are mimicked by a butterfly social parasite, *Science*, 323(5915), 782–785.

Bargmann, C. I. (2006) Comparative chemosensation from receptors to ecology, *Nature*, 444(7117), 295–301.

Barth, F. G. (2002) *A spider's world: Senses and behavior*. Berlin: Springer.

Barth, F. (2015) A spider's tactile hairs, *Scholarpedia*, 10(3), 7267.

Barth, F. G., and Höller, A. (1999) Dynamics of arthropod filiform hairs. V. The response of spider trichobothria to natural stimuli, *Philosophical Transactions of the Royal Society B: Biological Sciences*, 354(1380), 183–192.

Barton, B. T., et al. (2018) Testing the AC/DC hypothesis: Rock and roll is noise pollution and weakens a trophic cascade, *Ecology and Evolution*, 8(15), 7649–7656.

Basolo, A. L. (1990) Female preference predates the evolution of the sword in swordtail fish, *Science*, 250(4982), 808–810.

Bates, A. E., et al. (2010) Deep-sea hydrothermal vent animals seek cool fluids in a highly variable thermal environment, *Nature Communications*, 1(1), 14.

Bates, L. A., et al. (2007) Elephants classify human ethnic groups by odor and garment color, *Current Biology*, 17(22), 1938–1942.

Bates, L. A., et al. (2008) African elephants have expectations about the locations of out-of-sight family members, *Biology Letters*, 4(1), 34–36.

Bateson, P. (1991) Assessment of pain in animals, *Animal Behaviour*, 42(5), 827–839.

Bauer, G. B., et al. (2012) Tactile discrimination of textures by Florida manatees (*Trichechus manatus latirostris*), *Marine Mammal Science*, 28(4), E456–E471.

Bauer, G. B., Reep, R. L., and Marshall, C. D. (2018) The tactile senses of marine mammals, *International Journal of Comparative Psychology*, 31.

Baxi, K. N., Dorries, K. M., and Eisthen, H. L. (2006) Is the vomeronasal system really specialized for detecting pheromones?, *Trends in Neurosciences*, 29(1), 1–7.

Bedore, C. N., and Kajiura, S. M. (2013) Bioelectric fields of marine organisms: Voltage and frequency contributions to detectability by electroreceptive predators, *Physiological and Biochemical Zoology*, 86(3), 298–311.

Bedore, C. N., Kajiura, S. M., and Johnsen, S. (2015) Freezing behaviour facilitates bioelectric crypsis in cuttlefish faced with predation risk, *Proceedings of the Royal Society B: Biological Sciences*, 282(1820), 20151886.

Benoit-Bird, K. J., and Au, W. W. L. (2009a) Cooperative prey herding by the pelagic dolphin, *Stenella longirostris*, *Journal of the Acoustical Society of America*, 125(1), 125–137.

Benoit-Bird, K. J., and Au, W. W. L. (2009b) Phonation behavior of cooperatively foraging spinner dolphins, *Journal of the Acoustical Society of America*, 125(1), 539–546.

Bernal, X. E., Rand, A. S., and Ryan, M. J. (2006) Acoustic preferences and localization performance of blood-sucking flies (*Corethrella Coquillett*) to túngara frog calls, *Behavioral Ecology*, 17(5), 709–715.

Beston, H. (2003) *The outermost house: A year of life on the great beach of Cape Cod*. New York: Holt Paperbacks.

Bianco, G., Ilieva, M., and Åkesson, S. (2019) Magnetic storms disrupt nocturnal migratory activity in songbirds, *Biology Letters*, 15(3), 20180918.

Bingman, V. P., et al. (2017) Importance of the antenniform legs, but not vision, for homing by the neotropical whip spider *Paraphrynus laevifrons*, *Journal of Experimental Biology*, 220(Pt 5), 885–890.

Birkhead, T. (2013) *Bird sense: What it's like to be a bird*. New York: Bloomsbury.

Bisoffi, Z., et al. (2013) *Strongyloides stercoralis*: A plea for action, *PLOS Neglected Tropical Diseases*, 7(5), e2214.

Bjørge, M. H., et al. (2011) Behavioural changes following intraperitoneal vaccination in Atlantic salmon (*Salmo salar*), *Applied Animal Behaviour Science*, 133(1), 127–135.

Blackledge, T. A., Kuntner, M., and Agnarsson, I. (2011) The form and function of spider orb webs, in Casas, J. (ed), *Advances in insect physiology*, 175–262. Amsterdam: Elsevier.

Blackwall, J. (1830) Mr Murray's paper on the aerial spider, *Magazine of Natural History and Journal of Zoology, Botany, Mineralogy, Geology, and Meteorology*, 2, 116–413.

Blakemore, R. (1975) Magnetotactic bacteria, *Science*, 190(4212), 377–379.

Bleicher, S. S., et al. (2018) Divergent behavior amid convergent evolution: A case of four desert rodents learning to respond to known and novel vipers, *PLOS One*, 13(8), e0200672.

Blickley, J. L., et al. (2012) Experimental chronic noise is related to elevated fecal corticosteroid metabolites in lekking male greater sage-grouse (*Centrocercus urophasianus*), *PLOS One*, 7(11), e50462.

Bok, M. J., et al. (2014) Biological sunscreens tune polychromatic ultraviolet vision in mantis shrimp, *Current Biology*, 24(14), 1636–1642.

Bok, M. J., Capa, M., and Nilsson, D.-E. (2016) Here, there and everywhere: The radiolar eyes of fan worms (Annelida, Sabellidae), *Integrative and Comparative Biology*, 56(5), 784–795.

Boles, L. C., and Lohmann, K. J. (2003) True navigation and magnetic maps in spiny lobsters, *Nature*, 421(6918), 60–63.

Bonadonna, F., et al. (2006) Evidence that blue petrel, *Halobaena caerulea*, fledglings can detect and orient to dimethyl sulfide, *Journal of Experimental Biology*, 209(11), 2165–2169.

Boonman, A., et al. (2013) It's not black or white: On the range of vision and echolocation in echolocating bats, *Frontiers in Physiology*, 4, 248.

Boonman, A., Bumrungsri, S., and Yovel, Y. (2014) Nonecholocating fruit bats produce biosonar clicks with their wings, *Current Biology*, 24(24), 2962–2967.

Boström, J. E., et al. (2016) Ultra-rapid vision in birds, *PLOS One*, 11(3), e0151099.

Bottesch, M., et al. (2016) A magnetic compass that might help coral reef fish larvae return to their natal reef, *Current Biology*, 26(24), R1266–R1267.

Braithwaite, V. (2010) *Do fish feel pain?* New York: Oxford University Press.

Braithwaite, V., and Droege, P. (2016) Why human pain can't tell us whether fish feel pain, *Animal Sentience*, 3(3).

Braude, S., et al. (2021) Surprisingly long survival of premature conclusions about naked mole-rat biology, *Biological Reviews*, 96(2), 376–393.

Brill, R. L., et al. (1992) Target detection, shape discrimination, and signal characteristics of an echolocating false killer whale (*Pseudorca crassidens*), *Journal of the Acoustical Society of America*, 92(3), 1324–1330.

Brinkløv, S., Elemans, C. P. H., and Ratcliffe, J. M. (2017) Oilbirds produce echolocation signals beyond their best hearing range and adjust signal design to natural light conditions, *Royal Society Open Science*, 4(5), 170255.

Brinkløv, S., Fenton, M. B., and Ratcliffe, J. M. (2013) Echolocation in oilbirds and swiftlets, *Frontiers in Physiology*, 4, 123.

Brinkløv, S., Kalko, E. K. V., and Surlykke, A. (2009) Intense echolocation calls from two "whispering" bats, *Artibeus jamaicensis* and *Macrophyllum macrophyllum* (Phyllostomidae), *Journal of Experimental Biology*, 212(Pt 1), 11–20.

Brinkløv, S., and Warrant, E. (2017) Oilbirds, *Current Biology*, 27(21), R1145–R1147.

Briscoe, A. D., et al. (2010) Positive selection of a duplicated UV-sensitive visual pigment coincides with wing pigment evolution in *Heliconius* butterflies, *Proceedings of the National Academy of Sciences*, 107(8), 3628–3633.

Broom, D. (2001) Evolution of pain, *Vlaams Diergeneeskundig Tijdschrift*, 70, 17–21.

Brothers, J. R., and Lohmann, K. J. (2018) Evidence that magnetic navigation and geomagnetic imprinting shape spatial genetic variation in sea turtles, *Current Biology*, 28(8), 1325–1329.e2.

Brown, F. A. (1962) Responses of the planarian, dugesia, and the protozoan, paramecium, to very weak horizontal magnetic fields, *Biological Bulletin*, 123(2), 264–281.

Brown, F. A., Webb, H. M., and Barnwell, F. H. (1964) A compass directional phenomenon in mud-snails and its relation to magnetism, *Biological Bulletin*, 127(2), 206–220.

Brown, R. E., and Fedde, M. R. (1993) Airflow sensors in the avian wing, *Journal of Experimental Biology*, 179(1), 13–30.

Brownell, P., and Farley, R. D. (1979a) Detection of vibrations in sand by tarsal sense organs of the nocturnal scorpion, *Paruroctonus mesaensis, Journal of Comparative Physiology A*, 131(1), 23–30.

Brownell, P., and Farley, R. D. (1979b) Orientation to vibrations in sand by the nocturnal scorpion, *Paruroctonus mesaensis*: Mechanism of target localization, *Journal of Comparative Physiology A*, 131(1), 31–38.

Brownell, P., and Farley, R. D. (1979c) Prey-localizing behaviour of the nocturnal desert scorpion, *Paruroctonus mesaensis*: Orientation to substrate vibrations, *Animal Behaviour*, 27(Pt 1), 185–193.

Brownell, P. H. (1984) Prey detection by the sand scorpion, *Scientific American*, 251(6), 86–97.

Brumm, H. (2004) The impact of environmental noise on song amplitude in a territorial bird, *Journal of Animal Ecology*, 73(3), 434–440.

Brunetta, L., and Craig, C. L. (2012) *Spider silk: Evolution and 400 million years of spinning, waiting, snagging, and mating*. New Haven, CT: Yale University Press.

Bryant, A. S., et al. (2018) A critical role for thermosensation in host seeking by skin-penetrating nematodes, *Current Biology*, 28(14), 2338–2347.e6.

Bryant, A. S., and Hallem, E. A. (2018) Temperature-dependent behaviors of parasitic helminths, *Neuroscience Letters*, 687, 290–303.

Bullock, T. H. (1969) Species differences in effect of electroreceptor input on electric organ pacemakers and other aspects of behavior in electric fish, *Brain, Behavior and Evolution*, 2(2), 102–118.

Bullock, T. H., Behrend, K., and Heiligenberg, W. (1975) Comparison of the jamming avoidance responses in Gymnotoid and Gymnarchid electric fish: A case of convergent evolution of behavior and its sensory basis, *Journal of Comparative Physiology*, 103(1), 97–121.

Bullock, T. H., and Diecke, F. P. J. (1956) Properties of an infra-red receptor, *Journal of Physiology*, 134(1), 47–87.

Bush, N. E., Solla, S. A., and Hartmann, M. J. (2016) Whisking mechanics and active sensing, *Current Opinion in Neurobiology*, 40, 178–188.

Buxton, R. T., et al. (2017) Noise pollution is pervasive in U.S. protected areas, *Science*, 356(6337), 531–533.

Cadena, V., et al. (2013) Evaporative respiratory cooling augments pit organ thermal detection in rattlesnakes, *Journal of Comparative Physiology A*, 199(12), 1093–1104.

Caldwell, M. S., McDaniel, J. G., and Warkentin, K. M. (2010) Is it safe? Red-eyed treefrog embryos assessing predation risk use two features of rain vibrations to avoid false alarms, *Animal Behaviour*, 79(2), 255–260.

Calma, J. (2020) The pandemic turned the volume down on ocean noise pollution, *The Verge*. Available at: www.theverge.com/22166314/covid-19-pandemic-ocean-noise-pollution.

Caprio, J. (1975) High sensitivity of catfish taste receptors to amino acids, *Comparative Biochemistry and Physiology Part A: Physiology*, 52(1), 247–251.

Caprio, J., et al. (1993) The taste system of the channel catfish: From biophysics to behavior, *Trends in Neurosciences*, 16(5), 192–197.

Caputi, A. A. (2017) Active electroreception in weakly electric fish, in Sherman, S. M. (ed), *Oxford research encyclopedia of neuroscience*. New York: Oxford University Press. Available at: DOI: 10.1093/acrefore/9780190264086.013.106.

Caputi, A. A., et al. (2013) On the haptic nature of the active electric sense of fish, *Brain Research*, 1536, 27–43.

Caputi, Á. A., Aguilera, P. A., and Pereira, A. C. (2011) Active electric imaging: Body-object interplay and object's "electric texture," *PLOS One*, 6(8), e22793.

Caras, M. L. (2013) Estrogenic modulation of auditory processing: A vertebrate comparison, *Frontiers in Neuroendocrinology*, 34(4), 285–299.

Carlson, B. A. (2002) Electric signaling behavior and the mechanisms of electric organ discharge production in mormyrid fish, *Journal of Physiology-Paris*, 96(5), 405–419.

Carlson, B. A., et al. (eds), (2019) *Electroreception: Fundamental insights from comparative approaches*. Cham: Springer.

Carlson, B. A., and Arnegard, M. E. (2011) Neural innovations and the diversification of African weakly electric fishes, *Communicative & Integrative Biology*, 4(6), 720–725.

Carlson, B. A., and Sisneros, J. A. (2019) A brief history of electrogenesis and electroreception in fishes, in Carlson, B. A., et al. (eds), *Electroreception: Fundamental insights from comparative approaches*, 1–23. Cham: Springer.

Caro, T. M. (2016) *Zebra stripes*. Chicago: University of Chicago Press.

Caro, T., et al. (2019) Benefits of zebra stripes: Behaviour of tabanid flies around zebras and horses, *PLOS One*, 14(2), e0210831.

Carpenter, C. W., et al. (2018) Human ability to discriminate surface chemistry by touch, *Materials Horizons*, 5(1), 70–77.

Carr, A. (1995) Notes on the behavioral ecology of sea turtles, in Bjorndal, K. A. (ed), *Biology and conservation of sea turtles*, rev. ed., 19–26. Washington, DC: Smithsonian Institution Press.

Carr, A. L., and Salgado, V. L. (2019) Ticks home in on body heat: A new understanding of Haller's organ and repellent action, *PLOS One*, 14(8), e0221659.

Carr, C. E., and Christensen-Dalsgaard, J. (2015) Sound localization strategies in three predators, *Brain, Behavior and Evolution*, 86(1), 17–27.

Carr, C. E., and Christensen-Dalsgaard, J. (2016) Evolutionary trends in directional hearing, *Current Opinion in Neurobiology*, 40, 111–117.

Carr, T. D., et al. (2017) A new tyrannosaur with evidence for anagenesis and crocodile-like facial sensory system, *Scientific Reports*, 7(1), 44942.

Carrete, M., et al. (2012) Mortality at wind-farms is positively related to large-scale distribution and aggregation in griffon vultures, *Biological Conservation*, 145(1), 102–108.

Carvalho, L. S., et al. (2017) The genetic and evolutionary drives behind primate color vision, *Frontiers in Ecology and Evolution*, 5, 34.

Casas, J., and Dangles, O. (2010) Physical ecology of fluid flow sensing in arthropods, *Annual Review of Entomology*, 55(1), 505–520.

Casas, J., and Steinmann, T. (2014) Predator-induced flow disturbances alert prey, from the onset of an attack, *Proceedings of the Royal Society B: Biological Sciences*, 281(1790), 20141083.

Catania, K. C. (1995a) Magnified cortex in star-nosed moles, *Nature*, 375(6531), 453–454.

Catania, K. C. (1995b) Structure and innervation of the sensory organs on the snout of the star-nosed mole, *Journal of Comparative Neurology*, 351(4), 536–548.

Catania, K. C. (2006) Olfaction: Underwater "sniffing" by semi-aquatic mammals, *Nature*, 444(7122), 1024–1025.

Catania, K. C. (2008) Worm grunting, fiddling, and charming—Humans unknowingly mimic a predator to harvest bait, *PLOS One*, 3(10), e3472.

Catania, K. C. (2011) The sense of touch in the star-nosed mole: From mechanoreceptors to the brain, *Philosophical Transactions of the Royal Society B: Biological Sciences*, 366(1581), 3016–3025.

Catania, K. C. (2016) Leaping eels electrify threats, supporting Humboldt's account of a battle with horses, *Proceedings of the National Academy of Sciences*, 113(25), 6979–6984.

Catania, K. C. (2019) The astonishing behavior of electric eels, *Frontiers in Integrative Neuroscience*, 13, 23.

Catania, K. C., et al. (1993) Nose stars and brain stripes, *Nature*, 364(6437), 493.

Catania, K. C., and Kaas, J. H. (1997a) Somatosensory fovea in the star-nosed mole: Behavioral use of the star in relation to innervation patterns and cortical representation, *Journal of Comparative Neurology*, 387(2), 215–233.

Catania, K. C., and Kaas, J. H. (1997b) The mole nose instructs the brain, *Somatosensory & Motor Research*, 14(1), 56–58.

Catania, K. C., Northcutt, R. G., and Kaas, J. H. (1999) The development of a biological novelty: A different way to make appendages as revealed in the snout of the star-nosed mole *Condylura cristata*, *Journal of Experimental Biology*, 202(Pt 20), 2719–2726.

Catania, K. C., and Remple, F. E. (2004) Tactile foveation in the star-nosed mole, *Brain, Behavior and Evolution*, 63(1), 1–12.

Catania, K. C., and Remple, F. E. (2005) Asymptotic prey profitability drives star-nosed moles to the foraging speed limit, *Nature*, 433(7025), 519–522.

Catania, K. C., and Remple, M. S. (2002) Somatosensory cortex dominated by the representation of teeth in the naked mole-rat brain, *Proceedings of the National Academy of Sciences*, 99(8), 5692–5697.

Caves, E. M., Brandley, N. C., and Johnsen, S. (2018) Visual acuity and the evolution of signals, *Trends in Ecology & Evolution*, 33(5), 358–372.

Ceballos, G., Ehrlich, P. R., and Dirzo, R. (2017) Biological annihilation via the ongoing sixth mass extinction signaled by vertebrate population losses and declines, *Proceedings of the National Academy of Sciences*, 114(30), E6089–E6096.

Chappuis, C. J., et al. (2013) Water vapour and heat combine to elicit biting and biting persistence in tsetse, *Parasites & Vectors*, 6(1), 240.

Chatigny, F. (2019) The controversy on fish pain: A veterinarian's perspective, *Journal of Applied Animal Welfare Science*, 22(4), 400–410.

Chen, P.-J., et al. (2016) Extreme spectral richness in the eye of the common bluebottle butterfly, *Graphium sarpedon*, *Frontiers in Ecology and Evolution*, 4, 12.

Chen, Q., et al. (2012) Reduced performance of prey targeting in pit vipers with contralaterally occluded infrared and visual senses, *PLOS One*, 7(5), e34989.

Chernetsov, N., Kishkinev, D., and Mouritsen, H. (2008) A long-distance avian migrant compensates for longitudinal displacement during spring migration, *Current Biology*, 18(3), 188–190.

Chiou, T.-H., et al. (2008) Circular polarization vision in a stomatopod crustacean, *Current Biology*, 18(6), 429–434.

Chiszar, D., et al. (1983) Strike-induced chemosensory searching by rattlesnakes: The role of envenomation-related chemical cues in the post-strike environment, in Müller-Schwarze, D., and Silverstein, R. M. (eds), *Chemical signals in vertebrates*, 3:1–24. Boston: Springer.

Chiszar, D., et al. (1999) Discrimination between envenomated and nonenvenomated prey by western diamondback rattlesnakes (*Crotalus atrox*): Chemosensory consequences of venom, *Copeia*, 1999(3), 640–648.

Chiszar, D., Walters, A., and Smith, H. M. (2008) Rattlesnake preference for envenomated prey: Species specificity, *Journal of Herpetology*, 42(4), 764–767.

Chittka, L. (1997) Bee color vision is optimal for coding flower color, but flower colors are not optimal for being coded—why?, *Israel Journal of Plant Sciences*, 45(2–3), 115–127.

Chittka, L., and Menzel, R. (1992) The evolutionary adaptation of flower colours and the insect pollinators' colour vision, *Journal of Comparative Physiology A*, 171(2), 171–181.

Chittka, L., and Niven, J. (2009) Are bigger brains better?, *Current Biology*, 19(21), R995–R1008.

Chiu, C., and Moss, C. F. (2008) When echolocating bats do not echolocate, *Communicative & Integrative Biology*, 1(2), 161–162.

Chiu, C., Xian, W., and Moss, C. F. (2008) Flying in silence: Echolocating bats cease vocalizing to avoid sonar jamming, *Proceedings of the National Academy of Sciences*, 105(35), 13116–13121.

Chiu, C., Xian, W., and Moss, C. F. (2009) Adaptive echolocation behavior in bats for the analysis of auditory scenes, *Journal of Experimental Biology*, 212(9), 1392–1404.

Cinzano, P., Falchi, F., and Elvidge, C. D. (2001) The first world atlas of the artificial night sky brightness, *Monthly Notices of the Royal Astronomical Society*, 328(3), 689–707.

Clark, C. J., LePiane, K., and Liu, L. (2020) Evolution and ecology of silent flight in owls and other flying vertebrates, *Integrative Organismal Biology*, 2(1), obaa001.

Clark, C. W., and Gagnon, G. C. (2004) Low-frequency vocal behaviors of baleen whales in the North Atlantic: Insights from IUSS detections, locations and tracking from 1992 to 1996, *Journal of Underwater Acoustics*, 52, 609–640.

Clark, G. A., and de Cruz, J. B. (1989) Functional interpretation of protruding filoplumes in oscines, *The Condor*, 91(4), 962–965.

Clark, R. (2004) Timber rattlesnakes (*Crotalus horridus*) use chemical cues to select ambush sites, *Journal of Chemical Ecology*, 30(3), 607–617.

Clark, R., and Ramirez, G. (2011) Rosy boas (*Lichanura trivirgata*) use chemical cues to identify female mice (*Mus musculus*) with litters of dependent young, *Herpetological Journal*, 21(3), 187–191.

Clarke, D., et al. (2013) Detection and learning of floral electric fields by bumblebees, *Science*, 340(6128), 66–69.

Clarke, D., Morley, E., and Robert, D. (2017) The bee, the flower, and the electric field: Electric ecology and aerial electroreception, *Journal of Comparative Physiology A*, 203(9), 737–748.

Cocroft, R. (1999) Offspring-parent communication in a subsocial treehopper (Hemiptera: Membracidae: *Umbonia crassicornis*), *Behaviour*, 136(1), 1–21.

Cocroft, R. B. (2011) The public world of insect vibrational communication, *Molecular Ecology*, 20(10), 2041–2043.

Cocroft, R. B., and Rodríguez, R. L. (2005) The behavioral ecology of insect vibrational communication, *BioScience*, 55(4), 323–334.

Cohen, K. E., et al. (2020) Knowing when to stick: Touch receptors found in the remora adhesive disc, *Royal Society Open Science*, 7(1), 190990.

Cohen, K. L., Seid, M. A., and Warkentin, K. M. (2016) How embryos escape from danger: The mechanism of rapid, plastic hatching in red-eyed treefrogs, *Journal of Experimental Biology*, 219(12), 1875–1883.

Cokl, A., and Virant-Doberlet, M. (2003) Communication with substrate-borne signals in small plant-dwelling insects, *Annual Review of Entomology*, 48, 29–50.

Cole, J. (2016) *Losing touch: A man without his body*. Oxford: Oxford University Press.

Collin, S. P. (2019) Electroreception in vertebrates and invertebrates, in Choe, J. C. (ed), *Encyclopedia of animal behavior*, 2nd ed., 120–131. Amsterdam: Elsevier.

Collin, S. P., et al. (2009) The evolution of early vertebrate photoreceptors, *Philosophical Transactions of the Royal Society B: Biological Sciences*, 364(1531), 2925–2940.

Collins, C. E., Hendrickson, A., and Kaas, J. H. (2005) Overview of the visual system of Tarsius, *The Anatomical Record: Part A, Discoveries in Molecular, Cellular, and Evolutionary Biology*, 287(1), 1013–1025.

Colour Blind Awareness (n.d.) Living with Colour Vision Deficiency, Colour Blind Awareness. Available at: www.colourblindawareness.org/colour-blindness/living-with-colour-vision-deficiency/.

Conner, W. E., and Corcoran, A. J. (2012) Sound strategies: The 65-million-year-old battle between bats and insects, *Annual Review of Entomology*, 57(1), 21–39.

Corbet, S. A., Beament, J., and Eisikowitch, D. (1982) Are electrostatic forces involved in pollen transfer?, *Plant, Cell & Environment*, 5(2), 125–129.

Corcoran, A. J., et al. (2011) How do tiger moths jam bat sonar?, *Journal of Experimental Biology*, 214(14), 2416–2425.

Corcoran, A. J., Barber, J. R., and Conner, W. E. (2009) Tiger moth jams bat sonar, *Science*, 325(5938), 325–327.

Corcoran, A. J., and Moss, C. F. (2017) Sensing in a noisy world: Lessons from auditory specialists, echolocating bats, *Journal of Experimental Biology*, 220(24), 4554–4566.

Corfas, R. A., and Vosshall, L. B. (2015) The cation channel TRPA1 tunes mosquito thermotaxis to host temperatures, *eLife*, 4, e11750.

Costa, D. (1993) The secret life of marine mammals: Novel tools for studying their behavior and biology at sea, *Oceanography*, 6(3), 120–128.

Costa, D., and Kooyman, G. (2011) Oxygen consumption, thermoregulation, and the effect of fur oiling and washing on the sea otter, *Enhydra lutris, Canadian Journal of Zoology*, 60(11), 2761–2767.

Cowart, L. (2021) *Hurts so good: The science and culture of pain on purpose*. New York: PublicAffairs.

Cox, J. J., et al. (2006) An SCN9A channelopathy causes congenital inability to experience pain, *Nature*, 444(7121), 894–898.

Crampton, W. G. R. (2019) Electroreception, electrogenesis and electric signal evolution, *Journal of Fish Biology*, 95(1), 92–134.

Cranford, T. W., Amundin, M., and Norris, K. S. (1996) Functional morphology and homology in the odontocete nasal complex: Implications for sound generation, *Journal of Morphology*, 228(3), 223–285.

Crapse, T. B., and Sommer, M. A. (2008) Corollary discharge across the animal kingdom, *Nature Reviews Neuroscience*, 9(8), 587–600.

Craven, B. A., Paterson, E. G., and Settles, G. S. (2010) The fluid dynamics of canine olfaction: Unique nasal airflow patterns as an explanation of macrosmia, *Journal of the Royal Society Interface*, 7(47), 933–943.

Crish, C., Crish, S., and Comer, C. (2015) Tactile sensing in the naked mole rat, *Scholarpedia*, 10(3), 7164.

Cronin, T. W. (2018) A different view: Sensory drive in the polarized-light realm, *Current Zoology*, 64(4), 513–523.

Cronin, T. W., et al. (2014) *Visual Ecology*. Princeton, NJ: Princeton University Press.

Cronin, T. W., and Bok, M. J. (2016) Photoreception and vision in the ultraviolet, *Journal of Experimental Biology*, 219(18), 2790–2801.

Cronin, T. W., and Marshall, N. J. (1989a) A retina with at least ten spectral types of photoreceptors in a mantis shrimp, *Nature*, 339(6220), 137–140.

Cronin, T. W., and Marshall, N. J. (1989b) Multiple spectral classes of photoreceptors in the retinas of gonodactyloid stomatopod crustaceans, *Journal of Comparative Physiology A*, 166(2), 261–275.

Cronin, T. W., Marshall, N. J., and Caldwell, R. L. (2017) Stomatopod vision, in Sherman, S. M. (ed), *Oxford research encyclopedia of neuroscience*. New York: Oxford University Press. Available at: oxfordre.com/neuroscience/view/10.1093/acrefore/9780190264086.001.0001/acrefore-9780190264086-e-157.

Cronon, W. (1996) The trouble with wilderness; Or, getting back to the wrong nature, *Environmental History*, 1(1), 7–28.

Crook, R. J. (2021) Behavioral and neurophysiological evidence suggests affective pain experience in octopus, *iScience*, 24(3), 102229.

Crook, R. J., et al. (2011) Peripheral injury induces long-term sensitization of defensive responses to visual and tactile stimuli in the squid *Loligo pealeii*, Lesueur 1821, *Journal of Experimental Biology*, 214(19), 3173–3185.

Crook, R. J., et al. (2014) Nociceptive sensitization reduces predation risk, *Current Biology*, 24(10), 1121–1125.

Crook, R. J., Hanlon, R. T., and Walters, E. T. (2013) Squid have nociceptors that display widespread long-term sensitization and spontaneous activity after bodily injury, *Journal of Neuroscience*, 33(24), 10021–10026.

Crook, R. J., and Walters, E. T. (2014) Neuroethology: Self-recognition helps octopuses avoid entanglement, *Current Biology*, 24(11), R520–R521.

Cross, F. R., et al. (2020) Arthropod intelligence? The case for Portia, *Frontiers in Psychology*, 11.

Crowe-Riddell, J. M., Simões, B. F., et al. (2019) Phototactic tails: Evolution and molecular basis of a novel sensory trait in sea snakes, *Molecular Ecology*, 28(8), 2013–2028.

Crowe-Riddell, J. M., Williams, R., et al. (2019) Ultrastructural evidence of a mechanosensory function of scale organs (sensilla) in sea snakes (Hydrophiinae), *Royal Society Open Science*, 6(4), 182022.

Cullen, K. E. (2004) Sensory signals during active versus passive movement, *Current Opinion in Neurobiology*, 14(6), 698–706.

Cummings, M. E., Rosenthal, G. G., and Ryan, M. J. (2003) A private ultraviolet channel in visual communication, *Proceedings of the Royal Society B: Biological Sciences*, 270(1518), 897–904.

Cunningham, S., et al. (2010) Bill morphology of ibises suggests a remote-tactile sensory system for prey detection, *The Auk*, 127(2), 308–316.

Cunningham, S., Castro, I., and Alley, M. (2007) A new prey-detection mechanism for kiwi (*Apteryx* spp.) suggests convergent evolution between paleognathous and neognathous birds, *Journal of Anatomy*, 211(4), 493–502.

Cunningham, S. J., Alley, M. R., and Castro, I. (2011) Facial bristle feather histology and morphology in New Zealand birds: Implications for function, *Journal of Morphology*, 272(1), 118–128.

Cuthill, I. C., et al. (2017) The biology of color, *Science*, 357(6350), eaan0221.

Czaczkes, T. J., et al. (2018) Reduced light avoidance in spiders from populations in light-polluted urban environments, *Naturwissenschaften*, 105(11–12), 64.

Czech-Damal, N. U., et al. (2012) Electroreception in the Guiana dolphin (*Sotalia guianensis*), *Proceedings of the Royal Society B: Biological Sciences*, 279(1729), 663–668.

Czech-Damal, N. U., et al. (2013) Passive electroreception in aquatic mammals, *Journal of Comparative Physiology A*, 199(6), 555–563.

Daan, S., Barnes, B. M., and Strijkstra, A. M. (1991) Warming up for sleep? Ground squirrels sleep during arousals from hibernation, *Neuroscience Letters*, 128(2), 265–268.

Daly, I., et al. (2016) Dynamic polarization vision in mantis shrimps, *Nature Communications*, 7, 12140.

Daly, I. M., et al. (2018) Complex gaze stabilization in mantis shrimp, *Proceedings of the Royal Society B: Biological Sciences*, 285(1878), 20180594.

Dangles, O., Casas, J., and Coolen, I. (2006) Textbook cricket goes to the field: The ecological scene of the neuroethological play, *Journal of Experimental Biology*, 209(3), 393–398.

Darwin, C. (1871) *The descent of man, and selection in relation to sex*. London: J. Murray.

Darwin, C. (1890) *The formation of vegetable mould, through the action of worms, with observations on their habits*. New York: D. Appleton and Company.

Darwin, C. (1958) *The origin of species by means of natural selection*. New York: Signet.

De Brito Sanchez, M. G., et al. (2014) The tarsal taste of honey bees: Behavioral and electrophysiological analyses, *Frontiers in Behavioral Neuroscience*, 8.

Degen, T., et al. (2016) Street lighting: Sex-independent impacts on moth movement, *Journal of Animal Ecology*, 85(5), 1352–1360.

DeGennaro, M., et al. (2013) *Orco* mutant mosquitoes lose strong preference for humans and are not repelled by volatile DEET, *Nature*, 498(7455), 487–491.

Dehnhardt, G., et al. (2001) Hydrodynamic trail-following in harbor seals (*Phoca vitulina*), *Science*, 293(5527), 102–104.

Dehnhardt, G., Mauck, B., and Hyvärinen, H. (1998) Ambient temperature does not affect the tactile sensitivity of mystacial vibrissae in harbour seals, *Journal of Experimental Biology*, 201(22), 3023–3029.

Dennis, E. J., Goldman, O. V., and Vosshall, L. B. (2019) *Aedes aegypti* mosquitoes use their legs to sense DEET on contact, *Current Biology*, 29(9), 1551–1556.e5.

Derryberry, E. P., et al. (2020) Singing in a silent spring: Birds respond to a half-century soundscape reversion during the COVID-19 shutdown, *Science*, 370(6516), 575–579.

DeRuiter, S. L., et al. (2013) First direct measurements of behavioural responses by Cuvier's beaked whales to mid-frequency active sonar, *Biology Letters*, 9(4), 20130223.

De Santana, C. D., et al. (2019) Unexpected species diversity in electric eels with a description of the strongest living bioelectricity generator, *Nature Communications*, 10(1), 4000.

D'Estries, M. (2019) This bat-friendly town turned the night red, *Treehugger*. Available at: www.treehugger.com/worlds-first-bat-friendly-town-turns-night-red-4868381.

D'Ettorre, P. (2016) Genomic and brain expansion provide ants with refined sense of smell, *Proceedings of the National Academy of Sciences*, 113(49), 13947–13949.

Deutschlander, M. E., Borland, S. C., and Phillips, J. B. (1999) Extraocular magnetic compass in newts, *Nature*, 400(6742), 324–325.

Diderot, D. (1749) Lettre sur les aveugles à l'usage de ceux qui voient. Available at: www .google.com/books/edition/Lettre_sur_les_aveugles/W30HAAAAQAAJ?hl=en&gbpv=1.

Dijkgraaf, S. (1963) The functioning and significance of the lateral-line organs, *Biological Reviews*, 38(1), 51–105.

Dijkgraaf, S. (1989) A short personal review of the history of lateral line research, in Coombs, S., Görner, P., and Münz, H. (eds), *The mechanosensory lateral line*, 7–14. New York: Springer.

Dijkgraaf, S., and Kalmijn, A. J. (1962) Verhaltensversuche zur Funktion der Lorenzinischen Ampullen, *Naturwissenschaften*, 49, 400.

Dinets, V. (2016) No cortex, no cry, *Animal Sentience*, 1(3).

Di Silvestro, R. (2012) Spider-Man vs the real deal: Spider powers, National Wildlife Foundation blog. Available at: blog.nwf.org/2012/06/spiderman-vs-the-real-deal-spider -powers/.

Dominoni, D. M., et al. (2020) Why conservation biology can benefit from sensory ecology, *Nature Ecology & Evolution*, 4(4), 502–511.

Dominy, N. J., and Lucas, P. W. (2001) Ecological importance of trichromatic vision to primates, *Nature*, 410(6826), 363–366.

Dominy, N. J., Svenning, J.-C., and Li, W.-H. (2003) Historical contingency in the evolution of primate color vision, *Journal of Human Evolution*, 44(1), 25–45.

Dooling, R. J., et al. (2002) Auditory temporal resolution in birds: Discrimination of harmonic complexes, *Journal of the Acoustical Society of America*, 112(2), 748–759.

Dooling, R. J., Lohr, B., and Dent, M. L. (2000) Hearing in birds and reptiles, in Dooling, R. J., Fay, R. R., and Popper, A. N. (eds), *Comparative hearing: Birds and reptiles*, 308–359. New York: Springer.

Dooling, R. J., and Prior, N. H. (2017) Do we hear what birds hear in birdsong?, *Animal Behaviour*, 124, 283–289.

Douglas, R. H., and Jeffery, G. (2014) The spectral transmission of ocular media suggests ultraviolet sensitivity is widespread among mammals, *Proceedings of the Royal Society B: Biological Sciences*, 281(1780), 20132995.

Dreyer, D., et al. (2018) The Earth's magnetic field and visual landmarks steer migratory flight behavior in the nocturnal Australian bogong moth, *Current Biology*, 28(13), 2160–2166.e5.

Du, W.-G., et al. (2011) Behavioral thermoregulation by turtle embryos, *Proceedings of the National Academy of Sciences*, 108(23), 9513–9515.

Duarte, C. M., et al. (2021) The soundscape of the Anthropocene ocean, *Science*, 371(6529), eaba4658.

Dunlop, R., and Laming, P. (2005) Mechanoreceptive and nociceptive responses in the central nervous system of goldfish (*Carassius auratus*) and trout (*Oncorhynchus mykiss*), *Journal of Pain*, 6(9), 561–568.

Dunning, D. C., and Roeder, K. D. (1965) Moth sounds and the insect-catching behavior of bats, *Science*, 147(3654), 173–174.

Duque, F. G., Rodríguez-Saltos, C. A., and Wilczynski, W. (2018) High-frequency vocalizations in Andean hummingbirds, *Current Biology*, 28, R909–R930.

Duranton, C., and Horowitz, A. (2019) Let me sniff! Nosework induces positive judgment bias in pet dogs, *Applied Animal Behaviour Science*, 211, 61–66.

Durso, A. (2013) Non-toxic venoms?, *Life is short, but snakes are long* (blog). Available at: snakesarelong.blogspot.com/2013/03/non-toxic-venoms.html.

Dusenbery, D. B. (1992) *Sensory ecology: How organisms acquire and respond to information.* New York: W. H. Freeman.

Dusenbery, M. (2018) *Doing harm: The truth about how bad medicine and lazy science leave women dismissed, misdiagnosed, and sick.* New York: HarperOne.

Eaton, J. (2014) When it comes to smell, the turkey vulture stands (nearly) alone, *Bay Nature.* Available at: baynature.org/article/comes-smell-turkey-vulture-stands-nearly -alone/.

Eaton, M. D. (2005) Human vision fails to distinguish widespread sexual dichromatism among sexually "monochromatic" birds, *Proceedings of the National Academy of Sciences,* 102(31), 10942–10946.

Ebert, J., and Westhoff, G. (2006) Behavioural examination of the infrared sensitivity of rattle-snakes (*Crotalus atrox*), *Journal of Comparative Physiology A,* 192(9), 941–947.

Edelman, N. B., et al. (2015) No evidence for intracellular magnetite in putative vertebrate magnetoreceptors identified by magnetic screening, *Proceedings of the National Academy of Sciences,* 112(1), 262–267.

Eder, S. H. K., et al. (2012) Magnetic characterization of isolated candidate vertebrate magne-toreceptor cells, *Proceedings of the National Academy of Sciences,* 109(30), 12022–12027.

Einwich, A., et al. (2020) A novel isoform of cryptochrome 4 (Cry4b) is expressed in the retina of a night-migratory songbird, *Scientific Reports,* 10(1), 15794.

Eisemann, C. H., et al. (1984) Do insects feel pain? A biological view, *Experientia,* 40(2), 164–167.

Eisenberg, J. F., and Gould, E. (1966) The behavior of *Solenodon paradoxus* in captivity with comments on the behavior of other insectivora, *Zoologica,* 51(4), 49–60.

Eisthen, H. L. (2002) Why are olfactory systems of different animals so similar?, *Brain, Behavior and Evolution,* 59(5–6), 273–293.

Elemans, C. P. H., et al. (2011) Superfast muscles set maximum call rate in echolocating bats, *Science,* 333(6051), 1885–1888.

Elwood, R. W. (2011) Pain and suffering in invertebrates?, *ILAR Journal,* 52(2), 175–184.

Elwood, R. W. (2019) Discrimination between nociceptive reflexes and more complex re-sponses consistent with pain in crustaceans, *Philosophical Transactions of the Royal Society B: Biological Sciences,* 374(1785), 20190368.

Elwood, R. W., and Appel, M. (2009) Pain experience in hermit crabs?, *Animal Behaviour,* 77(5), 1243–1246.

Embar, K., et al. (2018) Pit fights: Predators in evolutionarily independent communities, *Journal of Mammalogy,* 99(5), 1183–1188.

Emerling, C. A., and Springer, M. S. (2015) Genomic evidence for rod monochromacy in sloths and armadillos suggests early subterranean history for *Xenarthra, Proceedings of the Royal Society B: Biological Sciences,* 282(1800), 20142192.

Engels, S., et al. (2012) Night-migratory songbirds possess a magnetic compass in both eyes, *PLOS One,* 7(9), e43271.

Engels, S., et al. (2014) Anthropogenic electromagnetic noise disrupts magnetic compass orien-tation in a migratory bird, *Nature,* 509(7500), 353–356.

Erbe, C., et al. (2019) The effects of ship noise on marine mammals—A review, *Frontiers in Marine Science,* 6, 606.

Erbe, C., Dunlop, R., and Dolman, S. (2018) Effects of noise on marine mammals, in Slab-bekoorn, H., et al. (eds), *Effects of anthropogenic noise on animals,* 277–309. New York: Springer.

Eriksson, A., et al. (2012) Exploitation of insect vibrational signals reveals a new method of pest management, *PLOS One,* 7(3), e32954.

Etheredge, J. A., et al. (1999) Monarch butterflies (*Danaus plexippus* L.) use a magnetic compass for navigation, *Proceedings of the National Academy of Sciences,* 96(24), 13845–13846.

European Parliament, Council of the European Union (2010) Directive 2010/63/EU of the European Parliament and of the Council of 22 September 2010 on the protection of animals used for scientific purposes: Text with EEA relevance, L 276(20.10.2010), 33–79.

Evans, J. E., et al. (2012) Short-term physiological and behavioural effects of high- versus low-frequency fluorescent light on captive birds, *Animal Behaviour*, 83(1), 25–33.

Falchi, F., et al. (2016) The new world atlas of artificial night sky brightness, *Science Advances*, 2(6), e1600377.

Fedigan, L. M., et al. (2014) The heterozygote superiority hypothesis for polymorphic color vision is not supported by long-term fitness data from wild neotropical monkeys, *PLOS One*, 9(1), e84872.

Feller, K. D., et al. (2021) Surf and turf vision: Patterns and predictors of visual acuity in compound eye evolution, *Arthropod Structure & Development*, 60, 101002.

Fenton, M. B., et al. (eds), (2016) *Bat bioacoustics*. New York: Springer.

Fenton, M. B., Faure, P. A., and Ratcliffe, J. M. (2012) Evolution of high duty cycle echolocation in bats, *Journal of Experimental Biology*, 215(17), 2935–2944.

Fertin, A., and Casas, J. (2007) Orientation towards prey in antlions: Efficient use of wave propagation in sand, *Journal of Experimental Biology*, 210(19), 3337–3343.

Feynman, R. (1964) *The Feynman Lectures on Physics*, vol. 9, *Electricity in the Atmosphere*. Available at: www.feynmanlectures.caltech.edu/II_09.html.

Finger, S., and Piccolino, M. (2011) *The shocking history of electric fishes: From ancient epochs to the birth of modern neurophysiology*. New York: Oxford University Press.

Finkbeiner, S. D., et al. (2017) Ultraviolet and yellow reflectance but not fluorescence is important for visual discrimination of conspecifics by *Heliconius erato*, *Journal of Experimental Biology*, 220(7), 1267–1276.

Finneran, J. J. (2013) Dolphin "packet" use during long-range echolocation tasks, *Journal of the Acoustical Society of America*, 133(3), 1796–1810.

Firestein, S. (2005) A Nobel nose: The 2004 Nobel Prize in Physiology and Medicine, *Neuron*, 45(3), 333–338.

Fishbein, A. R., et al. (2020) Sound sequences in birdsong: How much do birds really care?, *Philosophical Transactions of the Royal Society B: Biological Sciences*, 375(1789), 20190044.

Fleissner, G., et al. (2003) Ultrastructural analysis of a putative magnetoreceptor in the beak of homing pigeons, *Journal of Comparative Neurology*, 458(4), 350–360.

Fleissner, G., et al. (2007) A novel concept of Fe-mineral-based magnetoreception: Histological and physicochemical data from the upper beak of homing pigeons, *Naturwissenschaften*, 94(8), 631–642.

Forbes, A. A., et al. (2018) Quantifying the unquantifiable: Why Hymenoptera, not Coleoptera, is the most speciose animal order, *BMC Ecology*, 18(1), 21.

Ford, N. B., and Low, J. R. (1984) Sex pheromone source location by garter snakes, *Journal of Chemical Ecology*, 10(8), 1193–1199.

Forel, A. (1874) *Les fourmis de la Suisse: Systématique, notices anatomiques et physiologiques, architecture, distribution géographique, nouvelles expériences et observations de moeurs*. Zurich: Druck von Zürcher & Furrer.

Fournier, J. P., et al. (2013) If a bird flies in the forest, does an insect hear it?, *Biology Letters*, 9(5), 20130319.

Fox, R., Lehmkuhle, S. W., and Westendorf, D. H. (1976) Falcon visual acuity, *Science*, 192(4236), 263–265.

Francis, C. D., et al. (2012) Noise pollution alters ecological services: Enhanced pollination and disrupted seed dispersal, *Proceedings of the Royal Society B: Biological Sciences*, 279(1739), 2727–2735.

Francis, C. D., et al. (2017) Acoustic environments matter: Synergistic benefits to humans and ecological communities, *Journal of Environmental Management*, 203(Pt 1), 245–254.

Fransson, T., et al. (2001) Magnetic cues trigger extensive refuelling, *Nature*, 414(6859), 35–36.

Friis, I., Sjulstok, E., and Solov'yov, I. A. (2017) Computational reconstruction reveals a candidate magnetic biocompass to be likely irrelevant for magnetoreception, *Scientific Reports,* 7(1), 13908.

Frisk, G. V. (2012) Noiseonomics: The relationship between ambient noise levels in the sea and global economic trends, *Scientific Reports,* 2(1), 437.

Fritsches, K. A., Brill, R. W., and Warrant, E. J. (2005) Warm eyes provide superior vision in swordfishes, *Current Biology,* 15(1), 55–58.

Fukutomi, M., and Carlson, B. A. (2020) A history of corollary discharge: Contributions of mormyrid weakly electric fish, *Frontiers in Integrative Neuroscience,* 14, 42.

Fullard, J. H., and Yack, J. E. (1993) The evolutionary biology of insect hearing, *Trends in Ecology & Evolution,* 8(7), 248–252.

Gagliardo, A., et al. (2013) Oceanic navigation in Cory's shearwaters: Evidence for a crucial role of olfactory cues for homing after displacement, *Journal of Experimental Biology,* 216(15), 2798–2805.

Gagnon, Y. L., et al. (2015) Circularly polarized light as a communication signal in mantis shrimps, *Current Biology,* 25(23), 3074–3078.

Gal, R., et al. (2014) Sensory arsenal on the stinger of the parasitoid jewel wasp and its possible role in identifying cockroach brains, *PLOS One,* 9(2), e89683.

Galambos, R., and Griffin, D. R. (1942) Obstacle avoidance by flying bats: The cries of bats, *Journal of Experimental Zoology,* 89(3), 475–490.

Gall, M. D., Salameh, T. S., and Lucas, J. R. (2013) Songbird frequency selectivity and temporal resolution vary with sex and season, *Proceedings of the Royal Society B: Biological Sciences,* 280(1751), 20122296.

Gall, M. D., and Wilczynski, W. (2015) Hearing conspecific vocal signals alters peripheral auditory sensitivity, *Proceedings of the Royal Society B: Biological Sciences,* 282(1808), 20150749.

Garcia-Larrea, L., and Bastuji, H. (2018) Pain and consciousness, *Progress in Neuro-Psychopharmacology and Biological Psychiatry,* 87(Pt B), 193–199.

Gardiner, J. M., et al. (2014) Multisensory integration and behavioral plasticity in sharks from different ecological niches, *PLOS One,* 9(4), e93036.

Garm, A., and Nilsson, D.-E. (2014) Visual navigation in starfish: First evidence for the use of vision and eyes in starfish, *Proceedings of the Royal Society B: Biological Sciences,* 281(1777), 20133011.

Garstang, M., et al. (1995) Atmospheric controls on elephant communication, *Journal of Experimental Biology,* 198(Pt 4), 939–951.

Gaspard, J. C., et al. (2017) Detection of hydrodynamic stimuli by the postcranial body of Florida manatees (*Trichechus manatus latirostris*), *Journal of Comparative Physiology A,* 203(2), 111–120.

Gaston, K. J. (2019) Nighttime ecology: The "nocturnal problem" revisited, *The American Naturalist,* 193(4), 481–502.

Gavelis, G. S., et al. (2015) Eye-like ocelloids are built from different endosymbiotically acquired components, *Nature,* 523(7559), 204–207.

Gehring, J., Kerlinger, P., and Manville, A. (2009) Communication towers, lights, and birds: Successful methods of reducing the frequency of avian collisions, *Ecological Applications,* 19(2), 505–514.

Gehring, W. J., and Wehner, R. (1995) Heat shock protein synthesis and thermotolerance in *Cataglyphis,* an ant from the Sahara desert, *Proceedings of the National Academy of Sciences,* 92(7), 2994–2998.

Geipel, I., et al. (2019) Bats actively use leaves as specular reflectors to detect acoustically camouflaged prey, *Current Biology,* 29(16), 2731–2736.e3.

Geipel, I., Jung, K., and Kalko, E. K. V. (2013) Perception of silent and motionless prey on vegetation by echolocation in the gleaning bat *Micronycteris microtis,* *Proceedings of the Royal Society B: Biological Sciences,* 280(1754), 20122830.

Geiser, F. (2013) Hibernation, *Current Biology,* 23(5), R188–R193.

Gentle, M. J., and Breward, J. (1986) The bill tip organ of the chicken (*Gallus gallus* var. *domesticus*), *Journal of Anatomy,* 145, 79–85.

Ghose, K., Moss, C. F., and Horiuchi, T. K. (2007) Flying big brown bats emit a beam with two lobes in the vertical plane, *Journal of the Acoustical Society of America,* 122(6), 3717–3724.

Gil, D., et al. (2015) Birds living near airports advance their dawn chorus and reduce overlap with aircraft noise, *Behavioral Ecology,* 26(2), 435–443.

Gill, A. B., et al. (2014) Marine renewable energy, electromagnetic (EM) fields and EM-sensitive animals, in Shields, M. A., and Payne, A. I. L. (eds), *Marine renewable energy technology and environmental interactions,* 61–79. Dordrecht: Springer.

Gläser, N., and Kröger, R. H. H. (2017) Variation in rhinarium temperature indicates sensory specializations in placental mammals, *Journal of Thermal Biology,* 67, 30–34.

Godfrey-Smith, P. (2016) *Other minds: The octopus, the sea, and the deep origins of consciousness.* New York: Farrar, Straus and Giroux.

Goerlitz, H. R., et al. (2010) An aerial-hawking bat uses stealth echolocation to counter moth hearing, *Current Biology,* 20(17), 1568–1572.

Goldberg, Y. P., et al. (2012) Human Mendelian pain disorders: A key to discovery and validation of novel analgesics, *Clinical Genetics,* 82(4), 367–373.

Goldbogen, J. A., et al. (2019) Extreme bradycardia and tachycardia in the world's largest animal, *Proceedings of the National Academy of Sciences,* 116(50), 25329–25332.

Gol'din, P. (2014) "Antlers inside": Are the skull structures of beaked whales (Cetacea: Ziphiidae) used for echoic imaging and visual display?, *Biological Journal of the Linnean Society,* 113(2), 510–515.

Goldsmith, T. H. (1980) Hummingbirds see near ultraviolet light, *Science,* 207(4432), 786–788.

Gonzalez-Bellido, P. T., Wardill, T. J., and Juusola, M. (2011) Compound eyes and retinal information processing in miniature dipteran species match their specific ecological demands, *Proceedings of the National Academy of Sciences,* 108(10), 4224–4229.

Göpfert, M. C., and Hennig, R. M. (2016) Hearing in insects, *Annual Review of Entomology,* 61, 257–276.

Göpfert, M. C., Surlykke, A., and Wasserthal, L. T. (2002) Tympanal and atympanal "mouth-ears" in hawkmoths (Sphingidae), *Proceedings of the Royal Academy B: Biological Sciences,* 269(1486), 89–95.

Gordon, T. A. C., et al. (2018) Habitat degradation negatively affects auditory settlement behavior of coral reef fishes, *Proceedings of the National Academy of Sciences,* 115(20), 5193–5198.

Gordon, T. A. C., et al. (2019) Acoustic enrichment can enhance fish community development on degraded coral reef habitat, *Nature Communications,* 10(1), 5414.

Gorham, P. W. (2013) Ballooning spiders: The case for electrostatic flight, arXiv:1309.4731.

Goris, R. C. (2011) Infrared organs of snakes: An integral part of vision, *Journal of Herpetology,* 45(1), 2–14.

Goté, J. T., et al. (2019) Growing tiny eyes: How juvenile jumping spiders retain high visual performance in the face of size limitations and developmental constraints, *Vision Research,* 160, 24–36.

Gould, E. (1965) Evidence for echolocation in the Tenrecidae of Madagascar, *Proceedings of the American Philosophical Society,* 109(6), 352–360.

Goutte, S., et al. (2017) Evidence of auditory insensitivity to vocalization frequencies in two frogs, *Scientific Reports,* 7(1), 12121.

Gracheva, E. O., et al. (2010) Molecular basis of infrared detection by snakes, *Nature,* 464(7291), 1006–1011.

Gracheva, E. O., et al. (2011) Ganglion-specific splicing of TRPV1 underlies infrared sensation in vampire bats, *Nature,* 476(7358), 88–91.

Gracheva, E. O., and Bagriantsev, S. N. (2015) Evolutionary adaptation to thermosensation, *Current Opinion in Neurobiology,* 34, 67–73.

Granger, J., et al. (2020) Gray whales strand more often on days with increased levels of atmospheric radio-frequency noise, *Current Biology,* 30(4), R155–R156.

Grant, R. A., Breakell, V., and Prescott, T. J. (2018) Whisker touch sensing guides locomotion in small, quadrupedal mammals, *Proceedings of the Royal Society B: Biological Sciences,* 285(1880), 20180592.

Grant, R. A., Sperber, A. L., and Prescott, T. J. (2012) The role of orienting in vibrissal touch sensing, *Frontiers in Behavioral Neuroscience,* 6, 39.

Grasso, F. W. (2014) The octopus with two brains: How are distributed and central representations integrated in the octopus central nervous system?, in Darmaillacq, A.-S., Dickel, L., and Mather, J. (eds), *Cephalopod cognition,* 94–122. Cambridge: Cambridge University Press.

Graziadei, P. P., and Gagne, H. T. (1976) Sensory innervation in the rim of the octopus sucker, *Journal of Morphology,* 150(3), 639–679.

Greenwood, V. (2012) The humans with super human vision, *Discover Magazine.* Available at: www.discovermagazine.com/mind/the-humans-with-super-human-vision.

Gregory, J. E., et al. (1989) Responses of electroreceptors in the snout of the echidna, *Journal of Physiology,* 414, 521–538.

Greif, S., et al. (2017) Acoustic mirrors as sensory traps for bats, *Science,* 357(6355), 1045–1047.

Griffin, D. R. (1944a) Echolocation by blind men, bats and radar, *Science,* 100(2609), 589–590.

Griffin, D. R. (1944b) The sensory basis of bird navigation, *The Quarterly Review of Biology,* 19(1), 15–31.

Griffin, D. R. (1953) Bat sounds under natural conditions, with evidence for echolocation of insect prey, *Journal of Experimental Zoology,* 123(3), 435–465.

Griffin, D. R. (1974) *Listening in the dark: The acoustic orientation of bats and men.* New York: Dover Publications.

Griffin, D. R. (2001) Return to the magic well: Echolocation behavior of bats and responses of insect prey, *BioScience,* 51(7), 555–556.

Griffin, D. R., and Galambos, R. (1941) The sensory basis of obstacle avoidance by flying bats, *Journal of Experimental Zoology,* 86(3), 481–506.

Griffin, D. R., Webster, F. A., and Michael, C. R. (1960) The echolocation of flying insects by bats, *Animal Behaviour,* 8(3), 141–154.

Grinnell, A. D. (1966) Mechanisms of overcoming interference in echolocating animals, in Busnel, R.-G. (ed), *Animal Sonar Systems: Biology and Bionics,* 1, 451–480.

Grinnell, A .D., Gould, E., and Fenton, M. B. (2016) A history of the study of echolocation, in Fenton, M. B., et al. (eds), *Bat bioacoustics,* 1–24. New York: Springer.

Grinnell, A. D., and Griffin, D. R. (1958) The sensitivity of echolocation in bats, *Biological Bulletin,* 114(1), 10–22.

Gross, K., Pasinelli, G., and Kunc, H. P. (2010) Behavioral plasticity allows short-term adjustment to a novel environment, *The American Naturalist,* 176(4), 456–464.

Grüsser, O.-J. (1994) Early concepts on efference copy and reafference, *Behavioral and Brain Sciences,* 17(2), 262–265.

Gu, J.-J., et al. (2012) Wing stridulation in a Jurassic katydid (Insecta, Orthoptera) produced low-pitched musical calls to attract females, *Proceedings of the National Academy of Sciences,* 109(10), 3868–3873.

Günther, R. H., O'Connell-Rodwell, C. E., and Klemperer, S. L. (2004) Seismic waves from elephant vocalizations: A possible communication mode?, *Geophysical Research Letters,* 31(11).

Gutnick, T., et al. (2011) *Octopus vulgaris* uses visual information to determine the location of its arm, *Current Biology,* 21(6), 460–462.

Hagedorn, M. (2004) Essay: The lure of field research on electric fish, in von der Emde, G., Mogdans, J., and Kapoor, B. G. (eds), *The senses of fish: Adaptations for the reception of natural stimuli,* 362–368. Dordrecht: Springer.

Hagedorn, M., and Heiligenberg, W. (1985) Court and spark: Electric signals in the courtship and mating of gymnotoid fish, *Animal Behaviour,* 33(1), 254–265.

Hager, F. A., and Kirchner, W. H. (2013) Vibrational long-distance communication in the termites *Macrotermes natalensis* and *Odontotermes* sp., *Journal of Experimental Biology,* 216(17), 3249–3256.

Hager, F. A., and Krausa, K. (2019) Acacia ants respond to plant-borne vibrations caused by mammalian browsers, *Current Biology*, 29(5), 717–725.e3.

Halfwerk, W., et al. (2019) Adaptive changes in sexual signalling in response to urbanization, *Nature Ecology & Evolution*, 3(3), 374–380.

Hamel, J. A., and Cocroft, R. B. (2012) Negative feedback from maternal signals reduces false alarms by collectively signalling offspring, *Proceedings of the Royal Society B: Biological Sciences*, 279(1743), 3820–3826.

Han, C. S., and Jablonski, P. G. (2010) Male water striders attract predators to intimidate females into copulation, *Nature Communications*, 1(1), 52.

Hanke, F. D., and Kelber, A. (2020) The eye of the common octopus (*Octopus vulgaris*), *Frontiers in Physiology*, 10, 1637.

Hanke, W., et al. (2010) Harbor seal vibrissa morphology suppresses vortex-induced vibrations, *Journal of Experimental Biology*, 213(15), 2665–2672.

Hanke, W., and Dehnhardt, G. (2015) Vibrissal touch in pinnipeds, *Scholarpedia*, 10(3), 6828.

Hanke, W., Römer, R., and Dehnhardt, G. (2006) Visual fields and eye movements in a harbor seal (*Phoca vitulina*), *Vision Research*, 46(17), 2804–2814.

Hardy, A. R., and Hale, M. E. (2020) Sensing the structural characteristics of surfaces: Texture encoding by a bottom-dwelling fish, *Journal of Experimental Biology*, 223(21), jeb227280.

Harley, H. E., Roitblat, H. L., and Nachtigall, P. E. (1996) Object representation in the bottlenose dolphin (*Tursiops truncatus*): Integration of visual and echoic information, *Journal of Experimental Psychology: Animal Behavior Processes*, 22(2), 164–174.

Hart, N. S., et al. (2011) Microspectrophotometric evidence for cone monochromacy in sharks, *Naturwissenschaften*, 98(3), 193–201.

Hartline, P. H., Kass, L., and Loop, M. S. (1978) Merging of modalities in the optic tectum: Infrared and visual integration in rattlesnakes, *Science*, 199(4334), 1225–1229.

Hartzell, P. L., et al. (2011) Distribution and phylogeny of glacier ice worms (*Mesenchytraeus solifugus* and *Mesenchytraeus solifugus rainierensis*), *Canadian Journal of Zoology*, 83(9), 1206–1213.

Haspel, G., et al. (2012) By the teeth of their skin, cavefish find their way, *Current Biology*, 22(16), R629–R630.

Haynes, K. F., et al. (2002) Aggressive chemical mimicry of moth pheromones by a bolas spider: How does this specialist predator attract more than one species of prey?, *Chemoecology*, 12(2), 99–105.

Healy, K., et al. (2013) Metabolic rate and body size are linked with perception of temporal information, *Animal Behaviour*, 86(4), 685–696.

Heffner, H. E. (1983) Hearing in large and small dogs: Absolute thresholds and size of the tympanic membrane, *Behavioral Neuroscience*, 97(2), 310–318.

Heffner, H. E., and Heffner, R. S. (2018) The evolution of mammalian hearing, in *To the ear and back again—Advances in auditory biophysics: Proceedings of the 13th Mechanics of Hearing Workshop*, St. Catharines, Canada, 130001. Available at: aip.scitation.org/doi/abs/10.1063/1.5038516.

Heffner, R. S., and Heffner, H. E. (1985) Hearing range of the domestic cat, *Hearing Research*, 19(1), 85–88.

Hein, C. M., et al. (2011) Robins have a magnetic compass in both eyes, *Nature*, 471(7340), E1.

Heinrich, B. (1993) *The hot-blooded insects: Strategies and mechanisms of thermoregulation*. Berlin: Springer.

Henninger, J., et al. (2018) Statistics of natural communication signals observed in the wild identify important yet neglected stimulus regimes in weakly electric fish, *Journal of Neuroscience*, 38(24), 5456–5465.

Henry, K. S., et al. (2011) Songbirds tradeoff auditory frequency resolution and temporal resolution, *Journal of Comparative Physiology A*, 197(4), 351–359.

Henson, O. W. (1965) The activity and function of the middle-ear muscles in echo-locating bats, *Journal of Physiology*, 180(4), 871–887.

Hepper, P. G. (1988) The discrimination of human odour by the dog, *Perception*, 17(4), 549–554.

Hepper, P. G., and Wells, D. L. (2005) How many footsteps do dogs need to determine the direction of an odour trail?, *Chemical Senses*, 30(4), 291–298.

Herberstein, M. E., Heiling, A. M., and Cheng, K. (2009) Evidence for UV-based sensory exploitation in Australian but not European crab spiders, *Evolutionary Ecology*, 23(4), 621–634.

Heyers, D., et al. (2007) A visual pathway links brain structures active during magnetic compass orientation in migratory birds, *PLOS One*, 2(9), e937.

Hildebrand, J. (2005) Impacts of anthropogenic sound, in Reynolds, J. E., et al. (eds), *Marine mammal research: Conservation beyond crisis*, 101–124. Baltimore: Johns Hopkins University Press.

Hill, P. S. M. (2008) *Vibrational communication in animals*. Cambridge, MA: Harvard University Press.

Hill, P. S. M. (2009) How do animals use substrate-borne vibrations as an information source?, *Naturwissenschaften*, 96(12), 1355–1371.

Hill, P. S. M. (2014) Stretching the paradigm or building a new? Development of a cohesive language for vibrational communication, in Cocroft, R. B., et al. (eds), *Studying vibrational communication*, 13–30. Berlin: Springer.

Hill, P. S. M., and Wessel, A. (2016) Biotremology, *Current Biology*, 26(5), R187–R191.

Hines, H. M., et al. (2011) Wing patterning gene redefines the mimetic history of *Heliconius* butterflies, *Proceedings of the National Academy of Sciences*, 108(49), 19666–19671.

Hiramatsu, C., et al. (2017) Experimental evidence that primate trichromacy is well suited for detecting primate social colour signals, *Proceedings of the Royal Society B: Biological Sciences*, 284(1856), 20162458.

Hiryu, S., et al. (2005) Doppler-shift compensation in the Taiwanese leaf-nosed bat (*Hipposideros terasensis*) recorded with a telemetry microphone system during flight, *Journal of the Acoustical Society of America*, 118(6), 3927–3933.

Hochner, B. (2012) An embodied view of octopus neurobiology, *Current Biology*, 22(20), R887–R892.

Hochner, B. (2013) How nervous systems evolve in relation to their embodiment: What we can learn from octopuses and other molluscs, *Brain, Behavior and Evolution*, 82(1), 19–30.

Hochstoeger, T., et al. (2020) The biophysical, molecular, and anatomical landscape of pigeon CRY4: A candidate light-based quantal magnetosensor, *Science Advances*, 6(33), eabb9110.

Hofer, B. (1908) Studien über die Hautsinnesorgane der Fische. I. Die Funktion der Seitenorgane bei den Fischen, *Berichte aus der Kgl. Bayerischen Biologischen Versuchsstation in München*, 1, 115–164.

Hoffstaetter, L. J., Bagriantsev, S. N., and Gracheva, E. O. (2018) TRPs et al.: A molecular toolkit for thermosensory adaptations, *Pflügers Archiv—European Journal of Physiology*, 470(5), 745–759.

Holderied, M. W., and von Helversen, O. (2003) Echolocation range and wingbeat period match in aerial-hawking bats, *Proceedings of the Royal Society B: Biological Sciences*, 270(1530), 2293–2299.

Holland, R. A., et al. (2006) Navigation: Bat orientation using Earth's magnetic field, *Nature*, 444(7120), 702.

Holy, T. E., and Guo, Z. (2005) Ultrasonic songs of male mice, *PLOS Biology*, 3(12), e386.

Hopkins, C., and Bass, A. (1981) Temporal coding of species recognition signals in an electric fish, *Science*, 212(4490), 85–87.

Hopkins, C. D. (1981) On the diversity of electric signals in a community of mormyrid electric fish in West Africa, *American Zoologist*, 21(1), 211–222.

Hopkins, C. D. (2005) Passive electrolocation and the sensory guidance of oriented behavior, in Bullock, T. H., et al. (eds), *Electroreception*, 264–289. New York: Springer.

Hopkins, C. D. (2009) Electrical perception and communication, in Squire, L. R. (ed), *Encyclopedia of neuroscience*, 813–831. Amsterdam: Elsevier.

Hore, P. J., and Mouritsen, H. (2016) The radical-pair mechanism of magnetoreception, *Annual Review of Biophysics*, 45(1), 299–344.

Horowitz, A. (2010) *Inside of a dog: What dogs see, smell, and know*. London: Simon & Schuster UK.

Horowitz, A. (2016) *Being a dog: Following the dog into a world of smell*. New York: Scribner.

Horowitz, A., and Franks, B. (2020) What smells? Gauging attention to olfaction in canine cognition research, *Animal Cognition*, 23(1), 11–18.

Horváth, G., et al. (2009) Polarized light pollution: A new kind of ecological photopollution, *Frontiers in Ecology and the Environment*, 7(6), 317–325.

Horwitz, J. (2015) *War of the whales: A true story*. New York: Simon & Schuster.

Hughes, A. (1977) The topography of vision in mammals of contrasting life style: Comparative optics and retinal organisation, in Crescitelli, F. (ed), *The visual system in vertebrates*, 613–756. New York: Springer.

Hughes, H. C. (2001) *Sensory exotica: A world beyond human experience*. Cambridge, MA: MIT Press.

Hulgard, K., et al. (2016) Big brown bats (*Eptesicus fuscus*) emit intense search calls and fly in stereotyped flight paths as they forage in the wild, *Journal of Experimental Biology*, 219(3), 334–340.

Hunt, S., et al. (1998) Blue tits are ultraviolet tits, *Proceedings of the Royal Society B: Biological Sciences*, 265(1395), 451–455.

Hurst, J., et al. (eds), (2008) *Chemical signals in vertebrates 11*. New York: Springer.

Ibrahim, N., et al. (2014) Semiaquatic adaptations in a giant predatory dinosaur, *Science*, 345(6204), 1613–1616.

Ikinamo (2011) Simroid dental training humanoid robot communicates with trainee dentists #DigInfo. [Video] Available at: www.youtube.com/watch?v=C47NHADFQSo.

Inger, R., et al. (2014) Potential biological and ecological effects of flickering artificial light, *PLOS One*, 9(5), e98631.

Inman, M. (2013) Why the mantis shrimp is my new favorite animal, *The Oatmeal*. Available at: theoatmeal.com/comics/mantis_shrimp.

Irwin, W. P., Horner, A. J., and Lohmann, K. J. (2004) Magnetic field distortions produced by protective cages around sea turtle nests: Unintended consequences for orientation and navigation?, *Biological Conservation*, 118(1), 117–120.

Ivanov, M. P. (2004) Dolphin's echolocation signals in a complicated acoustic environment, *Acoustical Physics*, 50(4), 469–479.

Jacobs, G. H. (1984) Within-species variations in visual capacity among squirrel monkeys (*Saimiri sciureus*): Color vision, *Vision Research*, 24(10), 1267–1277.

Jacobs, G. H., and Neitz, J. (1987) Inheritance of color vision in a New World monkey (*Saimiri sciureus*), *Proceedings of the National Academy of Sciences*, 84(8), 2545–2549.

Jacobs, G. H., Neitz, J., and Deegan, J. F. (1991) Retinal receptors in rodents maximally sensitive to ultraviolet light, *Nature*, 353(6345), 655–656.

Jacobs, L. F. (2012) From chemotaxis to the cognitive map: The function of olfaction, *Proceedings of the National Academy of Sciences*, 109(Suppl. 1), 10693–10700.

Jakob, E. M., et al. (2018) Lateral eyes direct principal eyes as jumping spiders track objects, *Current Biology*, 28(18), R1092–R1093.

Jakobsen, L., Ratcliffe, J. M., and Surlykke, A. (2013) Convergent acoustic field of view in echolocating bats, *Nature*, 493(7430), 93–96.

Japyassú, H. F., and Laland, K. N. (2017) Extended spider cognition, *Animal Cognition*, 20(3), 375–395.

Jechow, A., and Hölker, F. (2020) Evidence that reduced air and road traffic decreased artificial night-time skyglow during COVID-19 lockdown in Berlin, Germany, *Remote Sensing*, 12(20), 3412.

Jiang, P., et al. (2012) Major taste loss in carnivorous mammals, *Proceedings of the National Academy of Sciences*, 109(13), 4956–4961.

Johnsen, S. (2012) *The optics of life: A biologist's guide to light in nature.* Princeton, NJ: Princeton University Press.

Johnsen, S. (2014) Hide and seek in the open sea: Pelagic camouflage and visual countermeasures, *Annual Review of Marine Science,* 6(1), 369–392.

Johnsen, S. (2017) Open questions: We don't really know anything, do we? Open questions in sensory biology, *BMC Biology,* 15, art. 43.

Johnsen, S., and Lohmann, K. J. (2005) The physics and neurobiology of magnetoreception, *Nature Reviews Neuroscience,* 6(9), 703–712.

Johnsen, S., Lohmann, K. J., and Warrant, E. J. (2020) Animal navigation: A noisy magnetic sense?, *Journal of Experimental Biology,* 223(18), jeb164921.

Johnsen, S., and Widder, E. (2019) Mission logs: June 20, Here be monsters: We filmed a giant squid in America's backyard, *NOAA Ocean Exploration.* Available at: oceanexplorer.noaa .gov/explorations/19biolum/logs/jun20/jun20.html.

Johnson, M., et al. (2004) Beaked whales echolocate on prey, *Proceedings of the Royal Society B: Biological Sciences,* 271(Suppl. 6), S383–S386.

Johnson, M., Aguilar de Soto, N., and Madsen, P. (2009) Studying the behaviour and sensory ecology of marine mammals using acoustic recording tags: A review, *Marine Ecology Progress Series,* 395, 55–73.

Johnson, R. N., et al. (2018) Adaptation and conservation insights from the koala genome, *Nature Genetics,* 50(8), 1102–1111.

Jones, G., and Teeling, E. (2006) The evolution of echolocation in bats, *Trends in Ecology & Evolution,* 21(3), 149–156.

Jordan, G., et al. (2010) The dimensionality of color vision in carriers of anomalous trichromacy, *Journal of Vision,* 10(8), 12.

Jordan, G., and Mollon, J. (2019) Tetrachromacy: The mysterious case of extra-ordinary color vision, *Current Opinion in Behavioral Sciences,* 30, 130–134.

Jordt, S.-E., and Julius, D. (2002) Molecular basis for species-specific sensitivity to "hot" chili peppers, *Cell,* 108(3), 421–430.

Josberger, E. E., et al. (2016) Proton conductivity in ampullae of Lorenzini jelly, *Science Advances,* 2(5), e1600112.

Jung, J., et al. (2019) How do red-eyed treefrog embryos sense motion in predator attacks? Assessing the role of vestibular mechanoreception, *Journal of Experimental Biology,* 222(21), jeb206052.

Jung, K., Kalko, E. K. V., and von Helversen, O. (2007) Echolocation calls in Central American emballonurid bats: Signal design and call frequency alternation, *Journal of Zoology,* 272(2), 125–137.

Kajiura, S. M. (2001) Head morphology and electrosensory pore distribution of carcharhinid and sphyrnid sharks, *Environmental Biology of Fishes,* 61(2), 125–133.

Kajiura, S. M. (2003) Electroreception in neonatal bonnethead sharks, *Sphyrna tiburo, Marine Biology,* 143(3), 603–611.

Kajiura, S. M., and Holland, K. N. (2002) Electroreception in juvenile scalloped hammerhead and sandbar sharks, *Journal of Experimental Biology,* 205(23), 3609–3621.

Kalberer, N. M., Reisenman, C. E., and Hildebrand, J. G. (2010) Male moths bearing transplanted female antennae express characteristically female behaviour and central neural activity, *Journal of Experimental Biology,* 213(8), 1272–1280.

Kalka, M. B., Smith, A. R., and Kalko, E. K. V. (2008) Bats limit arthropods and herbivory in a tropical forest, *Science,* 320(5872), 71.

Kalmijn, A. J. (1971) The electric sense of sharks and rays, *Journal of Experimental Biology,* 55(2), 371–383.

Kalmijn, A. J. (1974) The detection of electric fields from inanimate and animate sources other than electric organs, in Fessard, A. (ed), *Electroreceptors and other specialized receptors in lower vertebrates,* 147–200. Berlin: Springer.

Kalmijn, A. J. (1982) Electric and magnetic field detection in elasmobranch fishes, *Science*, 218(4575), 916–918.

Kaminski, J., et al. (2019) Evolution of facial muscle anatomy in dogs, *Proceedings of the National Academy of Sciences*, 116(29), 14677–14681.

Kane, S. A., Van Beveren, D., and Dakin, R. (2018) Biomechanics of the peafowl's crest reveals frequencies tuned to social displays, *PLOS One*, 13(11), e0207247.

Kant, I. (2007) *Anthropology, history, and education*. Cambridge: Cambridge University Press.

Kapoor, M. (2020) The only catfish native to the western U.S. is running out of water, *High Country News*. Available at: www.hcn.org/issues/52.7/fish-the-only-catfish-native-to-the -western-u-s-is-running-out-of-water.

Kardong, K. V., and Berkhoudt, H. (1999) Rattlesnake hunting behavior: Correlations between plasticity of predatory performance and neuroanatomy, *Brain, Behavior and Evolution*, 53(1), 20–28.

Kardong, K. V., and Mackessy, S. P. (1991) The strike behavior of a congenitally blind rattlesnake, *Journal of Herpetology*, 25(2), 208–211.

Kasumyan, A. O. (2019) The taste system in fishes and the effects of environmental variables, *Journal of Fish Biology*, 95(1), 155–178.

Katz, H. K., et al. (2015) Eye movements in chameleons are not truly independent—Evidence from simultaneous monocular tracking of two targets, *Journal of Experimental Biology*, 218(13), 2097–2105.

Kavaliers, M. (1988) Evolutionary and comparative aspects of nociception, *Brain Research Bulletin*, 21(6), 923–931.

Kawahara, A. Y., et al. (2019) Phylogenomics reveals the evolutionary timing and pattern of butterflies and moths, *Proceedings of the National Academy of Sciences*, 116(45), 22657–22663.

Kelber, A., Balkenius, A., and Warrant, E. J. (2002) Scotopic colour vision in nocturnal hawkmoths, *Nature*, 419(6910), 922–925.

Kelber, A., Vorobyev, M., and Osorio, D. (2003) Animal colour vision—Behavioural tests and physiological concepts, *Biological Reviews of the Cambridge Philosophical Society*, 78(1), 81–118.

Keller, A., et al. (2007) Genetic variation in a human odorant receptor alters odour perception, *Nature*, 449(7161), 468–472.

Keller, A., and Vosshall, L. B. (2004a) A psychophysical test of the vibration theory of olfaction, *Nature Neuroscience*, 7(4), 337–338.

Keller, A., and Vosshall, L. B. (2004b) Human olfactory psychophysics, *Current Biology*, 14(20), R875–R878.

Kempster, R. M., Hart, N. S., and Collin, S. P. (2013) Survival of the stillest: Predator avoidance in shark embryos, *PLOS One*, 8(1), e52551.

Ketten, D. R. (1997) Structure and function in whale ears, *Bioacoustics*, 8(1–2), 103–135.

Key, B. (2016) Why fish do not feel pain, *Animal Sentience*, 1(3).

Key, F. M., et al. (2018) Human local adaptation of the TRPM8 cold receptor along a latitudinal cline, *PLOS Genetics*, 14(5), e1007298.

Kick, S., and Simmons, J. (1984) Automatic gain control in the bat's sonar receiver and the neuroethology of echolocation, *Journal of Neuroscience*, 4(11), 2725–2737.

Kimchi, T., Etienne, A. S., and Terkel, J. (2004) A subterranean mammal uses the magnetic compass for path integration, *Proceedings of the National Academy of Sciences*, 101(4), 1105–1109.

King, J. E., Becker, R. F., and Markee, J. E. (1964) Studies on olfactory discrimination in dogs: (3) Ability to detect human odour trace, *Animal Behaviour*, 12(2), 311–315.

Kingston, A. C. N., et al. (2015) Visual phototransduction components in cephalopod chromatophores suggest dermal photoreception, *Journal of Experimental Biology*, 218(10), 1596–1602.

Kirschfeld, K. (1976) The resolution of lens and compound eyes, in Zettler, F., and Weiler, R. (eds), *Neural principles in vision*, 354–370. Berlin: Springer.

Kirschvink, J., et al. (1997) Measurement of the threshold sensitivity of honeybees to weak, extremely low-frequency magnetic fields, *Journal of Experimental Biology*, 200(Pt 9), 1363–1368.

Kish, D. (1995) Echolocation: How humans can "see" without sight. Unpublished master's thesis, California State University.

Kish, D. (2015) How I use sonar to navigate the world. TED Talk. Available at: www.ted.com/talks/daniel_kish_how_i_use_sonar_to_navigate_the_world.

Klärner, D., and Barth, F. G. (1982) Vibratory signals and prey capture in orb-weaving spiders (*Zygiella x-notata, Nephila clavipes;* Araneidae), *Journal of Comparative Physiology*, 148(4), 445–455.

Klopsch, C., Kuhlmann, H. C., and Barth, F. G. (2012) Airflow elicits a spider's jump towards airborne prey. I. Airflow around a flying blowfly, *Journal of the Royal Society Interface*, 9(75), 2591–2602.

Klopsch, C., Kuhlmann, H. C., and Barth, F. G. (2013) Airflow elicits a spider's jump towards airborne prey. II. Flow characteristics guiding behaviour, *Journal of the Royal Society Interface*, 10(82), 20120820.

Knop, E., et al. (2017) Artificial light at night as a new threat to pollination, *Nature*, 548(7666), 206–209.

Knudsen, E. I., Blasdel, G. G., and Konishi, M. (1979) Sound localization by the barn owl (*Tyto alba*) measured with the search coil technique, *Journal of Comparative Physiology A*, 133(1), 1–11.

Kober, R., and Schnitzler, H. (1990) Information in sonar echoes of fluttering insects available for echolocating bats, *Journal of the Acoustical Society of America*, 87(2), 882–896.

Kojima, S. (1990) Comparison of auditory functions in the chimpanzee and human, *Folia Primatologica*, 55(2), 62–72.

Kolbert, E. (2014) *The sixth extinction: An unnatural history.* New York: Henry Holt.

Konishi, M. (1969) Time resolution by single auditory neurones in birds, *Nature*, 222(5193), 566–567.

Konishi, M. (1973) Locatable and nonlocatable acoustic signals for barn owls, *The American Naturalist*, 107(958), 775–785.

Konishi, M. (2012) How the owl tracks its prey, *American Scientist*, 100(6), 494.

Koselj, K., Schnitzler, H.-U., and Siemers, B. M. (2011) Horseshoe bats make adaptive prey-selection decisions, informed by echo cues, *Proceedings of the Royal Society B: Biological Sciences*, 278(1721), 3034–3041.

Koshitaka, H., et al. (2008) Tetrachromacy in a butterfly that has eight varieties of spectral receptors, *Proceedings of the Royal Society B: Biological Sciences*, 275(1637), 947–954.

Kothari, N. B., et al. (2014) Timing matters: Sonar call groups facilitate target localization in bats, *Frontiers in Physiology*, 5, 168.

Krestel, D., et al. (1984) Behavioral determination of olfactory thresholds to amyl acetate in dogs, *Neuroscience and Biobehavioral Reviews*, 8(2), 169–174.

Kröger, R. H. H., and Goiricelaya, A. B. (2017) Rhinarium temperature dynamics in domestic dogs, *Journal of Thermal Biology*, 70, 15–19.

Krumm, B., et al. (2017) Barn owls have ageless ears, *Proceedings of the Royal Society B: Biological Sciences*, 284(1863), 20171584.

Kuhn, R. A., et al. (2010) Hair density in the Eurasian otter *Lutra lutra* and the sea otter *Enhydra lutris, Acta Theriologica*, 55(3), 211–222.

Kuna, V. M., and Nábělek, J. L. (2021) Seismic crustal imaging using fin whale songs, *Science*, 371(6530), 731–735.

Kunc, H., et al. (2014) Anthropogenic noise affects behavior across sensory modalities, *The American Naturalist*, 184 (4), E93–E100.

Kürten, L., and Schmidt, U. (1982) Thermoperception in the common vampire bat (*Desmodus rotundus*), *Journal of Comparative Physiology A*, 146(2), 223–228.

Kwon, D. (2019) Watcher of whales: A profile of Roger Payne. *The Scientist*. Available at: www.the-scientist.com/profile/watcher-of-whales--a-profile-of-roger-payne-66610.

Kyba, C. C. M., et al. (2017) Artificially lit surface of Earth at night increasing in radiance and extent, *Science Advances*, 3(11), e1701528.

Land, M. F. (1966) A multilayer interference reflector in the eye of the scallop, *Pecten maximus*, *Journal of Experimental Biology*, 45(3), 433–447.

Land, M. F. (1969a) Movements of the retinae of jumping spiders (Salticidae: Dendryphantinae) in response to visual stimuli, *Journal of Experimental Biology*, 51(2), 471–493.

Land, M. F. (1969b) Structure of the retinae of the principal eyes of jumping spiders (Salticidae: Dendryphantinae) in relation to visual optics, *Journal of Experimental Biology*, 51(2), 443–470.

Land, M. F. (2003) The spatial resolution of the pinhole eyes of giant clams (*Tridacna maxima*), *Proceedings of the Royal Society B: Biological Sciences*, 270(1511), 185–188.

Land, M. F. (2018) *Eyes to see: The astonishing variety of vision in nature.* Oxford: Oxford University Press.

Land, M. F., et al. (1990) The eye-movements of the mantis shrimp *Odontodactylus scyllarus* (Crustacea: Stomatopoda), *Journal of Comparative Physiology A*, 167(2), 155–166.

Landler, L., et al. (2018) Comment on "Magnetosensitive neurons mediate geomagnetic orientation in *Caenorhabditis elegans*," *eLife*, 7, e30187.

Landolfa, M. A., and Barth, F. G. (1996) Vibrations in the orb web of the spider *Nephila clavipes*: Cues for discrimination and orientation, *Journal of Comparative Physiology A*, 179(4), 493–508.

Lane, K. A., Lucas, K. M., and Yack, J. E. (2008) Hearing in a diurnal, mute butterfly, *Morpho peleides* (Papilionoidea, Nymphalidae), *Journal of Comparative Neurology*, 508(5), 677–686.

Laska, M. (2017) Human and animal olfactory capabilities compared, in Buettner, A. (ed), *Springer handbook of odor*, 81–82. New York: Springer.

Laughlin, S. B., and Weckström, M. (1993) Fast and slow photoreceptors—A comparative study of the functional diversity of coding and conductances in the Diptera, *Journal of Comparative Physiology A*, 172(5), 593–609.

Laursen, W. J., et al. (2016) Low-cost functional plasticity of TRPV1 supports heat tolerance in squirrels and camels, *Proceedings of the National Academy of Sciences*, 113(40), 11342–11347.

LaVinka, P. C., and Park, T. J. (2012) Blunted behavioral and C Fos responses to acidic fumes in the African naked mole-rat, *PLOS One*, 7(9), e45060.

Lavoué, S., et al. (2012) Comparable ages for the independent origins of electrogenesis in African and South American weakly electric fishes, *PLOS One*, 7(5), e36287.

Lawson, S. L., et al. (2018) Relative salience of syllable structure and syllable order in zebra finch song, *Animal Cognition*, 21(4), 467–480.

Lazzari, C. R. (2009) Orientation towards hosts in haematophagous insects, in Simpson, S., and Casas, J. (eds), *Advances in insect physiology*, vol. 37, 1–58. Amsterdam: Elsevier.

Lecocq, T., et al. (2020) Global quieting of high-frequency seismic noise due to COVID-19 pandemic lockdown measures, *Science*, 369(6509), 1338–1343.

Lee-Johnson, C. P., and Carnegie, D. A. (2010) Mobile robot navigation modulated by artificial emotions, *IEEE Transactions on Systems, Man, and Cybernetics, Part B (Cybernetics)*, 40(2), 469–480.

Legendre, F., Marting, P. R., and Cocroft, R. B. (2012) Competitive masking of vibrational signals during mate searching in a treehopper, *Animal Behaviour*, 83(2), 361–368.

Leitch, D. B., and Catania, K. C. (2012) Structure, innervation and response properties of integumentary sensory organs in crocodilians, *Journal of Experimental Biology*, 215(23), 4217–4230.

Lenoir, A., et al. (2001) Chemical ecology and social parasitism in ants, *Annual Review of Entomology*, 46(1), 573–599.

Leonard, M. L., and Horn, A. G. (2008) Does ambient noise affect growth and begging call structure in nestling birds?, *Behavioral Ecology*, 19(3), 502–507.

Leonhardt, S. D., et al. (2016) Ecology and evolution of communication in social insects, *Cell*, 164(6), 1277–1287.

Levy, G., and Hochner, B. (2017) Embodied organization of *Octopus vulgaris* morphology, vision, and locomotion, *Frontiers in Physiology,* 8, 164.

Lewin, G., Lu, Y., and Park, T. (2004) A plethora of painful molecules, *Current Opinion in Neurobiology,* 14(4), 443–449.

Lewis, E. R., et al. (2006) Preliminary evidence for the use of microseismic cues for navigation by the Namib golden mole, *Journal of the Acoustical Society of America,* 119(2), 1260–1268.

Lewis, J. (2014) Active electroreception: Signals, sensing, and behavior, in Evans, D. H., Claiborne, J. B., and Currie, S. (eds), *The physiology of fishes,* 4th ed., 373–388. Boca Raton, FL: CRC Press.

Li, F. (2013) Taste perception: From the tongue to the testis, *Molecular Human Reproduction,* 19(6), 349–360.

Li, L., et al. (2015) Multifunctionality of chiton biomineralized armor with an integrated visual system, *Science,* 350(6263), 952–956.

Lind, O., et al. (2013) Ultraviolet sensitivity and colour vision in raptor foraging, *Journal of Experimental Biology,* 216(Pt 10), 1819–1826.

Linsley, E. G. (1943) Attraction of *Melanophila* beetles by fire and smoke, *Journal of Economic Entomology,* 36(2), 341–342.

Linsley, E. G., and Hurd, P. D. (1957) *Melanophila* beetles at cement plants in Southern California (Coleoptera, Buprestidae), *Coleopterists Bulletin,* 11(1/2), 9–11.

Lissmann, H. W. (1951) Continuous electrical signals from the tail of a fish, *Gymnarchus niloticus* Cuv., *Nature,* 167(4240), 201–202.

Lissmann, H. W. (1958) On the function and evolution of electric organs in fish, *Journal of Experimental Biology,* 35(1), 156–191.

Lissmann, H. W., and Machin, K. E. (1958) The mechanism of object location in *Gymnarchus niloticus* and similar fish, *Journal of Experimental Biology,* 35(2), 451–486.

Liu, M. Z., and Vosshall, L. B. (2019) General visual and contingent thermal cues interact to elicit attraction in female *Aedes aegypti* mosquitoes, *Current Biology,* 29(13), 2250–2257.e4.

Liu, Z., et al. (2014) Repeated functional convergent effects of NaV1.7 on acid insensitivity in hibernating mammals, *Proceedings of the Royal Society B: Biological Sciences,* 281(1776), 20132950.

Lloyd, E., et al. (2018) Evolutionary shift towards lateral line dependent prey capture behavior in the blind Mexican cavefish, *Developmental Biology,* 441(2), 328–337.

Lohmann, K. J. (1991) Magnetic orientation by hatchling loggerhead sea turtles (*Caretta caretta*), *Journal of Experimental Biology,* 155, 37–49.

Lohmann, K., et al. (1995) Magnetic orientation of spiny lobsters in the ocean: Experiments with undersea coil systems, *Journal of Experimental Biology,* 198(Pt 10), 2041–2048.

Lohmann, K. J., et al. (2001) Regional magnetic fields as navigational markers for sea turtles, *Science,* 294(5541), 364–366.

Lohmann, K. J., et al. (2004) Geomagnetic map used in sea-turtle navigation, *Nature,* 428(6986), 909–910.

Lohmann, K., and Lohmann, C. (1994) Detection of magnetic inclination angle by sea turtles: A possible mechanism for determining latitude, *Journal of Experimental Biology,* 194(1), 23–32.

Lohmann, K. J., and Lohmann, C. M. F. (1996) Detection of magnetic field intensity by sea turtles, *Nature,* 380(6569), 59–61.

Lohmann, K. J., and Lohmann, C. M. F. (2019) There and back again: Natal homing by magnetic navigation in sea turtles and salmon, *Journal of Experimental Biology,* 222(Suppl. 1), jeb184077.

Lohmann, K. J., Putman, N. F., and Lohmann, C. M. F. (2008) Geomagnetic imprinting: A unifying hypothesis of long-distance natal homing in salmon and sea turtles, *Proceedings of the National Academy of Sciences,* 105(49), 19096–19101.

Longcore, T. (2018) Hazard or hope? LEDs and wildlife, *LED Professional Review,* 70, 52–57.

Longcore, T., et al. (2012) An estimate of avian mortality at communication towers in the United States and Canada, *PLOS One*, 7(4), e34025.

Longcore, T., and Rich, C. (2016) *Artificial night lighting and protected lands: Ecological effects and management approaches*. Natural Resource Report 2017/1493.

Lu, P., et al. (2017) Extraoral bitter taste receptors in health and disease, *Journal of General Physiology*, 149(2), 181–197.

Lubbock, J. (1881) Observations on ants, bees, and wasps.—Part VIII, *Journal of the Linnean Society of London, Zoology*, 15(87), 362–387.

Lucas, J., et al. (2002) A comparative study of avian auditory brainstem responses: Correlations with phylogeny and vocal complexity, and seasonal effects, *Journal of Comparative Physiology A*, 188(11–12), 981–992.

Lucas, J. R., et al. (2007) Seasonal variation in avian auditory evoked responses to tones: A comparative analysis of Carolina chickadees, tufted titmice, and white-breasted nuthatches, *Journal of Comparative Physiology A*, 193(2), 201–215.

Ludeman, D. A., et al. (2014) Evolutionary origins of sensation in metazoans: Functional evidence for a new sensory organ in sponges, *BMC Evolutionary Biology*, 14(1), 3.

Maan, M. E., and Cummings, M. E. (2012) Poison frog colors are honest signals of toxicity, particularly for bird predators, *The American Naturalist*, 179(1), E1–E14.

Macpherson, F. (2011) Individuating the senses, in Macpherson, F. (ed), *The senses: Classic and contemporary philosophical perspectives*, 3–43. Oxford: Oxford University Press.

Madhav, M. S., et al. (2018) High-resolution behavioral mapping of electric fishes in Amazonian habitats, *Scientific Reports*, 8(1), 5830.

Madsen, P. T., et al. (2002) Sperm whale sound production studied with ultrasound time/depth-recording tags, *Journal of Experimental Biology*, 205(Pt 13), 1899–1906.

Madsen, P. T., et al. (2013) Echolocation in Blainville's beaked whales (*Mesoplodon densirostris*), *Journal of Comparative Physiology A*, 199(6), 451–469.

Madsen, P. T., and Surlykke, A. (2014) Echolocation in air and water, in Surlykke, A., et al. (eds), *Biosonar*, 257–304. New York: Springer.

Majid, A. (2015) Cultural factors shape olfactory language, *Trends in Cognitive Sciences*, 19(11), 629–630.

Majid, A., et al. (2017) What makes a better smeller?, *Perception*, 46(3–4), 406–430.

Majid, A., and Kruspe, N. (2018) Hunter-gatherer olfaction is special, *Current Biology*, 28(3), 409–413.e2.

Malakoff, D. (2010) A push for quieter ships, *Science*, 328(5985), 1502–1503.

Mancuso, K., et al. (2009) Gene therapy for red-green colour blindness in adult primates, *Nature*, 461(7625), 784–787.

Marder, E., and Bucher, D. (2007) Understanding circuit dynamics using the stomatogastric nervous system of lobsters and crabs, *Annual Review of Physiology*, 69(1), 291–316.

Marshall, C. D., et al. (1998) Prehensile use of perioral bristles during feeding and associated behaviors of the Florida manatee (*Trichechus manatus latirostris*), *Marine Mammal Science*, 14(2), 274–289.

Marshall, C. D., Clark, L. A., and Reep, R. L. (1998) The muscular hydrostat of the Florida manatee (*Trichechus manatus latirostris*): A functional morphological model of perioral bristle use, *Marine Mammal Science*, 14(2), 290–303.

Marshall, J., and Arikawa, K. (2014) Unconventional colour vision, *Current Biology*, 24(24), R1150–R1154.

Marshall, J., Carleton, K. L., and Cronin, T. (2015) Colour vision in marine organisms, *Current Opinions in Neurobiology*, 34, 86–94.

Marshall, J., and Oberwinkler, J. (1999) The colourful world of the mantis shrimp, *Nature*, 401(6756), 873–874.

Marshall, N. J. (1988) A unique colour and polarization vision system in mantis shrimps, *Nature*, 333(6173), 557–560.

Marshall, N. J., et al. (2019a) Colours and colour vision in reef fishes: Past, present and future research directions, *Journal of Fish Biology*, 95(1), 5–38.

Marshall, N. J., et al. (2019b) Polarisation signals: A new currency for communication, *Journal of Experimental Biology*, 222(3), jeb134213.

Marshall, N. J., Land, M. F., and Cronin, T. W. (2014) Shrimps that pay attention: Saccadic eye movements in stomatopod crustaceans, *Philosophical Transactions of the Royal Society B: Biological Sciences*, 369(1636), 20130042.

Martin, G. R. (2012) Through birds' eyes: Insights into avian sensory ecology, *Journal of Ornithology*, 153(Suppl. 1), 23–48.

Martin, G. R., Portugal, S. J., and Murn, C. P. (2012) Visual fields, foraging and collision vulnerability in *Gyps* vultures, *Ibis*, 154(3), 626–631.

Martinez, V., et al. (2020) Antlions are sensitive to subnanometer amplitude vibrations carried by sand substrates, *Journal of Comparative Physiology A*, 206(5), 783–791.

Masland, R. H. (2017) Vision: Two speeds in the retina, *Current Biology*, 27(8), R303–R305.

Mason, A. C., Oshinsky, M. L., and Hoy, R. R. (2001) Hyperacute directional hearing in a microscale auditory system, *Nature*, 410(6829), 686–690.

Mason, M. J. (2003) Bone conduction and seismic sensitivity in golden moles (Chrysochloridae), *Journal of Zoology*, 260(4), 405–413.

Mason, M. J., and Narins, P. M. (2002) Seismic sensitivity in the desert golden mole (*Eremitalpa granti*): A review, *Journal of Comparative Psychology*, 116(2), 158–163.

Mass, A. M., and Supin, A. Y. (1995) Ganglion cell topography of the retina in the bottlenosed dolphin, *Tursiops truncatus*, *Brain, Behavior and Evolution*, 45(5), 257–265.

Mass, A. M., and Supin, A. Y. (2007) Adaptive features of aquatic mammals' eye, *The Anatomical Record*, 290(6), 701–715.

Masters, W. M. (1984) Vibrations in the orbwebs of *Nuctenea sclopetaria* (Araneidae). I. Transmission through the web, *Behavioral Ecology and Sociobiology*, 15(3), 207–215.

Matos-Cruz, V., et al. (2017) Molecular prerequisites for diminished cold sensitivity in ground squirrels and hamsters, *Cell Reports*, 21(12), 3329–3337.

Maximov, V. V. (2000) Environmental factors which may have led to the appearance of colour vision, *Philosophical Transactions of the Royal Society B: Biological Sciences*, 355(1401), 1239–1242.

McArthur, C., et al. (2019) Plant volatiles are a salient cue for foraging mammals: Elephants target preferred plants despite background plant odour, *Animal Behaviour*, 155, 199–216.

McBride, C. S. (2016) Genes and odors underlying the recent evolution of mosquito preference for humans, *Current Biology*, 26(1), R41–R46.

McBride, C. S., et al. (2014) Evolution of mosquito preference for humans linked to an odorant receptor, *Nature*, 515(7526), 222–227.

McCulloch, K. J., Osorio, D., and Briscoe, A. D. (2016) Sexual dimorphism in the compound eye of *Heliconius erato*: A nymphalid butterfly with at least five spectral classes of photoreceptor, *Journal of Experimental Biology*, 219(15), 2377–2387.

McGann, J. P. (2017) Poor human olfaction is a 19th-century myth, *Science*, 356(6338), eaam7263.

McGregor, P. K., and Westby, G. M. (1992) Discrimination of individually characteristic electric organ discharges by a weakly electric fish, *Animal Behaviour*, 43(6), 977–986.

McKemy, D. D. (2007) Temperature sensing across species, *Pflügers Archiv—European Journal of Physiology*, 454(5), 777–791.

McKenzie, S. K., and Kronauer, D. J. C. (2018) The genomic architecture and molecular evolution of ant odorant receptors, *Genome Research*, 28(11), 1757–1765.

McMeniman, C. J., et al. (2014) Multimodal integration of carbon dioxide and other sensory cues drives mosquito attraction to humans, *Cell*, 156(5), 1060–1071.

Meister, M. (2016) Physical limits to magnetogenetics, *eLife*, 5, e17210.

Melin, A. D., et al. (2007) Effects of colour vision phenotype on insect capture by a free-ranging population of white-faced capuchins, *Cebus capucinus*, *Animal Behaviour*, 73(1), 205–214.

Melin, A. D., et al. (2016) Zebra stripes through the eyes of their predators, zebras, and humans, *PLOS One*, 11(1), e0145679.

Melin, A. D., et al. (2017) Trichromacy increases fruit intake rates of wild capuchins (*Cebus capucinus imitator*), *Proceedings of the National Academy of Sciences*, 114(39), 10402–10407.

Melo, N., et al. (2021) The irritant receptor TRPA1 mediates the mosquito repellent effect of catnip, *Current Biology*, 31(9), 1988–1994.e5.

Mencinger-Vračko, B., and Devetak, D. (2008) Orientation of the pit-building antlion larva *Euroleon* (Neuroptera, Myrmeleontidae) to the direction of substrate vibrations caused by prey, *Zoology*, 111(1), 2–8.

Menda, G., et al. (2019) The long and short of hearing in the mosquito *Aedes aegypti*, *Current Biology*, 29(4), 709–714.e4.

Merkel, F. W., and Fromme, H. G. (1958) Untersuchungen über das Orientierungsvermögen nächtlich ziehender Rotkehlchen, *Naturwissenschaften*, 45(2), 499–500.

Merker, B. (2005) The liabilities of mobility: A selection pressure for the transition to consciousness in animal evolution, *Consciousness and Cognition*, 14(1), 89–114.

Mettam, J. J., et al. (2011) The efficacy of three types of analgesic drugs in reducing pain in the rainbow trout, *Oncorhynchus mykiss*, *Applied Animal Behaviour Science*, 133(3), 265–274.

Meyer-Rochow, V. B. (1978) The eyes of mesopelagic crustaceans. II. *Streetsia challengeri* (amphipoda), *Cell and Tissue Research*, 186(2), 337–349.

Mhatre, N., Sivalinghem, S., and Mason, A. C. (2018) Posture controls mechanical tuning in the black widow spider mechanosensory system, bioRxiv. Available at: biorxiv.org/lookup/doi/10.1101/484238.

Middendorff, A. T. (1855) *Die Isepiptesen Russlands: Grundlagen zur Erforschung der Zugzeiten und Zugrichtungen der Vögel Russlands*. St. Petersburg: Academie impériale des Sciences.

Miles, R. N., Robert, D., and Hoy, R. R. (1995) Mechanically coupled ears for directional hearing in the parasitoid fly *Ormia ochracea*, *Journal of the Acoustical Society of America*, 98(6), 3059–3070.

Miller, A. K., Hensman, M. C., et al. (2015) African elephants (*Loxodonta africana*) can detect TNT using olfaction: Implications for biosensor application, *Applied Animal Behaviour Science*, 171, 177–183.

Miller, A. K., Maritz, B., et al. (2015) An ambusher's arsenal: Chemical crypsis in the puff adder (*Bitis arietans*), *Proceedings of the Royal Society B: Biological Sciences*, 282(1821), 20152182.

Miller, P. J. O., Kvadsheim, P. H., et al. (2015) First indications that northern bottlenose whales are sensitive to behavioural disturbance from anthropogenic noise, *Royal Society Open Science*, 2(6), 140484.

Millsopp, S., and Laming, P. (2008) Trade-offs between feeding and shock avoidance in goldfish (*Carassius auratus*), *Applied Animal Behaviour Science*, 113(1), 247–254.

Mitchinson, B., et al. (2011) Active vibrissal sensing in rodents and marsupials, *Philosophical Transactions of the Royal Society B: Biological Sciences*, 366(1581), 3037–3048.

Mitkus, M., et al. (2018) Raptor vision, in Sherman, S. M. (ed), *Oxford research encyclopedia of neuroscience*. Oxford: Oxford University Press.

Mitra, O., et al. (2009) Grunting for worms: Seismic vibrations cause *Diplocardia* earthworms to emerge from the soil, *Biology Letters*, 5(1), 16–19.

Moayedi, Y., Nakatani, M., and Lumpkin, E. (2015) Mammalian mechanoreception, *Scholarpedia*, 10(3), 7265.

Modrell, M. S., et al. (2011) Electrosensory ampullary organs are derived from lateral line placodes in bony fishes, *Nature Communications*, 2(1), 496.

Mogdans, J. (2019) Sensory ecology of the fish lateral-line system: Morphological and physiological adaptations for the perception of hydrodynamic stimuli, *Journal of Fish Biology*, 95(1), 53–72.

Møhl, B., et al. (2003) The monopulsed nature of sperm whale clicks, *Journal of the Acoustical Society of America*, 114(2), 1143–1154.

Moir, H. M., Jackson, J. C., and Windmill, J. F. C. (2013) Extremely high frequency sensitivity in a "simple" ear, *Biology Letters*, 9(4), 20130241.

Mollon, J. D. (1989) "Tho' she kneel'd in that place where they grew . . .": The uses and origins of primate colour vision, *Journal of Experimental Biology*, 146, 21–38.

Monnin, T., et al. (2002) Pretender punishment induced by chemical signalling in a queenless ant, *Nature*, 419(6902), 61–65.

Montague, M. J., Danek-Gontard, M., and Kunc, H. P. (2013) Phenotypic plasticity affects the response of a sexually selected trait to anthropogenic noise, *Behavioral Ecology*, 24(2), 343–348.

Montealegre-Z, F., et al. (2012) Convergent evolution between insect and mammalian audition, *Science*, 338(6109), 968–971.

Monterey Bay Aquarium (2016) Say hello to Selka!, Monterey Bay Aquarium. Available at: montereybayaquarium.tumblr.com/post/149326681398/say-hello-to-selka.

Montgomery, J., Bleckmann, H., and Coombs, S. (2013) Sensory ecology and neuroethology of the lateral line, in Coombs, S., et al. (eds), *The lateral line system*, 121–150. New York: Springer.

Montgomery, J. C., and Saunders, A. J. (1985) Functional morphology of the piper *Hyporhamphus ihi* with reference to the role of the lateral line in feeding, *Proceedings of the Royal Society B: Biological Sciences*, 224(1235), 197–208.

Mooney, T. A., Yamato, M., and Branstetter, B. K. (2012) Hearing in cetaceans: From natural history to experimental biology, *Advances in marine biology*, 63, 197–246.

Moore, B., et al. (2017) Structure and function of regional specializations in the vertebrate retina, in Kaas, J. H., and Streidter, G. (eds), *Evolution of nervous systems*, 351–372. Oxford, UK: Academic Press.

Moran, D., Softley, R., and Warrant, E. J. (2015) The energetic cost of vision and the evolution of eyeless Mexican cavefish, *Science Advances*, 1(8), e1500363.

Moreau, C. S., et al. (2006) Phylogeny of the ants: Diversification in the age of angiosperms, *Science*, 312(5770), 101–104.

Morehouse, N. (2020) Spider vision, *Current Biology*, 30(17), R975–R980.

Moreira, L. A. A., et al. (2019) Platyrrhine color signals: New horizons to pursue, *Evolutionary Anthropology: Issues, News, and Reviews*, 28(5), 236–248.

Morley, E. L., and Robert, D. (2018) Electric fields elicit ballooning in spiders, *Current Biology*, 28(14), 2324–2330.e2.

Mortimer, B. (2017) Biotremology: Do physical constraints limit the propagation of vibrational information?, *Animal Behaviour*, 130, 165–174.

Mortimer, B., et al. (2014) The speed of sound in silk: Linking material performance to biological function, *Advanced Materials*, 26(30), 5179–5183.

Mortimer, B., et al. (2016) Tuning the instrument: Sonic properties in the spider's web, *Journal of the Royal Society Interface*, 13(122), 20160341.

Mortimer, J. A., and Portier, K. M. (1989) Reproductive homing and internesting behavior of the green turtle (*Chelonia mydas*) at Ascension Island, South Atlantic Ocean, *Copeia*, 1989(4), 962–977.

Moss, C. F. (2018) Auditory mechanisms of echolocation in bats, in Sherman, S. M. (ed), *Oxford research encyclopedia of neuroscience*. Oxford: Oxford University Press.

Moss, C. F., et al. (2006) Active listening for spatial orientation in a complex auditory scene, *PLOS Biology*, 4(4), e79.

Moss, C. F., Chiu, C., and Surlykke, A. (2011) Adaptive vocal behavior drives perception by echolocation in bats, *Current Opinion in Neurobiology*, 21(4), 645–652.

Moss, C. F., and Schnitzler, H.-U. (1995) Behavioral studies of auditory information processing, in Popper, A. N., and Fay, R. R. (eds), *Hearing by bats*, 87–145. New York: Springer.

Moss, C. F., and Surlykke, A. (2010) Probing the natural scene by echolocation in bats, *Frontiers in Behavioral Neuroscience*, 4, 33.

Moss, C. J. (2000) *Elephant memories: Thirteen years in the life of an elephant family*. Chicago: University of Chicago Press.

Mouritsen, H. (2018) Long-distance navigation and magnetoreception in migratory animals, *Nature*, 558(7708), 50–59.

Mouritsen, H., et al. (2005) Night-vision brain area in migratory songbirds, *Proceedings of the National Academy of Sciences*, 102(23), 8339–8344.

Mourlam, M. J., and Orliac, M. J. (2017) Infrasonic and ultrasonic hearing evolved after the emergence of modern whales, *Current Biology*, 27(12), 1776–1781.e9.

Mugan, U., and MacIver, M. A. (2019) The shift from life in water to life on land advantaged planning in visually-guided behavior, bioRxiv, 585760.

Müller, P., and Robert, D. (2002) Death comes suddenly to the unprepared: Singing crickets, call fragmentation, and parasitoid flies, *Behavioral Ecology*, 13(5), 598–606.

Murchy, K. A., et al. (2019) Impacts of noise on the behavior and physiology of marine invertebrates: A meta-analysis, *Proceedings of Meetings on Acoustics*, 37(1), 040002.

Murphy, C. T., Reichmuth, C., and Mann, D. (2015) Vibrissal sensitivity in a harbor seal (*Phoca vitulina*), *Journal of Experimental Biology*, 218(15), 2463–2471.

Murray, R. W. (1960) Electrical sensitivity of the ampullæ of Lorenzini, *Nature*, 187(4741), 957.

Nachtigall, P. E. (2016) Biosonar and sound localization in dolphins, in Sherman, S. M. (ed), *Oxford research encyclopedia of neuroscience*. New York: Oxford University Press.

Nachtigall, P. E., and Supin, A. Y. (2008) A false killer whale adjusts its hearing when it echolocates, *Journal of Experimental Biology*, 211(11), 1714–1718.

Nagel, T. (1974) What is it like to be a bat?, *The Philosophical Review*, 83(4), 435–450.

Nakano, R., et al. (2009) Moths are not silent, but whisper ultrasonic courtship songs, *Journal of Experimental Biology*, 212(24), 4072–4078.

Nakano, R., et al. (2010) To females of a noctuid moth, male courtship songs are nothing more than bat echolocation calls, *Biology Letters*, 6(5), 582–584.

Nakata, K. (2010) Attention focusing in a sit-and-wait forager: A spider controls its prey-detection ability in different web sectors by adjusting thread tension, *Proceedings of the Royal Society B: Biological Sciences*, 277(1678), 29–33.

Nakata, K. (2013) Spatial learning affects thread tension control in orb-web spiders, *Biology Letters*, 9(4), 20130052.

Narins, P. M., and Lewis, E. R. (1984) The vertebrate ear as an exquisite seismic sensor, *Journal of the Acoustical Society of America*, 76(5), 1384–1387.

Narins, P. M., Stoeger, A. S., and O'Connell-Rodwell, C. (2016) Infrasonic and seismic communication in the vertebrates with special emphasis on the Afrotheria: An update and future directions, in Suthers, R. A., et al. (eds), *Vertebrate sound production and acoustic communication*, 191–227. Cham: Springer.

Necker, R. (1985) Observations on the function of a slowly-adapting mechanoreceptor associated with filoplumes in the feathered skin of pigeons, *Journal of Comparative Physiology A*, 156(3), 391–394.

Neil, T. R., et al. (2020) Moth wings are acoustic metamaterials, *Proceedings of the National Academy of Sciences*, 117(49), 31134–31141.

Neitz, J., Carroll, J., and Neitz, M. (2001) Color vision: Almost reason enough for having eyes, *Optics & Photonics News*, 12(1), 26–33.

Neitz, J., Geist, T., and Jacobs, G. H. (1989) Color vision in the dog, *Visual Neuroscience*, 3(2), 119–125.

Nesher, N., et al. (2014) Self-recognition mechanism between skin and suckers prevents octopus arms from interfering with each other, *Current Biology*, 24(11), 1271–1275.

Neumeyer, C. (1992) Tetrachromatic color vision in goldfish: Evidence from color mixture experiments, *Journal of Comparative Physiology A*, 171(5), 639–649.

Neunuebel, J. P., et al. (2015) Female mice ultrasonically interact with males during courtship displays, *eLife*, 4, e06203.

Nevitt, G. (2000) Olfactory foraging by Antarctic procellariiform seabirds: Life at high Reynolds numbers, *Biological Bulletin*, 198(2), 245–253.

Nevitt, G. A. (2008) Sensory ecology on the high seas: The odor world of the procellariiform seabirds, *Journal of Experimental Biology*, 211(11), 1706–1713.

Nevitt, G. A., and Bonadonna, F. (2005) Sensitivity to dimethyl sulphide suggests a mechanism for olfactory navigation by seabirds, *Biology Letters*, 1(3), 303–305.

Nevitt, G. A., and Hagelin, J. C. (2009) Symposium overview: Olfaction in birds: A dedication to the pioneering spirit of Bernice Wenzel and Betsy Bang, *Annals of the New York Academy of Sciences*, 1170(1), 424–427.

Nevitt, G. A., Losekoot, M., and Weimerskirch, H. (2008) Evidence for olfactory search in wandering albatross, *Diomedea exulans*, *Proceedings of the National Academy of Sciences*, 105(12), 4576–4581.

Nevitt, G. A., Veit, R. R., and Kareiva, P. (1995) Dimethyl sulphide as a foraging cue for Antarctic procellariiform seabirds, *Nature*, 376(6542), 680–682.

Newman, E. A., and Hartline, P. H. (1982) The infrared "vision" of snakes, *Scientific American*, 246(3), 116–127.

Nicolson, A. (2018) *The seabird's cry*. New York: Henry Holt.

Niesterok, B., et al. (2017) Hydrodynamic detection and localization of artificial flatfish breathing currents by harbour seals (*Phoca vitulina*), *Journal of Experimental Biology*, 220(2), 174–185.

Niimura, Y., Matsui, A., and Touhara, K. (2014) Extreme expansion of the olfactory receptor gene repertoire in African elephants and evolutionary dynamics of orthologous gene groups in 13 placental mammals, *Genome Research*, 24(9), 1485–1496.

Nilsson, D.-E. (2009) The evolution of eyes and visually guided behaviour, *Philosophical Transactions of the Royal Society B: Biological Sciences*, 364(1531), 2833–2847.

Nilsson, D.-E., et al. (2012) A unique advantage for giant eyes in giant squid, *Current Biology*, 22(8), 683–688.

Nilsson, D.-E., and Pelger, S. (1994) A pessimistic estimate of the time required for an eye to evolve, *Proceedings of the Royal Society B: Biological Sciences*, 256(1345), 53–58.

Nilsson, G. (1996) Brain and body oxygen requirements of *Gnathonemus petersii*, a fish with an exceptionally large brain, *Journal of Experimental Biology*, 199(3), 603–607.

Nimpf, S., et al. (2019) A putative mechanism for magnetoreception by electromagnetic induction in the pigeon inner ear, *Current Biology*, 29(23), 4052–4059.e4.

Niven, J. E., and Laughlin, S. B. (2008) Energy limitation as a selective pressure on the evolution of sensory systems, *Journal of Experimental Biology*, 211(Pt 11), 1792–1804.

Noble, G. K., and Schmidt, A. (1937) The structure and function of the facial and labial pits of snakes, *Proceedings of the American Philosophical Society*, 77(3), 263–288.

Noirot, E. (1966) Ultra-sounds in young rodents. I. Changes with age in albino mice, *Animal Behaviour*, 14(4), 459–462.

Noirot, I. C., et al. (2009) Presence of aromatase and estrogen receptor alpha in the inner ear of zebra finches, *Hearing Research*, 252(1–2), 49–55.

Nordmann, G. C., Hochstoeger, T., and Keays, D. A. (2017) Magnetoreception—A sense without a receptor, *PLOS Biology*, 15(10), e2003234.

Norman, L. J., and Thaler, L. (2019) Retinotopic-like maps of spatial sound in primary "visual" cortex of blind human echolocators, *Proceedings of the Royal Society B: Biological Sciences*, 286(1912), 20191910.

Norris, K. S., et al. (1961) An experimental demonstration of echolocation behavior in the porpoise, *Tursiops truncatus* (Montagu), *Biological Bulletin*, 120(2), 163–176.

Ntelezos, A., Guarato, F., and Windmill, J. F. C. (2016) The anti-bat strategy of ultrasound absorption: The wings of nocturnal moths (Bombycoidea: Saturniidae) absorb more ultrasound than the wings of diurnal moths (Chalcosiinae: Zygaenoidea: Zygaenidae), *Biology Open*, 6(1), 109–117.

O'Carroll, D. C., and Warrant, E. J. (2017) Vision in dim light: Highlights and challenges, *Philosophical Transactions of the Royal Society B: Biological Sciences*, 372(1717), 20160062.

O'Connell, C. (2008) *The elephant's secret sense: The hidden life of the wild herds of Africa.* Chicago: University of Chicago Press.

O'Connell, C. E., Arnason, B. T., and Hart, L. A. (1997) Seismic transmission of elephant vocalizations and movement, *Journal of the Acoustical Society of America*, 102(5), 3124.

O'Connell-Rodwell, C. E., et al. (2006) Wild elephant (*Loxodonta africana*) breeding herds respond to artificially transmitted seismic stimuli, *Behavioral Ecology and Sociobiology*, 59(6), 842–850.

O'Connell-Rodwell, C. E., et al. (2007) Wild African elephants (*Loxodonta africana*) discriminate between familiar and unfamiliar conspecific seismic alarm calls, *Journal of the Acoustical Society of America*, 122(2), 823–830.

O'Connell-Rodwell, C. E., Hart, L. A., and Arnason, B. T. (2001) Exploring the potential use of seismic waves as a communication channel by elephants and other large mammals, *American Zoologist*, 41(5), 1157–1170.

Olson, C. R., et al. (2018) Black Jacobin hummingbirds vocalize above the known hearing range of birds, *Current Biology*, 28(5), R204–R205.

Osorio, D., and Vorobyev, M. (1996) Colour vision as an adaptation to frugivory in primates, *Proceedings of the Royal Society B: Biological Sciences*, 263(1370), 593–599.

Osorio, D., and Vorobyev, M. (2008) A review of the evolution of animal colour vision and visual communication signals, *Vision Research*, 48(20), 2042–2051.

Ossiannilsson, F. (1949) Insect drummers, a study on the morphology and function of the sound-producing organ of Swedish *Homoptera auchenorrhyncha*, with notes on their sound-production. Dissertation, Entomologika sällskapet i Lund.

Owen, M. A., et al. (2015) An experimental investigation of chemical communication in the polar bear: Scent communication in polar bears, *Journal of Zoology*, 295(1), 36–43.

Owens, A. C. S., et al. (2020) Light pollution is a driver of insect declines, *Biological Conservation*, 241, 108259.

Owens, G. L., et al. (2012) In the four-eyed fish (*Anableps anableps*), the regions of the retina exposed to aquatic and aerial light do not express the same set of opsin genes, *Biology Letters*, 8(1), 86–89.

Pack, A., and Herman, L. (1995) Sensory integration in the bottlenosed dolphin: Immediate recognition of complex shapes across the senses of echolocation and vision, *Journal of the Acoustical Society of America*, 98, 722–33.

Page, R. A., and Ryan, M. J. (2008) The effect of signal complexity on localization performance in bats that localize frog calls, *Animal Behaviour*, 76(3), 761–769.

Pain, S. (2001) Stench warfare, *New Scientist*. Available at: www.newscientist.com/article/mg17122984-600-stench-warfare/.

Palmer, B. A., et al. (2017) The image-forming mirror in the eye of the scallop, *Science*, 358(6367), 1172–1175.

Panksepp, J., and Burgdorf, J. (2000) 50-kHz chirping (laughter?) in response to conditioned and unconditioned tickle-induced reward in rats: Effects of social housing and genetic variables, *Behavioural Brain Research*, 115(1), 25–38.

Park, T. J., et al. (2008) Selective inflammatory pain insensitivity in the African naked mole-rat (*Heterocephalus glaber*), *PLOS Biology*, 6(1), e13.

Park, T. J., et al. (2017) Fructose-driven glycolysis supports anoxia resistance in the naked mole-rat, *Science*, 356(6335), 307–311.

Park, T. J., Lewin, G. R., and Buffenstein, R. (2010) Naked mole rats: Their extraordinary sensory world, in Breed, M., and Moore, J. (eds), *Encyclopedia of animal behavior*, 505–512. Amsterdam: Elsevier.

Parker, A. (2004) *In the blink of an eye: How vision sparked the big bang of evolution.* New York: Basic Books.

Partridge, B. L., and Pitcher, T. J. (1980) The sensory basis of fish schools: Relative roles of lateral line and vision, *Journal of Comparative Physiology*, 135(4), 315–325.

Partridge, J. C., et al. (2014) Reflecting optics in the diverticular eye of a deep-sea barreleye fish (*Rhynchohyalus natalensis*), *Proceedings of the Royal Society B: Biological Sciences*, 281(1782), 20133223.

Patek, S. N., Korff, W. L., and Caldwell, R. L. (2004) Deadly strike mechanism of a mantis shrimp, *Nature*, 428(6985), 819–820.

Patton, P., Windsor, S., and Coombs, S. (2010) Active wall following by Mexican blind cavefish (*Astyanax mexicanus*), *Journal of Comparative Physiology A*, 196(11), 853–867.

Paul, S. C., and Stevens, M. (2020) Horse vision and obstacle visibility in horseracing, *Applied Animal Behaviour Science*, 222, 104882.

Paulin, M. G. (1995) Electroreception and the compass sense of sharks, *Journal of Theoretical Biology*, 174(3), 325–339.

Payne, K. (1999) *Silent thunder: In the presence of elephants*. London: Penguin.

Payne, K. B., Langbauer, W. R., and Thomas, E. M. (1986) Infrasonic calls of the Asian elephant (*Elephas maximus*), *Behavioral Ecology and Sociobiology*, 18(4), 297–301.

Payne, R. S. (1971) Acoustic location of prey by barn owls (*Tyto alba*), *Journal of Experimental Biology*, 54(3), 535–573.

Payne, R. S., and McVay, S. (1971) Songs of humpback whales, *Science*, 173(3997), 585–597.

Payne, R., and Webb, D. (1971) Orientation by means of long range acoustic signaling in baleen whales, *Annals of the New York Academy of Sciences*, 188(1 Orientation), 110–141.

Peichl, L. (2005) Diversity of mammalian photoreceptor properties: Adaptations to habitat and lifestyle?, *The Anatomical Record Part A: Discoveries in Molecular, Cellular, and Evolutionary Biology*, 287A(1), 1001–1012.

Peichl, L., Behrmann, G., and Kröger, R. H. (2001) For whales and seals the ocean is not blue: A visual pigment loss in marine mammals, *The European Journal of Neuroscience*, 13(8), 1520–1528.

Perry, M. W., and Desplan, C. (2016) Love spots, *Current Biology*, 26(12), R484–R485.

Persons, W. S., and Currie, P. J. (2015) Bristles before down: A new perspective on the functional origin of feathers, *Evolution: International Journal of Organic Evolution*, 69(4), 857–862.

Pettigrew, J. D., Manger, P. R., and Fine, S. L. B. (1998) The sensory world of the platypus, *Philosophical Transactions of the Royal Society B: Biological Sciences*, 353(1372), 1199–1210.

Phillips, J. N., et al. (2019) Background noise disrupts host-parasitoid interactions, *Royal Society Open Science*, 6(9), 190867.

Phippen, J. W. (2016) "Kill every buffalo you can! Every buffalo dead is an Indian gone," *The Atlantic*. Available at: www.theatlantic.com/national/archive/2016/05/the-buffalo-killers/482349/.

Picciani, N., et al. (2018) Prolific origination of eyes in Cnidaria with co-option of non-visual opsins, *Current Biology*, 28(15), 2413–2419.e4.

Piersma, T., et al. (1995) Holling's functional response model as a tool to link the food-finding mechanism of a probing shorebird with its spatial distribution, *Journal of Animal Ecology*, 64(4), 493–504.

Piersma, T., et al. (1998) A new pressure sensory mechanism for prey detection in birds: The use of principles of seabed dynamics?, *Proceedings of the Royal Society B: Biological Sciences*, 265(1404), 1377–1383.

Pihlström, H., et al. (2005) Scaling of mammalian ethmoid bones can predict olfactory organ size and performance, *Proceedings of the Royal Society B: Biological Sciences*, 272(1566), 957–962.

Pitcher, T. J., Partridge, B. L., and Wardle, C. S. (1976) A blind fish can school, *Science*, 194(4268), 963–965.

Plachetzki, D. C., Fong, C. R., and Oakley, T. H. (2012) Cnidocyte discharge is regulated by light and opsin-mediated phototransduction, *BMC Biology*, 10(1), 17.

Plotnik, J. M., et al. (2019) Elephants have a nose for quantity, *Proceedings of the National Academy of Sciences*, 116(25), 12566–12571.

Pointer, M. R., and Attridge, G. G. (1998) The number of discernible colours, *Color Research & Application*, 23(1), 52–54.

Polajnar, J., et al. (2015) Manipulating behaviour with substrate-borne vibrations—Potential for insect pest control, *Pest Management Science*, 71(1), 15–23.

Polilov, A. A. (2012) The smallest insects evolve anucleate neurons, *Arthropod Structure & Development*, 41(1), 29–34.

Pollack, L. (2012) Historical series: Magnetic sense of birds. Available at: www.ks.uiuc.edu/History/magnetoreception/.

Poole, J. H., et al. (1988) The social contexts of some very low frequency calls of African elephants, *Behavioral Ecology and Sociobiology*, 22(6), 385–392.

Popper, A. N., et al. (2004) Response of clupeid fish to ultrasound: A review, *ICES Journal of Marine Science*, 61(7), 1057–1061.

Porter, J., et al. (2007) Mechanisms of scent-tracking in humans, *Nature Neuroscience*, 10(1), 27–29.

Porter, M. L., et al. (2012) Shedding new light on opsin evolution, *Proceedings of the Royal Society B: Biological Sciences*, 279(1726), 3–14.

Porter, M. L., and Sumner-Rooney, L. (2018) Evolution in the dark: Unifying our understanding of eye loss, *Integrative and Comparative Biology*, 58(3), 367–371.

Potier, S., et al. (2017) Eye size, fovea, and foraging ecology in accipitriform raptors, *Brain, Behavior and Evolution*, 90(3), 232–242.

Poulet, J. F. A., and Hedwig, B. (2003) A corollary discharge mechanism modulates central auditory processing in singing crickets, *Journal of Neurophysiology*, 89(3), 1528–1540.

Poulson, S. J., et al. (2020) Naked mole-rats lack cold sensitivity before and after nerve injury, *Molecular Pain*, 16, 1744806920955103.

Prescott, T. J., Diamond, M. E., and Wing, A. M. (2011) Active touch sensing, *Philosophical Transactions of the Royal Society B: Biological Sciences*, 366(1581), 2989–2995.

Prescott, T. J., and Dürr, V. (2015) The world of touch, *Scholarpedia*, 10(4), 32688.

Prescott, T. J., Mitchinson, B., and Grant, R. (2011) Vibrissal behavior and function, *Scholarpedia*, 6(10), 6642.

Primack, R. B. (1982) Ultraviolet patterns in flowers, or flowers as viewed by insects, *Arnoldia*, 42(3), 139–146.

Prior, N. H., et al. (2018) Acoustic fine structure may encode biologically relevant information for zebra finches, *Scientific Reports*, 8(1), 6212.

Proske, U., and Gregory, E. (2003) Electrolocation in the platypus—Some speculations, *Comparative Biochemistry and Physiology Part A: Molecular & Integrative Physiology*, 136(4), 821–825.

Proust, M. (1993) *In search of lost time*, volume 5. Translated by C. K. Scott Moncrieff and Terence Kilmartin. New York: Modern Library.

Putman, N. F., et al. (2013) Evidence for geomagnetic imprinting as a homing mechanism in Pacific salmon, *Current Biology*, 23(4), 312–316.

Pye, D. (2004) Poem by David Pye: On the variety of hearing organs in insects, *Microscopic Research Techniques*, 63, 313–314.

Pyenson, N. D., et al. (2012) Discovery of a sensory organ that coordinates lunge feeding in rorqual whales, *Nature*, 485(7399), 498–501.

Pynn, L. K., and DeSouza, J. F. X. (2013) The function of efference copy signals: Implications for symptoms of schizophrenia, *Vision Research*, 76, 124–133.

Pytte, C. L., Ficken, M. S., Moiseff, A. (2004) Ultrasonic singing by the blue-throated hummingbird: A comparison between production and perception, *Journal of Comparative Physiology A*, 190(8), 665–673.

Qin, S., et al. (2016) A magnetic protein biocompass, *Nature Materials*, 15(2), 217–226.

Quignon, P., et al. (2012) Genetics of canine olfaction and receptor diversity, *Mammalian Genome*, 23(1–2), 132–143.

Raad, H., et al. (2016) Functional gustatory role of chemoreceptors in *Drosophila* wings, *Cell Reports*, 15(7), 1442–1454.

Radinsky, L. B. (1968) Evolution of somatic sensory specialization in otter brains, *Journal of Comparative Neurology*, 134(4), 495–505.

Ramey, E., et al. (2013) Desert-dwelling African elephants (*Loxodonta africana*) in Namibia dig wells to purify drinking water, *Pachyderm*, 53, 66–72.

Ramey, S. (2020) *The lady's handbook for her mysterious illness*. London: Fleet.

Ramsier, M. A., et al. (2012) Primate communication in the pure ultrasound, *Biology Letters*, 8(4), 508–511.

Rasmussen, L. E. L., et al. (1996) Insect pheromone in elephants, *Nature*, 379(6567), 684.

Rasmussen, L. E. L., and Krishnamurthy, V. (2000) How chemical signals integrate Asian elephant society: The known and the unknown, *Zoo Biology*, 19(5), 405–423.

Rasmussen, L. E. L., and Schulte, B. A. (1998) Chemical signals in the reproduction of Asian (*Elephas maximus*) and African (*Loxodonta africana*) elephants, *Animal Reproduction Science*, 53(1–4), 19–34.

Ratcliffe, J. M., et al. (2013) How the bat got its buzz, *Biology Letters*, 9(2), 20121031.

Ravaux, J., et al. (2013) Thermal limit for Metazoan life in question: In vivo heat tolerance of the Pompeii worm, *PLOS One*, 8(5), e64074.

Ravia, A., et al. (2020) A measure of smell enables the creation of olfactory metamers, *Nature*, 588(7836), 118–123.

Reep, R. L., Marshall, C. D., and Stoll, M. L. (2002) Tactile hairs on the postcranial body in Florida manatees: A mammalian lateral line?, *Brain, Behavior and Evolution*, 59(3), 141–154.

Reep, R., and Sarko, D. (2009) Tactile hair in manatees, *Scholarpedia*, 4(4), 6831.

Reilly, S. C., et al. (2008) Novel candidate genes identified in the brain during nociception in common carp (*Cyprinus carpio*) and rainbow trout (*Oncorhynchus mykiss*), *Neuroscience Letters*, 437(2), 135–138.

Reymond, L. (1985) Spatial visual acuity of the eagle *Aquila audax*: A behavioural, optical and anatomical investigation, *Vision Research*, 25(10), 1477–1491.

Reynolds, R. P., et al. (2010) Noise in a laboratory animal facility from the human and mouse perspectives, *Journal of the American Association for Laboratory Animal Science*, 49(5), 592–597.

Ridgway, S. H., and Au, W. W. L. (2009) Hearing and echolocation in dolphins, in Squire, L. R. (ed), *Encyclopedia of neuroscience*, 1031–1039. Amsterdam: Elsevier.

Riitters, K. H., and Wickham, J. D. (2003) How far to the nearest road?, *Frontiers in Ecology and the Environment*, 1(3), 125–129.

Ritz, T., Adem, S., and Schulten, K. (2000) A model for photoreceptor-based magneto-reception in birds, *Biophysical Journal*, 78(2), 707–718.

Robert, D., Amoroso, J., and Hoy, R. (1992) The evolutionary convergence of hearing in a parasitoid fly and its cricket host, *Science*, 258(5085), 1135–1137.

Robert, D., Mhatre, N., and McDonagh, T. (2010) The small and smart sensors of insect auditory systems, in *2010 Ninth IEEE Sensors Conference (SENSORS 2010)*, 2208–2211. Kona, HI: IEEE. Available at: ieeexplore.ieee.org/document/5690624/.

Roberts, S. A., et al. (2010) Darcin: A male pheromone that stimulates female memory and sexual attraction to an individual male's odour, *BMC Biology*, 8(1), 75.

Robinson, M. H., and Mirick, H. (1971) The predatory behavior of the golden-web spider *Nephila clavipes* (Araneae: Araneidae), *Psyche*, 78(3), 123–139.

Rogers, L. J. (2012) The two hemispheres of the avian brain: Their differing roles in perceptual processing and the expression of behavior, *Journal of Ornithology*, 153(1), 61–74.

Rolland, R. M., et al. (2012) Evidence that ship noise increases stress in right whales, *Proceedings of the Royal Society B: Biological Sciences*, 279(1737), 2363–2368.

Ros, M. (1935) Die Lippengruben der Pythonen als Temperaturorgane, *Jenaische Zeitschrift für Naturwissenschaft*, 70, 1–32.

Rose, J. D., et al. (2014) Can fish really feel pain?, *Fish and Fisheries*, 15(1), 97–133.

Rowe, A. H., et al. (2013) Voltage-gated sodium channel in grasshopper mice defends against bark scorpion toxin, *Science*, 342(6157), 441–446.

Rubin, J. J., et al. (2018) The evolution of anti-bat sensory illusions in moths, *Science Advances,* 4(7), eaar7428.

Ruck, P. (1958) A comparison of the electrical responses of compound eyes and dorsal ocelli in four insect species, *Journal of Insect Physiology,* 2(4), 261–274.

Rundus, A. S., et al. (2007) Ground squirrels use an infrared signal to deter rattlesnake predation, *Proceedings of the National Academy of Sciences,* 104(36), 14372–14376.

Ryan, M. J. (1980) Female mate choice in a neotropical frog, *Science,* 209(4455), 523–525.

Ryan, M. J. (2018) *A taste for the beautiful: The evolution of attraction.* Princeton, NJ: Princeton University Press.

Ryan, M. J., et al. (1990) Sexual selection for sensory exploitation in the frog *Physalaemus pustulosus, Nature,* 343(6253), 66–67.

Ryan, M. J., and Rand, A. S. (1993) Sexual selection and signal evolution: The ghost of biases past, *Philosophical Transactions of the Royal Society B: Biological Sciences,* 340(1292), 187–195.

Rycyk, A. M., et al. (2018) Manatee behavioral response to boats, *Marine Mammal Science,* 34(4), 924–962.

Ryerson, W. (2014) Why snakes flick their tongues: A fluid dynamics approach. Unpublished dissertation, University of Connecticut.

Sacks, O., and Wasserman, R. (1987) The case of the colorblind painter, *The New York Review of Books,* November 19. Available at: www.nybooks.com/articles/1987/11/19/the-case-of -the-colorblind-painter/.

Saito, C. A., et al. (2004) Alouatta trichromatic color vision—single-unit recording from retinal ganglion cells and microspectrophotometry, *Investigative Ophthalmology & Visual Science,* 45, 4276.

Salazar, V. L., Krahe, R., and Lewis, J. E. (2013) The energetics of electric organ discharge generation in gymnotiform weakly electric fish, *Journal of Experimental Biology,* 216(13), 2459–2468.

Sales, G. D. (2010) Ultrasonic calls of wild and wild-type rodents, in Brudzynski, S. (ed), *Handbook of behavioral neuroscience,* vol. 19, 77–88. Amsterdam: Elsevier.

Sanders, D., et al. (2021) A meta-analysis of biological impacts of artificial light at night, *Nature Ecology & Evolution,* 5(1), 74–81.

Sarko, D. K., Rice, F. L., and Reep, R. L. (2015) Elaboration and innervation of the vibrissal system in the rock hyrax (*Procavia capensis*), *Brain, Behavior and Evolution,* 85(3), 170–188.

Savoca, M. S., et al. (2016) Marine plastic debris emits a keystone infochemical for olfactory foraging seabirds, *Science Advances,* 2(11), e1600395.

Sawtell, N. B. (2017) Neural mechanisms for predicting the sensory consequences of behavior: Insights from electrosensory systems, *Annual Review of Physiology,* 79(1), 381–399.

Scanlan, M. M., et al. (2018) Magnetic map in nonanadromous Atlantic salmon, *Proceedings of the National Academy of Sciences,* 115(43), 10995–10999.

Schevill, W. E., and McBride, A. F. (1956) Evidence for echolocation by cetaceans, *Deep Sea Research,* 3(2), 153–154.

Schevill, W. E., Watkins, W. A., and Backus, R. H. (1964) The 20-cycle signals and *Balaenoptera* (fin whales), in Tavolga, W. N. (ed), *Marine bio-acoustics,* 147–152. Oxford: Pergamon Press.

Schiestl, F. P., et al. (2000) Sex pheromone mimicry in the early spider orchid (*Ophrys sphegodes*): Patterns of hydrocarbons as the key mechanism for pollination by sexual deception, *Journal of Comparative Physiology A,* 186(6), 567–574.

Schmitz, H., and Bleckmann, H. (1998) The photomechanic infrared receptor for the detection of forest fires in the beetle *Melanophila acuminata* (Coleoptera: Buprestidae), *Journal of Comparative Physiology A,* 182(5), 647–657.

Schmitz, H., and Bousack, H. (2012) Modelling a historic oil-tank fire allows an estimation of the sensitivity of the infrared receptors in pyrophilous *Melanophila* beetles, *PLOS One,* 7(5), e37627.

Schmitz, H., Schmitz, A., and Schneider, E. S. (2016) Matched filter properties of infrared receptors used for fire and heat detection in insects, in von der Emde, G., and Warrant, E. (eds), *The ecology of animal senses,* 207–234. Cham: Springer.

Schneider, E. R., et al. (2014) Neuronal mechanism for acute mechanosensitivity in tactile-foraging waterfowl, *Proceedings of the National Academy of Sciences,* 111(41), 14941–14946.

Schneider, E. R., et al. (2017) Molecular basis of tactile specialization in the duck bill, *Proceedings of the National Academy of Sciences,* 114(49), 13036–13041.

Schneider, E. R., et al. (2019) A cross-species analysis reveals a general role for Piezo2 in mechanosensory specialization of trigeminal ganglia from tactile specialist birds, *Cell Reports,* 26(8), 1979–1987.e3.

Schneider, E. S., Schmitz, A., and Schmitz, H. (2015) Concept of an active amplification mechanism in the infrared organ of pyrophilous *Melanophila* beetles, *Frontiers in Physiology,* 6, 391.

Schneider, W. T., et al. (2018) Vestigial singing behaviour persists after the evolutionary loss of song in crickets, *Biology Letters,* 14(2), 20170654.

Schneirla, T. C. (1944) A unique case of circular milling in ants, considered in relation to trail following and the general problem of orientation, *American Museum Novitates,* no. 1253.

Schnitzler, H.-U. (1967) Kompensation von Dopplereffekten bei Hufeisen-Fledermäusen, *Naturwissenschaften,* 54(19), 523.

Schnitzler, H.-U. (1973) Control of Doppler shift compensation in the greater horseshoe bat, *Rhinolophus ferrumequinum, Journal of Comparative Physiology,* 82(1), 79–92.

Schnitzler, H.-U., and Denzinger, A. (2011) Auditory fovea and Doppler shift compensation: Adaptations for flutter detection in echolocating bats using CF-FM signals, *Journal of Comparative Physiology A,* 197(5), 541–559.

Schnitzler, H.-U., and Kalko, E. K. V. (2001) Echolocation by insect-eating bats, *BioScience,* 51(7), 557–569.

Schraft, H. A., Bakken, G. S., and Clark, R. W. (2019) Infrared-sensing snakes select ambush orientation based on thermal backgrounds, *Scientific Reports,* 9(1), 3950.

Schraft, H. A., and Clark, R. W. (2019) Sensory basis of navigation in snakes: The relative importance of eyes and pit organs, *Animal Behaviour,* 147, 77–82.

Schraft, H. A., Goodman, C., and Clark, R. W. (2018) Do free-ranging rattlesnakes use thermal cues to evaluate prey?, *Journal of Comparative Physiology A,* 204(3), 295–303.

Schrope, M. (2013) Giant squid filmed in its natural environment, *Nature,* doi.org/10.1038/nature.2013.12202.

Schuergers, N., et al. (2016) Cyanobacteria use micro-optics to sense light direction, *eLife,* 5, e12620.

Schuller, G., and Pollak, G. (1979) Disproportionate frequency representation in the inferior colliculus of Doppler-compensating greater horseshoe bats: Evidence for an acoustic fovea, *Journal of Comparative Physiology,* 132(1), 47–54.

Schulten, K., Swenberg, C. E., and Weller, A. (1978) A biomagnetic sensory mechanism based on magnetic field modulated coherent electron spin motion, *Zeitschrift für Physikalische Chemie,* 111(1), 1–5.

Schumacher, S., et al. (2016) Cross-modal object recognition and dynamic weighting of sensory inputs in a fish, *Proceedings of the National Academy of Sciences,* 113(27), 7638–7643.

Schusterman, R. J., et al. (2000) Why pinnipeds don't echolocate, *Journal of the Acoustical Society of America,* 107(4), 2256–2264.

Schütz, S., et al. (1999) Insect antenna as a smoke detector, *Nature,* 398(6725), 298–299.

Schwenk, K. (1994) Why snakes have forked tongues, *Science,* 263(5153), 1573–1577.

Secor, S. M. (2008) Digestive physiology of the Burmese python: Broad regulation of integrated performance, *Journal of Experimental Biology,* 211(24), 3767–3774.

Seehausen, O., et al. (2008) Speciation through sensory drive in cichlid fish, *Nature,* 455(7213), 620–626.

Seehausen, O., van Alphen, J. J. M., and Witte, F. (1997) Cichlid fish diversity threatened by eutrophication that curbs sexual selection, *Science*, 277(5333), 1808–1811.

Seidou, M., et al. (1990) On the three visual pigments in the retina of the firefly squid, *Watasenia scintillans*, *Journal of Comparative Physiology A*, 166, 769–773.

Seneviratne, S. S., and Jones, I. L. (2008) Mechanosensory function for facial ornamentation in the whiskered auklet, a crevice-dwelling seabird, *Behavioral Ecology*, 19(4), 784–790.

Sengupta, P., and Garrity, P. (2013) Sensing temperature, *Current Biology*, 23(8), R304–R307.

Senzaki, M., et al. (2016) Traffic noise reduces foraging efficiency in wild owls, *Scientific Reports*, 6(1), 30602.

Sewell, G. D. (1970) Ultrasonic communication in rodents, *Nature*, 227(5256), 410.

Seyfarth, E.-A. (2002) Tactile body raising: Neuronal correlates of a "simple" behavior in spiders, in Toft, S., and Scharff, N. (eds), *European Arachnology 2000: Proceedings of the 19th European College of Arachnology*, 19–32. Aarhus: Aarhus University Press.

Shadwick, R. E., Potvin, J., and Goldbogen, J. A. (2019) Lunge feeding in rorqual whales, *Physiology*, 34(6), 409–418.

Shamble, P. S., et al. (2016) Airborne acoustic perception by a jumping spider, *Current Biology*, 26(21), 2913–2920.

Shan, L., et al. (2018) Lineage-specific evolution of bitter taste receptor genes in the giant and red pandas implies dietary adaptation, *Integrative Zoology*, 13(2), 152–159.

Shannon, G., et al. (2014) Road traffic noise modifies behaviour of a keystone species, *Animal Behaviour*, 94, 135–141.

Shannon, G., et al. (2016) A synthesis of two decades of research documenting the effects of noise on wildlife: Effects of anthropogenic noise on wildlife, *Biological Reviews*, 91(4), 982–1005.

Sharma, K. R., et al. (2015) Cuticular hydrocarbon pheromones for social behavior and their coding in the ant antenna, *Cell Reports*, 12(8), 1261–1271.

Shaw, J., et al. (2015) Magnetic particle-mediated magnetoreception, *Journal of the Royal Society Interface*, 12(110), 20150499.

Sherrington, C. S. (1903) Qualitative difference of spinal reflex corresponding with qualitative difference of cutaneous stimulus, *Journal of Physiology*, 30(1), 39–46.

Shimozawa, T., Murakami, J., and Kumagai, T. (2003) Cricket wind receptors: Thermal noise for the highest sensitivity known, in Barth, F. G., Humphrey, J. A. C., and Secomb, T. W. (eds), *Sensors and sensing in biology and engineering*, 145–157. Vienna: Springer.

Shine, R., et al. (2002) Antipredator responses of free-ranging pit vipers (*Gloydius shedaoensis*, Viperidae), *Copeia*, 2002(3), 843–850.

Shine, R., et al. (2003) Chemosensory cues allow courting male garter snakes to assess body length and body condition of potential mates, *Behavioral Ecology and Sociobiology*, 54(2), 162–166.

Sidebotham, J. (1877) Singing mice, *Nature*, 17(419), 29.

Siebeck, U. E., et al. (2010) A species of reef fish that uses ultraviolet patterns for covert face recognition, *Current Biology*, 20(5), 407–410.

Sieck, M. H., and Wenzel, B. M. (1969) Electrical activity of the olfactory bulb of the pigeon, *Electroencephalography and Clinical Neurophysiology*, 26(1), 62–69.

Siemers, B. M., et al. (2009) Why do shrews twitter? Communication or simple echo-based orientation, *Biology Letters*, 5(5), 593–596.

Silpe, J. E., and Bassler, B. L. (2019) A host-produced quorum-sensing autoinducer controls a phage lysis-lysogeny decision, *Cell*, 176(1–2), 268–280.e13.

Simmons, J. A., Ferragamo, M. J., and Moss, C. F. (1998) Echo-delay resolution in sonar images of the big brown bat, *Eptesicus fuscus*, *Proceedings of the National Academy of Sciences*, 95(21), 12647–12652.

Simmons, J. A., and Stein, R. A. (1980) Acoustic imaging in bat sonar: Echolocation signals and the evolution of echolocation, *Journal of Comparative Physiology*, 135(1), 61–84.

Simões, J. M., et al. (2021) Robustness and plasticity in *Drosophila* heat avoidance, *Nature Communications,* 12(1), 2044.

Simons, E. (2020) Backyard fly training and you, *Bay Nature.* Available at: baynature.org/article/lord-of-the-flies/.

Simpson, S. D., et al. (2016) Anthropogenic noise increases fish mortality by predation, *Nature Communications,* 7(1), 10544.

Sisneros, J. A. (2009) Adaptive hearing in the vocal plainfin midshipman fish: Getting in tune for the breeding season and implications for acoustic communication, *Integrative Zoology,* 4(1), 33–42.

Skedung, L., et al. (2013) Feeling small: Exploring the tactile perception limits, *Scientific Reports,* 3(1), 2617.

Slabbekoorn, H., and Peet, M. (2003) Birds sing at a higher pitch in urban noise, *Nature,* 424(6946), 267.

Smith, A. C., et al. (2003) The effect of colour vision status on the detection and selection of fruits by tamarins (*Saguinus* spp.), *Journal of Experimental Biology,* 206(18), 3159–3165.

Smith, B., et al. (2004) A survey of frog odorous secretions, their possible functions and phylogenetic significance, *Applied Herpetology,* 2, 47–82.

Smith, C. F., et al. (2009) The spatial and reproductive ecology of the copperhead (*Agkistrodon contortrix*) at the northeastern extreme of its range, *Herpetological Monographs,* 23(1), 45–73.

Smith, E. St. J., et al. (2011) The molecular basis of acid insensitivity in the African naked mole-rat, *Science,* 334(6062), 1557–1560.

Smith, E. St. J., Park, T. J., and Lewin, G. R. (2020) Independent evolution of pain insensitivity in African mole-rats: Origins and mechanisms, *Journal of Comparative Physiology A,* 206(3), 313–325.

Smith, F. A., et al. (2018) Body size downgrading of mammals over the late Quaternary, *Science,* 360(6386), 310–313.

Smith, L. M., et al. (2020) Impacts of COVID-19-related social distancing measures on personal environmental sound exposures, *Environmental Research Letters,* 15(10), 104094.

Sneddon, L. (2013) Do painful sensations and fear exist in fish?, in van der Kemp, T., and Lachance, M. (eds), *Animal suffering: From science to law,* 93–112. Toronto: Carswell.

Sneddon, L. U. (2018) Comparative physiology of nociception and pain, *Physiology,* 33(1), 63–73.

Sneddon, L. U. (2019) Evolution of nociception and pain: Evidence from fish models, *Philosophical Transactions of the Royal Society B: Biological Sciences,* 374(1785), 20190290.

Sneddon, L. U., et al. (2014) Defining and assessing animal pain, *Animal Behaviour,* 97, 201–212.

Sneddon, L. U., Braithwaite, V. A., and Gentle, M. J. (2003a) Do fishes have nociceptors? Evidence for the evolution of a vertebrate sensory system, *Proceedings of the Royal Society B: Biological Sciences,* 270(1520), 1115–1121.

Sneddon, L. U., Braithwaite, V. A., and Gentle, M. J. (2003b) Novel object test: Examining nociception and fear in the rainbow trout, *Journal of Pain,* 4(8), 431–440.

Snyder, J. B., et al. (2007) Omnidirectional sensory and motor volumes in electric fish, *PLOS Biology,* 5(11), e301.

Soares, D. (2002) An ancient sensory organ in crocodilians, *Nature,* 417(6886), 241–242.

Sobel, N., et al. (1999) The world smells different to each nostril, *Nature,* 402(6757), 35.

Solvi, C., Gutierrez Al-Khudhairy, S., and Chittka, L. (2020) Bumble bees display cross-modal object recognition between visual and tactile senses, *Science,* 367(6480), 910–912.

Speiser, D. I., and Johnsen, S. (2008a) Comparative morphology of the concave mirror eyes of scallops (Pectinoidea), *American Malacological Bulletin,* 26(1–2), 27–33.

Speiser, D. I., and Johnsen, S. (2008b) Scallops visually respond to the size and speed of virtual particles, *Journal of Experimental Biology,* 211(Pt 13), 2066–2070.

Sperry, R. W. (1950) Neural basis of the spontaneous optokinetic response produced by visual inversion, *Journal of Comparative and Physiological Psychology,* 43(6), 482–489.

Spoelstra, K., et al. (2017) Response of bats to light with different spectra: Light-shy and agile bat presence is affected by white and green, but not red light, *Proceedings of the Royal Society B: Biological Sciences*, 284(1855), 20170075.

Stack, D. W., et al. (2011) Reducing visitor noise levels at Muir Woods National Monument using experimental management, *Journal of the Acoustical Society of America*, 129(3), 1375–1380.

Stager, K. E. (1964) The role of olfaction in food location by the turkey vulture (*Cathartes aura*), *Contributions in Science*, 81, 1–63.

Stamp Dawkins, M. (2002) What are birds looking at? Head movements and eye use in chickens, *Animal Behaviour*, 63(5), 991–998.

Standing Bear, L. (2006) *Land of the spotted eagle*. Lincoln: Bison Books.

Stangl, F. B., et al. (2005) Comments on the predator-prey relationship of the Texas kangaroo rat (*Dipodomys elator*) and barn owl (*Tyto alba*), *The American Midland Naturalist*, 153(1), 135–141.

Stebbins, W. C. (1983) *The acoustic sense of animals*. Cambridge, MA: Harvard University Press.

Steen, J. B., et al. (1996) Olfaction in bird dogs during hunting, *Acta Physiologica Scandinavica*, 157(1), 115–119.

Sterbing-D'Angelo, S. J., et al. (2017) Functional role of airflow-sensing hairs on the bat wing, *Journal of Neurophysiology*, 117(2), 705–712.

Sterbing-D'Angelo, S. J., and Moss, C. F. (2014) Air flow sensing in bats, in Bleckmann, H., Mogdans, J., and Coombs, S. L. (eds), *Flow sensing in air and water*, 197–213. Berlin: Springer.

Stevens, M., and Cuthill, I. C. (2007) Hidden messages: Are ultraviolet signals a special channel in avian communication?, *BioScience*, 57(6), 501–507.

Stiehl, W. D., Lalla, L., and Breazeal, C. (2004) A "somatic alphabet" approach to "sensitive skin," in *Proceedings, ICRA '04, IEEE International Conference on Robotics and Automation, 2004*, 3, 2865–2870. New Orleans: IEEE.

Stoddard, M. C., et al. (2019) I see your false colours: How artificial stimuli appear to different animal viewers, *Interface Focus*, 9(1), 20180053.

Stoddard, M. C., et al. (2020) Wild hummingbirds discriminate nonspectral colors, *Proceedings of the National Academy of Sciences*, 117(26), 15112–15122.

Stokkan, K.-A., et al. (2013) Shifting mirrors: Adaptive changes in retinal reflections to winter darkness in Arctic reindeer, *Proceedings of the Royal Society B: Biological Sciences*, 280(1773), 20132451.

Stowasser, A., et al. (2010) Biological bifocal lenses with image separation, *Current Biology*, 20(16), 1482–1486.

Strauß, J., and Stumpner, A. (2015) Selective forces on origin, adaptation and reduction of tympanal ears in insects, *Journal of Comparative Physiology A*, 201(1), 155–169.

Strobel, S. M., et al. (2018) Active touch in sea otters: In-air and underwater texture discrimination thresholds and behavioral strategies for paws and vibrissae, *Journal of Experimental Biology*, 221(18), jeb181347.

Suga, N., and Schlegel, P. (1972) Neural attenuation of responses to emitted sounds in echolocating bats, *Science*, 177(4043), 82–84.

Sukhum, K. V., et al. (2016) The costs of a big brain: Extreme encephalization results in higher energetic demand and reduced hypoxia tolerance in weakly electric African fishes, *Proceedings of the Royal Society B: Biological Sciences*, 283(1845), 20162157.

Sullivan, J. J. (2013) One of us, *Lapham's Quarterly*. Available at: www.laphamsquarterly.org/animals/one-us.

Sumbre, G., et al. (2006) Octopuses use a human-like strategy to control precise point-to-point arm movements, *Current Biology*, 16(8), 767–772.

Sumner-Rooney, L., et al. (2018) Whole-body photoreceptor networks are independent of "lenses" in brittle stars, *Proceedings of the Royal Society B: Biological Sciences*, 285(1871), 20172590.

Sumner-Rooney, L. H., et al. (2014) Do chitons have a compass? Evidence for magnetic sensitivity in *Polyplacophora*, *Journal of Natural History*, 48(45–48), 3033–3045.

Sumner-Rooney, L. H., et al. (2020) Extraocular vision in a brittle star is mediated by chromatophore movement in response to ambient light, *Current Biology*, 30(2), 319–327.e4.

Supa, M., Cotzin, M., and Dallenbach, K. M. (1944) "Facial vision": The perception of obstacles by the blind, *The American Journal of Psychology*, 57(2), 133–183.

Suraci, J. P., et al. (2019) Fear of humans as apex predators has landscape-scale impacts from mountain lions to mice, *Ecology Letters*, 22(10), 1578–1586.

Surlykke, A., et al. (eds), (2014) *Biosonar*. New York: Springer.

Surlykke, A., and Kalko, E. K. V. (2008) Echolocating bats cry out loud to detect their prey, *PLOS One*, 3(4), e2036.

Surlykke, A., Simmons, J. A., and Moss, C. F. (2016) Perceiving the world through echolocation and vision, in Fenton, M. B., et al. (eds), *Bat bioacoustics*, 265–288. New York: Springer.

Suter, R. B. (1978) *Cyclosa turbinata* (Araneae, Araneidae): Prey discrimination via web-borne vibrations, *Behavioral Ecology and Sociobiology*, 3(3), 283–296.

Suthers, R. A. (1967) Comparative echolocation by fishing bats, *Journal of Mammalogy*, 48(1), 79–87.

Sutton, G. P., et al. (2016) Mechanosensory hairs in bumblebees (*Bombus terrestris*) detect weak electric fields, *Proceedings of the National Academy of Sciences*, 113(26), 7261–7265.

Swaddle, J. P., et al. (2015) A framework to assess evolutionary responses to anthropogenic light and sound, *Trends in Ecology & Evolution*, 30(9), 550–560.

Takeshita, F., and Murai, M. (2016) The vibrational signals that male fiddler crabs (*Uca lactea*) use to attract females into their burrows, *The Science of Nature*, 103, 49.

Tansley, K. (1965) *Vision in vertebrates*. London: Chapman and Hall.

Tautz, J., and Markl, H. (1978) Caterpillars detect flying wasps by hairs sensitive to airborne vibration, *Behavioral Ecology and Sociobiology*, 4(1), 101–110.

Tautz, J., and Rostás, M. (2008) Honeybee buzz attenuates plant damage by caterpillars, *Current Biology*, 18(24), R1125–R1126.

Taylor, C. J., and Yack, J. E. (2019) Hearing in caterpillars of the monarch butterfly (*Danaus plexippus*), *Journal of Experimental Biology*, 222(22), jeb211862.

Tedore, C., and Nilsson, D.-E. (2019) Avian UV vision enhances leaf surface contrasts in forest environments, *Nature Communications*, 10(1), 238.

Temple, S., et al. (2012) High-resolution polarisation vision in a cuttlefish, *Current Biology*, 22(4), R121–R122.

Ter Hofstede, H. M., and Ratcliffe, J. M. (2016) Evolutionary escalation: The bat-moth arms race, *Journal of Experimental Biology*, 219(11), 1589–1602.

Thaler, L., et al. (2017) Mouth-clicks used by blind expert human echolocators—Signal description and model based signal synthesis, *PLOS Computational Biology*, 13(8), e1005670.

Thaler, L., et al. (2020) The flexible action system: Click-based echolocation may replace certain visual functionality for adaptive walking, *Journal of Experimental Psychology: Human Perception and Performance*, 46(1), 21–35.

Thaler, L., Arnott, S. R., and Goodale, M. A. (2011) Neural correlates of natural human echolocation in early and late blind echolocation experts, *PLOS One*, 6(5), e20162.

Thaler, L., and Goodale, M. A. (2016) Echolocation in humans: An overview, *Wiley Interdisciplinary Reviews: Cognitive Science*, 7(6), 382–393.

Thoen, H. H., et al. (2014) A different form of color vision in mantis shrimp, *Science*, 343(6169), 411–413.

Thoma, V., et al. (2016) Functional dissociation in sweet taste receptor neurons between and within taste organs of *Drosophila*, *Nature Communications*, 7(1), 10678.

Thomas, K. N., Robison, B. H., and Johnsen, S. (2017) Two eyes for two purposes: In situ evidence for asymmetric vision in the cockeyed squids *Histioteuthis heteropsis* and *Stigmatoteuthis dofleini*, *Philosophical Transactions of the Royal Society B: Biological Sciences*, 372(1717), 20160069.

Thometz, N. M., et al. (2016) Trade-offs between energy maximization and parental care in a central place forager, the sea otter, *Behavioral Ecology*, 27(5), 1552–1566.

Thums, M., et al. (2013) Evidence for behavioural thermoregulation by the world's largest fish, *Journal of the Royal Society Interface,* 10(78), 20120477.

Tierney, K. B., et al. (2008) Salmon olfaction is impaired by an environmentally realistic pesticide mixture, *Environmental Science & Technology,* 42(13), 4996–5001.

Toda, Y., et al. (2021) Early origin of sweet perception in the songbird radiation, *Science,* 373(6551), 226–231.

Tracey, W. D. (2017) Nociception, *Current Biology,* 27(4), R129–R133.

Treiber, C. D., et al. (2012) Clusters of iron-rich cells in the upper beak of pigeons are macrophages not magnetosensitive neurons, *Nature,* 484(7394), 367–370.

Treisman, D. (2010) Ants and answers: A conversation with E. O. Wilson, *The New Yorker.* Available at: www.newyorker.com/books/page-turner/ants-and-answers-a-conversation -with-e-o-wilson.

Trible, W., et al. (2017) *Orco* mutagenesis causes loss of antennal lobe glomeruli and impaired social behavior in ants, *Cell,* 170(4), 727–735.e10.

Tricas, T. C., Michael, S. W., and Sisneros, J. A. (1995) Electrosensory optimization to conspecific phasic signals for mating, *Neuroscience Letters,* 202(1), 129–132.

Tsai, C.-C., et al. (2020) Physical and behavioral adaptations to prevent overheating of the living wings of butterflies, *Nature Communications,* 11(1), 551.

Tsujii, K., et al. (2018) Change in singing behavior of humpback whales caused by shipping noise, *PLOS One,* 13(10), e0204112.

Tumlinson, J. H., et al. (1971) Identification of the trail pheromone of a leaf-cutting ant, *Atta texana, Nature,* 234(5328), 348–349.

Turkel, W. J. (2013) *Spark from the deep: How shocking experiments with strongly electric fish powered scientific discovery.* Baltimore: Johns Hopkins University Press.

Tuthill, J. C., and Azim, E. (2018) Proprioception, *Current Biology,* 28(5), R194–R203.

Tuttle, M. D., and Ryan, M. J. (1981) Bat predation and the evolution of frog vocalizations in the neotropics, *Science,* 214(4521), 677–678.

Tyack, P. L. (1997) Studying how cetaceans use sound to explore their environment, in Owings, D. H., Beecher, M. D., and Thompson, N. S. (eds), *Perspectives in ethology,* vol. 12, 251–297. New York: Plenum Press.

Tyack, P. L., and Clark, C. W. (2000) Communication and acoustic behavior of dolphins and whales, in Au, W. W. L., Fay, R. R., and Popper, A. N. (eds), *Hearing by whales and dolphins,* 156–224. New York: Springer.

Tyler, N. J. C., et al. (2014) Ultraviolet vision may enhance the ability of reindeer to discriminate plants in snow, *Arctic,* 67(2), 159–166.

Uexküll, J. von (1909) *Umwelt und Innenwelt der Tiere.* Berlin: J. Springer.

Uexküll, J. von (2010) *A foray into the worlds of animals and humans: With a theory of meaning* (trans. J. D. O'Neil). Minneapolis: University of Minnesota Press.

Ulanovsky, N., and Moss, C. F. (2008) What the bat's voice tells the bat's brain, *Proceedings of the National Academy of Sciences,* 105(25), 8491–8498.

Ullrich-Luter, E. M., et al. (2011) Unique system of photoreceptors in sea urchin tube feet, *Proceedings of the National Academy of Sciences,* 108(20), 8367–8372.

Vaknin, Y., et al. (2000) The role of electrostatic forces in pollination, *Plant Systematics and Evolution,* 222(1), 133–142.

Van Buskirk, R. W., and Nevitt, G. A. (2008) The influence of developmental environment on the evolution of olfactory foraging behaviour in procellariiform seabirds, *Journal of Evolutionary Biology,* 21(1), 67–76.

Van der Horst, G., et al. (2011) Sperm structure and motility in the eusocial naked mole-rat, *Heterocephalus glaber:* A case of degenerative orthogenesis in the absence of sperm competition?, *BMC Evolutionary Biology,* 11(1), 351.

Van Doren, B. M., et al. (2017) High-intensity urban light installation dramatically alters nocturnal bird migration, *Proceedings of the National Academy of Sciences,* 114(42), 11175–11180.

Van Lenteren, J. C., et al. (2007) Structure and electrophysiological responses of gustatory organs on the ovipositor of the parasitoid *Leptopilina heterotoma*, *Arthropod Structure & Development*, 36(3), 271–276.

Van Staaden, M. J., et al. (2003) Serial hearing organs in the atympanate grasshopper *Bullacris membracioides* (Orthoptera, Pneumoridae), *Journal of Comparative Neurology*, 465(4), 579–592.

Veilleux, C. C., and Kirk, E. C. (2014) Visual acuity in mammals: Effects of eye size and ecology, *Brain, Behavior and Evolution*, 83(1), 43–53.

Vélez, A., Ryoo, D. Y., and Carlson, B. A. (2018) Sensory specializations of mormyrid fish are associated with species differences in electric signal localization behavior, *Brain, Behavior and Evolution*, 92(3–4), 125–141.

Vernaleo, B. A., and Dooling, R. J. (2011) Relative salience of envelope and fine structure cues in zebra finch song, *Journal of the Acoustical Society of America*, 129(5), 3373–3383.

Vidal-Gadea, A., et al. (2015) Magnetosensitive neurons mediate geomagnetic orientation in *Caenorhabditis elegans*, *eLife*, 4, e07493.

Viguier, C. (1882) Le sens de l'orientation et ses organes chez les animaux et chez l'homme, *Revue philosophique de la France et de l'étranger*, 14, 1–36.

Viitala, J., et al. (1995) Attraction of kestrels to vole scent marks visible in ultraviolet light, *Nature*, 373(6513), 425–427.

Vogt, R. G., and Riddiford, L. M. (1981) Pheromone binding and inactivation by moth antennae, *Nature*, 293(5828), 161–163.

Vollrath, F. (1979a) Behaviour of the kleptoparasitic spider *Argyrodes elevatus* (Araneae, theridiidae), *Animal Behaviour*, 27(Pt 2), 515–521.

Vollrath, F. (1979b) Vibrations: Their signal function for a spider kleptoparasite, *Science*, 205(4411), 1149–1151.

Von der Emde, G. (1990) Discrimination of objects through electrolocation in the weakly electric fish, *Gnathonemus petersii*, *Journal of Comparative Physiology A*, 167, 413–421.

Von der Emde, G. (1999) Active electrolocation of objects in weakly electric fish, *Journal of Experimental Biology*, 202, 1205–1215.

Von der Emde, G., et al. (1998) Electric fish measure distance in the dark, *Nature*, 395(6705), 890–894.

Von der Emde, G., and Ruhl, T. (2016) Matched filtering in African weakly electric fish: Two senses with complementary filters, in von der Emde, G., and Warrant, E. (eds), *The ecology of animal senses*, 237–263. Cham: Springer.

Von der Emde, G., and Schnitzler, H.-U. (1990) Classification of insects by echolocating greater horseshoe bats, *Journal of Comparative Physiology A*, 167(3), 423–430.

Von Dürckheim, K. E. M., et al. (2018) African elephants (*Loxodonta africana*) display remarkable olfactory acuity in human scent matching to sample performance, *Applied Animal Behaviour Science*, 200, 123–129.

Von Holst, E., and Mittelstaedt, H. (1950) Das reafferenzprinzip, *Naturwissenschaften*, 37(20), 464–476.

Wackermannová, M., Pinc, L., and Jebavý, L. (2016) Olfactory sensitivity in mammalian species, *Physiological Research*, 65(3), 369–390.

Walker, D. B., et al. (2006) Naturalistic quantification of canine olfactory sensitivity, *Applied Animal Behaviour Science*, 97(2–4), 241–254.

Walsh, C. M., Bautista, D. M., and Lumpkin, E. A. (2015) Mammalian touch catches up, *Current Opinion in Neurobiology*, 34, 133–139.

Wang, C. X., et al. (2019) Transduction of the geomagnetic field as evidenced from alpha-band activity in the human brain, *eNeuro*, 6(2), ENEURO.0483-18.2019.

Ward, J. (2013) Synesthesia, *Annual Review of Psychology*, 64(1), 49–75.

Wardill, T., et al. (2013) The miniature dipteran killer fly *Coenosia attenuata* exhibits adaptable aerial prey capture strategies, *Frontiers of Physiology Conference Abstract: International Conference on Invertebrate Vision*, doi:10.3389/conf.fphys.2013.25.00057.

Ware, H. E., et al. (2015) A phantom road experiment reveals traffic noise is an invisible source of habitat degradation, *Proceedings of the National Academy of Sciences*, 112(39), 12105–12109.

Warkentin, K. M. (1995) Adaptive plasticity in hatching age: A response to predation risk trade-offs, *Proceedings of the National Academy of Sciences*, 92(8), 3507–3510.

Warkentin, K. M. (2005) How do embryos assess risk? Vibrational cues in predator-induced hatching of red-eyed treefrogs, *Animal Behaviour*, 70(1), 59–71.

Warkentin, K. M. (2011) Environmentally cued hatching across taxa: Embryos respond to risk and opportunity, *Integrative and Comparative Biology*, 51(1), 14–25.

Warrant, E. J. (2017) The remarkable visual capacities of nocturnal insects: Vision at the limits with small eyes and tiny brains, *Philosophical Transactions of the Royal Society B: Biological Sciences*, 372(1717), 20160063.

Warrant, E. J., et al. (2004) Nocturnal vision and landmark orientation in a tropical halictid bee, *Current Biology*, 14(15), 1309–1318.

Warrant, E., et al. (2016) The Australian bogong moth *Agrotis infusa*: A long-distance nocturnal navigator, *Frontiers in Behavioral Neuroscience*, 10, 77.

Warrant, E. J., and Locket, N. A. (2004) Vision in the deep sea, *Biological Reviews of the Cambridge Philosophical Society*, 79(3), 671–712.

Watanabe, T. (1999) The influence of energetic state on the form of stabilimentum built by *Octonoba sybotides* (Araneae: Uloboridae), *Ethology*, 105(8), 719–725.

Watanabe, T. (2000) Web tuning of an orb-web spider, *Octonoba sybotides*, regulates prey-catching behaviour, *Proceedings of the Royal Society B: Biological Sciences*, 267(1443), 565–569.

Webb, B. (1996) A cricket robot, *Scientific American*. Available at: www.scientificamerican.com/article/a-cricket-robot/.

Webb, J. F. (2013) Morphological diversity, development, and evolution of the mechanosensory lateral line system, in Coombs, S., et al. (eds), *The lateral line system*, 17–72. New York: Springer.

Webster, D. B. (1962) A function of the enlarged middle-ear cavities of the kangaroo rat, *Dipodomys*, *Physiological Zoology*, 35(3), 248–255.

Webster, D. B., and Webster, M. (1971) Adaptive value of hearing and vision in kangaroo rat predator avoidance, *Brain, Behavior and Evolution*, 4(4), 310–322.

Webster, D. B., and Webster, M. (1980) Morphological adaptations of the ear in the rodent family heteromyidae, *American Zoologist*, 20(1), 247–254.

Weger, M., and Wagner, H. (2016) Morphological variations of leading-edge serrations in owls (*Strigiformes*), *PLOS One*, 11(3), e0149236.

Wehner, R. (1987) "Matched filters"—Neural models of the external world, *Journal of Comparative Physiology A*, 161(4), 511–531.

Weiss, T., et al. (2020) Human olfaction without apparent olfactory bulbs, *Neuron*, 105(1), 35–45.e5.

Wenzel, B. M., and Sieck, M. H. (1972) Olfactory perception and bulbar electrical activity in several avian species, *Physiology & Behavior*, 9(3), 287–293.

Wheeler, W. M. (1910) *Ants: Their structure, development and behavior.* New York: Columbia University Press.

Widder, E. (2019) The Medusa, NOAA Ocean Exploration. Available at: oceanexplorer.noaa.gov/explorations/19biolum/background/medusa/medusa.html.

Wieskotten, S., et al. (2010) Hydrodynamic determination of the moving direction of an artificial fin by a harbour seal (*Phoca vitulina*), *Journal of Experimental Biology*, 213(13), 2194–2200.

Wieskotten, S., et al. (2011) Hydrodynamic discrimination of wakes caused by objects of different size or shape in a harbour seal (*Phoca vitulina*), *Journal of Experimental Biology*, 214(11), 1922–1930.

Wignall, A. E., and Taylor, P. W. (2011) Assassin bug uses aggressive mimicry to lure spider prey, *Proceedings of the Royal Society B: Biological Sciences*, 278(1710), 1427–1433.

Wilcox, C., Van Sebille, E., and Hardesty, B. D. (2015) Threat of plastic pollution to seabirds is global, pervasive, and increasing, *Proceedings of the National Academy of Sciences*, 112(38), 11899–11904.

Wilcox, S. R., Jackson, R. R., and Gentile, K. (1996) Spiderweb smokescreens: Spider trickster uses background noise to mask stalking movements, *Animal Behaviour*, 51(2), 313–326.

Williams, C. J., et al. (2019) Analgesia for non-mammalian vertebrates, *Current Opinion in Physiology*, 11, 75–84.

Wilson, D. R., and Hare, J. F. (2004) Ground squirrel uses ultrasonic alarms, *Nature*, 430(6999), 523.

Wilson, E. O. (2015) Pheromones and other stimuli we humans don't get, with E. O. Wilson, *Big Think*. Available at: bigthink.com/videos/eo-wilson-on-the-world-of-pheromones.

Wilson, E. O., Durlach, N. I., and Roth, L. M. (1958) Chemical releasers of necrophoric behavior in ants, *Psyche*, 65(4), 108–114.

Wilson, S., and Moore, C. (2015) S1 somatotopic maps, *Scholarpedia*, 10(4), 8574.

Wiltschko, R., and Wiltschko, W. (2013) The magnetite-based receptors in the beak of birds and their role in avian navigation, *Journal of Comparative Physiology A*, 199(2), 89–98.

Wiltschko, R., and Wiltschko, W. (2019) Magnetoreception in birds, *Journal of the Royal Society Interface*, 16(158), 20190295.

Wiltschko, W. (1968) Über den Einfluß statischer Magnetfelder auf die Zugorientierung der Rotkehlchen (*Erithacus rubecula*), *Zeitschrift für Tierpsychologie*, 25(5), 537–558.

Wiltschko, W., et al. (2002) Lateralization of magnetic compass orientation in a migratory bird, *Nature*, 419(6906), 467–470.

Wiltschko, W., and Merkel, F. W. (1965) Orientierung zugunruhiger Rotkehlchen im statischen Magnetfeld, *Verhandlungen der Deutschen Zoologischen Gesellschaft in Jena*, 59, 362–367.

Windsor, D. A. (1998) Controversies in parasitology: Most of the species on Earth are parasites, *International Journal for Parasitology*, 28(12), 1939–1941.

Winklhofer, M., and Mouritsen, H. (2016) A room-temperature ferrimagnet made of metalloproteins?, bioRxiv, 094607.

Wisby, W. J., and Hasler, A. D. (1954) Effect of olfactory occlusion on migrating silver salmon (*O. kisutch*), *Journal of the Fisheries Research Board of Canada*, 11(4), 472–478.

Witherington, B., and Martin, R. E. (2003) Understanding, assessing, and resolving light-pollution problems on sea turtle nesting beaches, Florida Marine Research Institute Technical Report TR-2.

Witte, F., et al. (2013) Cichlid species diversity in naturally and anthropogenically turbid habitats of Lake Victoria, East Africa, *Aquatic Sciences*, 75(2), 169–183.

Woith, H., et al. (2018) Review: Can animals predict earthquakes?, *Bulletin of the Seismological Society of America*, 108(3A), 1031–1045.

Wolff, G. H., and Riffell, J. A. (2018) Olfaction, experience and neural mechanisms underlying mosquito host preference, *Journal of Experimental Biology*, 221(4), jeb157131.

Wu, C. H. (1984) Electric fish and the discovery of animal electricity, *American Scientist*, 72(6), 598–607.

Wu, L.-Q., and Dickman, J. D. (2012) Neural correlates of a magnetic sense, *Science*, 336(6084), 1054–1057.

Wueringer, B. E. (2012) Electroreception in elasmobranchs: Sawfish as a case study, *Brain, Behavior and Evolution*, 80(2), 97–107.

Wueringer, B. E., Squire, L., et al. (2012a) Electric field detection in sawfish and shovelnose rays, *PLOS One*, 7(7), e41605.

Wueringer, B. E., Squire, L., et al. (2012b) The function of the sawfish's saw, *Current Biology*, 22(5), R150–R151.

Wurtsbaugh, W. A., and Neverman, D. (1988) Post-feeding thermotaxis and daily vertical migration in a larval fish, *Nature*, 333(6176), 846–848.

Wyatt, T. (2015a) How animals communicate via pheromones, *American Scientist*, 103(2), 114.

Wyatt, T. D. (2015b) The search for human pheromones: The lost decades and the necessity of returning to first principles, *Proceedings of the Royal Society B: Biological Sciences*, 282(1804), 20142994.

Wynn, J., et al. (2020) Natal imprinting to the Earth's magnetic field in a pelagic seabird, *Current Biology*, 30(14), 2869–2873.e2.

Yadav, C. (2017) Invitation by vibration: Recruitment to feeding shelters in social caterpillars, *Behavioral Ecology and Sociobiology*, 71(3), 51.

Yager, D. D., and Hoy, R. R. (1986) The cyclopean ear: A new sense for the praying mantis, *Science*, 231(4739), 727–729.

Yanagawa, A., Guigue, A. M. A., and Marion-Poll, F. (2014) Hygienic grooming is induced by contact chemicals in *Drosophila melanogaster*, *Frontiers in Behavioral Neuroscience*, 8, 254.

Yarmolinsky, D. A., Zuker, C. S., and Ryba, N. J. P. (2009) Common sense about taste: From mammals to insects, *Cell*, 139(2), 234–244.

Yeates, L. C., Williams, T. M., and Fink, T. L. (2007) Diving and foraging energetics of the smallest marine mammal, the sea otter (*Enhydra lutris*), *Journal of Experimental Biology*, 210(11), 1960–1970.

Yong, E. (2020) America is trapped in a pandemic spiral, *The Atlantic*. Available at: www.theatlantic.com/health/archive/2020/09/pandemic-intuition-nightmare-spiral-winter/616204/.

Yoshizawa, M., et al. (2014) The sensitivity of lateral line receptors and their role in the behavior of Mexican blind cavefish (*Astyanax mexicanus*), *Journal of Experimental Biology*, 217(6), 886–895.

Yovel, Y., et al. (2009) The voice of bats: How greater mouse-eared bats recognize individuals based on their echolocation calls, *PLOS Computational Biology*, 5(6), e1000400.

Zagaeski, M., and Moss, C. F. (1994) Target surface texture discrimination by the echolocating bat, *Eptesicus fuscus*, *Journal of the Acoustical Society of America*, 95(5), 2881–2882.

Zapka, M., et al. (2009) Visual but not trigeminal mediation of magnetic compass information in a migratory bird, *Nature*, 461(7268), 1274–1277.

Zelenitsky, D. K., Therrien, F., and Kobayashi, Y. (2009) Olfactory acuity in theropods: Palaeobiological and evolutionary implications, *Proceedings of the Royal Society B: Biological Sciences*, 276(1657), 667–673.

Zimmer, C. (2012) Monet's ultraviolet eye, *Download the Universe*. Available at: www.downloadtheuniverse.com/dtu/2012/04/monets-ultraviolet-eye.html.

Zimmerman, A., Bai, L., and Ginty, D. D. (2014) The gentle touch receptors of mammalian skin, *Science*, 346(6212), 950–954.

Zimmermann, M. J. Y., et al. (2018) Zebrafish differentially process color across visual space to match natural scenes, *Current Biology*, 28(13), 2018–2032.e5.

Zions, M., et al. (2020) Nest carbon dioxide masks GABA-dependent seizure susceptibility in the naked mole-rat, *Current Biology*, 30(11), 2068–2077.e4.

Zippelius, H.-M. (1974) Ultraschall-Laute nestjunger Mäuse, *Behaviour*, 49(3–4), 197–204.

Zuk, M., Rotenberry, J. T., and Tinghitella, R. M. (2006) Silent night: Adaptive disappearance of a sexual signal in a parasitized population of field crickets, *Biology Letters*, 2(4), 521–524.

Zullo, L., et al. (2009) Nonsomatotopic organization of the higher motor centers in octopus, *Current Biology*, 19(19), 1632–1636.

Zupanc, G. K. H., and Bullock, T. H. (2005) From electrogenesis to electroreception: An overview, in Bullock, T. H., et al. (eds), *Electroreception*, 5–46. New York: Springer.

Insert Photo Credits

Page 6: top (barn owl): AHisgett; bottom (fly): treegrow
Page 7: top (frog): brian.gratwicke; bottom (finch): archer10 (Dennis)
Page 8: top (whale): greyloch; bottom (elephant): Kumaravel
Page 9: top (tarsier): berniedup; center (wax moth): Andy Reago & Chrissy McClarren; bottom (hummingbird): Bettina Arrigoni
Page 10: (bat): Jesse Barber
Page 11: (dolphins): J. D. Ebberly
Page 12: top (black ghost knifefish): blickwinkel / Alamy Stock Photo; center right (electric eel): chrisbb@prodigy.net; center left (glass knifefish): Charles & Clint; bottom (elephantfish): Imagebroker / Alamy Stock Photo
Page 13: top (shark ampullae): Albert kok; center (sawfish): Simon Fraser University; bottom (hammerhead): Numinosity by Gary J. Wood
Page 14: top (platypus): Klaus; bottom (bumblebee): wwarby
Page 15: top (moth): CSIRO; center (robin): tallpomlin; bottom (turtle): Dionysisa303
Page 16: (octopus): Joe Parks

Index

About the Author

ED YONG is an award-winning science writer on the staff of *The Atlantic,* where he won the Pulitzer Prize in Explanatory Reporting and the George Polk Award for Science Reporting, among other honors. His first book, *I Contain Multitudes,* was a *New York Times* bestseller and won numerous awards. His work has appeared in *The New Yorker, National Geographic, Wired, The New York Times, Scientific American,* and other publications. He lives in Washington, D.C.

Twitter: @edyong209

To inquire about booking Ed Yong for a speaking engagement, please contact the Penguin Random House Speakers Bureau at speakers@penguinrandomhouse.com.